An Introduction to

LASERS

THEORY AND APPLICATIONS

An Introduction to
LASERS
THEORY AND APPLICATIONS

Dr. M.N. AVADHANULU
M.Sc., Ph.D.
Head of the Department of Applied Physics
Kavikulguru Institute of Technology and Science
Ramtek, Nagpur

Dr. P.S. HEMNE
M.Sc., Ph.D.
Head, P.G. Department of Physics
Nevjabai Hitkarini College, Bramhapuri
(Nagpur University)

S. CHAND
PUBLISHING

S Chand And Company Limited
(ISO 9001 Certified Company)

S Chand And Company Limited
(ISO 9001 Certified Company)
Head Office: D-92, Sector–2, Noida – 201301, U.P. (India), Ph. 91-120-4682700
Registered Office: A-27, 2nd Floor, Mohan Co-operative Industrial Estate, New Delhi – 110 044, Phone: 011-49731800
www.**schandpublishing.com**; e-mail: **info@schandpublishing.com**

Marketing Offices:

Chennai	:	Ph: 23632120; chennai@schandpublishing.com
Guwahati	:	Ph: 2738811, 2735640; guwahati@schandpublishing.com
Hyderabad	:	Ph: 40186018; hyderabad@schandpublishing.com
Jalandhar	:	Ph: 4645630; jalandhar@schandpublishing.com
Kolkata	:	Ph: 23357458, 23353914; kolkata@schandpublishing.com
Lucknow	:	Ph: 4003633; lucknow@schandpublishing.com
Mumbai	:	Ph: 25000297; mumbai@schandpublishing.com
Patna	:	Ph: 2260011; patna@schandpublishing.com

First Edition 2001
Subsequent Editions and Reprints 2008, 2009 (Twice), 2010 (Twice)
Second Revised Edition 2012; Reprints 2013, 2015, 2017, 2018, 2019, 2022

Reprint 2024

ISBN: 978-81-219-2071-1 **Product Code:** H6LAS68PHYS10ENAB12O

PRINTED IN INDIA

By Vikas Publishing House Private Limited, Plot 20/4, Site-IV, Industrial Area Sahibabad, Ghaziabad – 201 010 and Published by S Chand And Company Limited, A-27, 2nd Floor, Mohan Co-operative Industrial Estate, New Delhi – 110 044.

PREFACE TO THE REVISED EDITION

We feel great pleasure to present the revised edition of the presently existing book, '***An Introduction to Lasers: Theory and Applications***.' Many more advance techniques are joining their hands in the enrichment of the topic 'LASER'. Accordingly, the syllabi of various universities in India and abroad are continuously revising. To keep pace with this, we are bound to fulfill the changing requirements demanded by our students' platform. We feel very happy and satisfied to present this revised edition at the hands of the students at the proper time.

Laser is one of the most revolutionary inventions of the 20th century (invention in 1958) and has proved to be a boon for the scientific research and development of technology. It is a light source but differs vastly from the common light sources and is not useful for illumination purposes as does the incandescent lamps. Lasers are more akin to radio and microwave transmitters which produce coherent, monochromatic and polarized beam of electromagnetic radiation. Because of the remarkable characteristics of directionality, coherence, monochromaticity and high intensity exhibited by laser beam, laser occupies a unique place in modern technology. It works as a '*Tool box*' in a wide variety of fields such as *material processing, laser printing, electronics industry, nuclear energy, entertainment, communications, medical diagnosis, surgery, ophthalmology, LIDAR, weapon monitoring in defence* and so many.

The generation of laser is based upon the application of quantum concept suggested by Planck. Hence, the book starts with the basic principles used in the development of quantum theory which are discussed in a simple manner for the beginners. *Chapter 1 is thoroughly revised.* History of quantum theory, being the backbone of the development of the theory of lasers, is deliberately incorporated here so as to get the students community acquainted with the foundation of the topic. This includes *Planck's quantum theory*, *Planck's radiation law*, *Photoelectric effect*, *Compton effect*, *the wave-particle duality of light*, *the de- Broglie hypothesis*, *wave velocity and group velocity*, *Heisenberg uncertainty principle*, *wave function* (ψ), *Schrödinger wave equation etc*. Some of the mathematical details given may not be necessary for initial appreciation of the subject but are described for the sake of consistency. The authors hope that they will be of use for such cross-section of readers who want to make use of this book as a text book. *Chapter 2 deals with the type of lasers*. Some of the lasers commonly used in practical applications are described in detail. Semiconductor diode laser is given special attention and the relevant theory and principles are discussed at length. *Chapter 3 covers the characteristics of a laser beam* which make it distinct from the light beam generated by common light sources. *Chapter 4 discusses the ways and means of modifying the laser output*. Single mode selection, giant pulse generation, and mode locking are briefly introduced. Chapter 5, the applications of lasers, is also elaborated in detail, as per the demand of the revised syllabi of some Technical Universities in India, particularly Gautam Budh Technical University (Formerly U.P. Technical University Lucknow). It is a vast subject and it is very difficult to do justice to it. However, we have tried to give adequate justice by accommodating and discussing various applications of lasers. *In this revised edition, we have added Exercises comprising of Questions and Numericals*, asked in the examinations of various universities, including Technical universities, at the end of each chapter. This will be helpful to the students to prepare for their examination.

The authors have referred number of books while preparing the present book and some of the books which influenced much are listed at the end of the book. The authors do not claim originality in content or presentation of the material included in this book. The authors record here their indebtedness to authors of those original works from where the information is liberally drawn.

The authors humbly invite suggestions from the users of this book for the improvement of the material.

We express sincere thanks to the Management team and Editorial staff of S. Chand & Company Ltd. for taking all pains to bring this book in the present form.

Nagpur **AUTHORS**

CONTENTS

CHAPTER 1

Basic Theory

1.1 INTRODUCTION

The word laser is an acronym for **L**ight **A**mplification by **S**timulated **E**mission of **R**adiation.

Laser is one of the outstanding inventions of the second half of the last century. Laser has become a valuable tool in a variety of fields starting with medicine to communications. Laser is a light source but it is very much different from many of traditional light sources. Laser is not used for illumination purposes as we use the other light sources. Lasers produce a highly directional and high intensity beam with a narrow frequency range than that available from the common types of light sources. They are more widely used as a high power electromagnetic beam rather than a light beam. The laser beams are used as a special type of drill bit to drill holes in hard materials, as a saw to cut thick metal sheets, as a phonograph needle for compact discs, as a knife during surgical operations, as target designators for military weapons and so on. Thus, laser is a high technology device affecting our lives in many ways.

1.2 A BRIEF HISTORY OF THE LASER

Lasers are in fact generators of light. They are based on the amplification of light by means of stimulated radiation of atoms or molecules. In 1917 Einstein predicted the possibility of such stimulated radiation.

In 1952, Ch. Townes, J. Gordon and H. Zeiger in U.S.A. and N. Basov and A. Prokhorov in USSR, independently suggested the principle of generating and amplifying microwave oscillations based on the concept of stimulated radiation. It lead to the invention of MASER (Microwave Amplification by Stimulated Emission of Radiation) in 1954. MASERS used a two-level system. In 1955, Basov and Prokhonov suggested use of three-level system.

In 1958, Townes and Schawlow and Basov and Prokhorov independently expressed their ideas about extending the maser concept to optical frequencies. They developed the concept of an optical amplifier surrounded by an optical mirror resonant cavity to allow for growth of the beam. Townes, Basov and Prokhorov both received Nobel Prizes for their work in this field.

In 1960, Theodore Maiman of Hughes Research Laboratories produced the first laser using a ruby crystal as the amplifier and a flashlamp as the energy source. The helical flashlamp surrounded a rod-shaped ruby crystal, and the optical cavity was formed by coating the flattened ends of the rudy rod with a highly reflecting material. An intense red beam was observed to emerge from the end of the rod when the flashlamp was fired.

The first gas laser was developed in 1961 by A. Javan, W. Bennett, and D. Harriott of Bell Laboratories, using a mixture of helium and neon gases. At the same laboratories, L.F. Johnson and K. Nassau demonstrated the first neodymium laser, which has since become one of the most reliable lasers available. This was followed in 1962 by the first semiconductor laser, demonstrated by R. Hall at the General Electric Research Laboratories. Table 1.1 gives the major landmarks in the development of lasers.

TABLE 1.1
MAJOR LANDMARKS IN THE DEVELOPMENT OF LASERS

Year	*Discoverer*	*Type of Laser/Principle*
1917	Albert Einstein	Stimulated Emission Process
1952	N.G. Basov, A.M. Prokhorov and Townes	Maser Principle
1954	Townes, Gordon, Zeiger	Maser
1958	Townes, Schawlow, Basov, Prokhorov	Laser Principle
1960	Theodore Maiman	Ruby Laser
1961	A. Javan, W. Bennett and D. Harriott	Helium-Neon Laser
1961	L.F. Johnson and K. Nassau	Neodymium Laser
1962	R.Hall	Semiconductor Laser
1963	C.K.N. Patel	Carbon dioxide Laser
1964	W. Bridges	Argon Ion Laser
1966	W. Silfvast, G.R. Fowles, and B.D. Hopkins	He-Cd Laser
1966	P.P. Sorokin and J.R. Lankard	Tunable Dye Laser
1975	J.J. Ewing and C. Brau	Excimer Laser
1976	J.M.J. Madey and coworkers	Free-electron Laser
1979	Walling and coworkers	Alexandrite Laser
1985	D.Mathews and coworkers	X-ray Laser

Difference between laser and ordinary light beam.

Laser light differs from ordinary light is following respects:

(*i*) Laser beam is highly monochromatic.

(*ii*) It travels as a highly concentrated parallel beam along a particular direction.

(*iii*) Laser beam is highly coherent with all the waves exactly in phase with each other and in the same state of polarization. The beam has both spatial and temporal coherence.

On the other hand-

(*i*) Ordinary light is not monochromatic. Light spectrum from a source may extend over a wide range of wavelength.

(*ii*) The light from an ordinary source travels and spreads in all directions.

(*iii*) Ordinary light is incoherent *i.e.*, there is a wide phase difference between the light observed at a point at different times and at different points in space.

Directionality, monochromacity, intensity and coherence are the peculiar properties of laser beam (see details in chapter 3).

1.3 INTERACTION OF LIGHT AND MATTER

Light is a form of radiant energy. Different hypotheses were put forward at different stages regarding the nature of light. Newton considered light as a stream of corpuscles while Huygens regarded light as made up of waves. Maxwell established that light belongs to the group of electromagnetic waves which propagate at a speed '*c*' in a vacuum. The frequency and the wavelength of the light wave are related to *c* through the expression

$$\boxed{c = \nu\lambda} \quad ...(1.1)$$

It is implicit in this visualization that light waves are continuous; are of infinite extension; and could carry any arbitrary amounts of energy.

When light is incident on a substance, there may occur reflection, transmission, absorption and scattering of light to varying degrees depending on the nature of the substance.

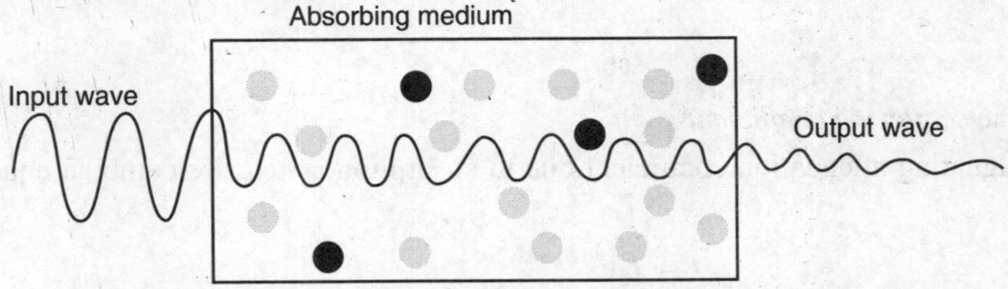

Fig. 1.1. Attenuation of a light wave in an absorbing medium.

A light wave incident on a transparent solid induces periodic oscillations of orbital electrons in step with its own frequency. The oscillating electrons act as point sources and emit waves in all directions. If the medium is homogeneous, the secondary waves destroy one another in all directions except in the direction of propagation of the incident wave. The resultant beam will have maximum intensity in that direction and constitutes the transmitted beam.

Part of the incident light energy gets transformed into the energy of motion of the atoms in the solid. It leads to a loss of energy and as a result, the light intensity decreases with distance in the solid as shown in Fig. 1.1. We say in this case that the medium absorbs light. The reduction in intensity of light with increasing length of propagation in a medium is called *absorption* or *attenuation* of light. We can describe the attenuation of light in an optical medium with the help of absorption coefficient α.

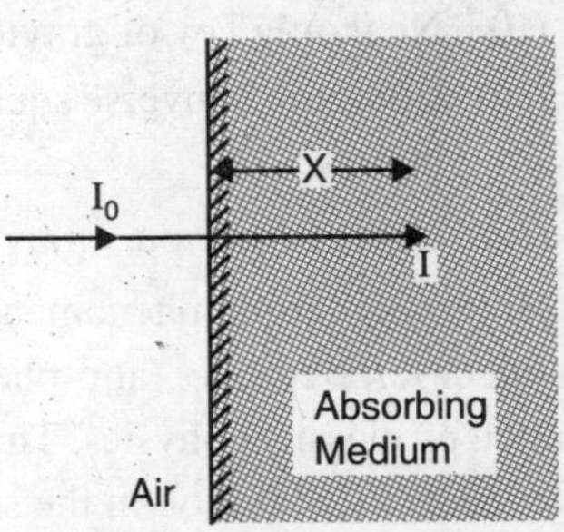

Fig. 1.2. Absorption of light in a medium.

When a beam of light passes through a thin section *dx* of a transparent material (Fig. 1.2), let the decrease in intensity of the light be *dI*. It is found that *dI* is proportional to the initial intensity of light, *I*, and to the thickness of the medium, *dx*. Thus,

$$dI = -\alpha I\, dx \qquad ...(1.2)$$

where α is the constant of proportionality. It is called the *coefficient of absorption* (the fraction of light absorbed per unit length) of the optical medium. The negative sign indicates that the intensity decreases with distance. We can rearrange Eq. (1.2) as

$$\frac{dI}{I} = -\alpha\, dx \qquad ...(1.3)$$

The total loss in light intensity after passing through a distance *x* in the medium is obtained by integrating Eq. (1.3). Thus, if I_0 is the value of intensity at $x = 0$, the value of the intensity *I* at a distance *x* is

$$\int_{I_0}^{I} \frac{dI}{I} = -\alpha \int_0^x dx$$

$$\text{In}\, \frac{I}{I_0} = \alpha x$$

or

$$\boxed{I = I_0 e^{-ax}} \qquad ...(1.4)$$

Eq. (1.4) shows that light intensity decreases exponentially with distance in the medium.

When a light beam encounters obstacles which have sizes smaller than a wavelength ($d << \lambda$), it is redirected into different directions. We say light is *scattered* by the obstacle. Scattering phenomenon also causes attenuation of energy because part of the light energy is lost from the original beam. Scattering effects are described by an exponential law that is much similar in form to Eq.(1.4).

$$I = I_0 e^{-\alpha_s x} \qquad ...(1.5)$$

where α_s is the *scattering coefficient*.

By designating attenuation coefficient due to absorption as α_a, we combine equations (1.4) and (1.5) as

$$I = I_0 e^{-(\alpha_0 + \alpha_s) x} \qquad ...(1.6)$$

Eq. (1.6) describes attenuation of light due to both absorption and scattering effects.

1.4 A BRIEF HISTORY OF QUANTUM THEORY

Phenomena such as motions of mechanical objects involving distances larger than about 10^{-6}m can be explained satisfactorily by applying the laws of classical theoretical physics which is based on the following basic laws:

(*i*) Newton's laws of motion,

(*ii*) Newton'a law of gravitational attraction,

(*iii*) Coulomb's inverse square law of attraction or repulsion between two electrically charged bodies,

(*iv*) Law of force on a moving chanrge in a magnetic field, *i.e.,* the Lorentz force.

However, certain phenomena such as spectral distribution of energy in black body radiation, photo-electric effect etc., and phenomena involving distances of the order of 10^{-10} could not be explained by classical physics. The failure of classical physics to explain above facts. particularly to expalin energy distibution in the spectrum of a block body led Max Planck to propose the *quantum hypothesis* in 1900 and this was the origin of quantum theory.

1.5 PLANCK'S QUANTUM THEORY

In 1900, Max Planck proposed that light consists of discrete bundles or chunks of energy. The amount of energy of each bundle is "*hv*". These bundles of radiant energy are called *quanta.* In 1905, Einstein refined the quantum hypothesis of Planck and provided theoretical justification for the features of photoelectric effect. He gave the name *photon* to the quantum of light energy.

A photon represents the minimum energy unit of light. It is localized in a small volume of space and remains localized as it moves away from the light source. Each photon carries an amount of energy proportional to the frequency of the light wave, as given by

$$\boxed{E = hv} \qquad ...(1.7)$$

where h is Planck's constant. It is obvious that the higher the frequency of a photon, the more energy it possesses. Thus UV light photons are more energetic than visible light photons.

In the quantum theory of radiation Max Planck assumed that the atoms in the walls of a black body behave like simple harmonic oscillators and each has a characteristic frequency of oscillation. In this theory, he made the following two radical assumptions:

(1) A simple harmonic oscillator can not have any arbitrary values of energy but only those vales of the total energy E that are given by the relation:

$$E = nh\nu \qquad ...(1.9a)$$

Where $n = 0, 1, 2, ...$, n is called the *quantum number*, ν is the frequency of oscillation and h is a universal constant called Planck's constant ($h = 6.626 \times 10^{-34}$ Js). In this relation $h\nu$ is the basic unit of energy and is called a *quantum of energy*. Thus, the relation shows that the total energy of an oscillator is *quantized.*

(2) As long as the oscillator has energy equal to one of the allowed values given by the relation $E = nh\nu$, it can not emit or absorb energy. Therefore, the oscillator is said to be in a stationary state or a quantum state of energy. *The emission or absorption of energy occurs only when the oscillator jumps from one energy state to another.* If the oscillator jumps down from a higher energy state of quantum number n_2 to a lower energy state of quantum number n_1, the energy *emitted* is given by

$$E_2 - E_1 = (n_2 - n_1)h\nu \qquad ...(1.9b)$$

if $n_2 - n_1$ = one unit, then $E_2 - E_1 = h\nu$

Similarly an oscillator *absorbs* a quantum $h\nu$ of energy when it jumps to its higher energy state.

According to Planck, the quantum theory is applicable only to the process of emission and absorption of radiant energy.

In 1905, Einstein extended Planck's quantum theory by assuming that a monochromatic radiation of frequency ν consists of a stream of photons each of energy $h\nu$ and the photons travel through space with the speed of light.

The interaction of light with matter is better explained using the concept of the photon rather than by the wave concept. When light interacts with matter, the energy exchange can take place only at certain discrete values for which the photon is the minimum energy unit that light can give or accept.

Photon energy is usually expressed in terms of electron-volts (eV). The photon energy (in eV) corresponding to light of wavelength λ is given by

$$E = \frac{12,400}{\lambda} \qquad ...(1.9)$$

where λ is in Angstrom unit.

The wave picture of light is classical and the photon picture is quantum mechanical.

The laser is inherently a quantum mechanical device because its operation depends on the existence of photons.

Example 1.1. *The He-Ni system is capable of lasing at several different IR wavelengths, the prominent one being 3.3913 μm. Determine the energy difference (in eV) between the upper and lower levels for this wavelength.*

Solution :

$$E = \frac{12400}{\lambda\,(\text{Å})}\text{ eV} = \frac{12400}{33913\text{ Å}}\text{ eV}$$

$$= \mathbf{0.37eV}$$

Example 1.2. *The CO_2 laser is one of the most powerful lasers. The energy difference between the two laser levels is 0.117 eV. Determine the frequency and wavelength of the radiation.*

Solution :

$$\lambda\,(\text{Å}) = \frac{12400}{E(\text{eV})} = \frac{12400}{0.117}$$

or
$$\lambda = 105983\ \text{Å}$$
$$= \mathbf{10.5\ \mu m}$$

$$\therefore \quad \nu = \frac{c}{\lambda} = \frac{3 \times 10^8\ m/s}{10.5 \times 10^{-6}\ m} = \mathbf{2.9 \times 10^{13}\ Hz}$$

Example 1.3. *A pulsed laser is constructed with a ruby crystal as the active element. The ruby rod contains typically a total of 3 × 10^{19} Cr^{3+} ions. if the laser emits light at 6943 Å wavelength, find (a) the energy of one emitted photon (in eV) and (b) the total energy available per laser pulse (assuming total population inversion).*

Solution : (*a*) The photon energy in eV is given by

$$E = \frac{12400}{\lambda\ (\text{Å})}\ \text{eV}$$

$$\therefore \quad E = \frac{12400}{6943\ \text{Å}} = \mathbf{1.79 eV}$$

(*b*) Energy per pulse = (Energy of one photon) × (Total number of photons)

= (Energy of one photon) × (Total number of atoms in the excited state)

$$E_T = (1.79\text{eV} \times 1.602 \times 10^{-19}\ \text{J/eV}) \times (3 \times 10^{19})$$
$$= \mathbf{8.6\ J}$$

1.6 PLANCK'S RADIATION LAW

On the basis of quantum theory, Planck obtained the formula for an average energy of ar oscillator is

$$E = \frac{h\nu}{e^{h\nu/kT} - 1} \qquad ...(1.10)$$

It can be shown that the number of oscillations or degrees of freedom per unit volume in the frequency range ν and ν + *d*ν and in given by

$$N(\nu)\ d\nu = \frac{8\ \pi\nu^2}{C^3}\ d\nu \qquad ...(1.11)$$

where *c* is the speed of light in vacuum.

Then assuming that the average value of energies of the various modes of oscillations in black body radiation is given by Eq. 1.10, Planck obtained the relation.

$$p_\nu\ d\nu = \frac{8\pi h\nu^3}{c^3} \cdot \frac{1}{e^{h\nu/kT} - 1} d\nu \qquad ...(1.12)$$

Where p_ν *d*ν is the energy per unit volume in the frequency range ν and ν + *d*ν and p_ν is the energy per unit volume per unit frequency range at frequency ν.

In terms of wavelength of radiation, the equation is expressed as

$$p_\lambda\ d\lambda = \frac{8\pi hc}{\lambda^5} \cdot \frac{1}{e^{hc/\lambda kT} - 1} d\lambda \qquad ...(1.13)$$

Eq. (1.12) and Eq. (1.13) are two forms of Planck's radiation law.

When the values of p_λ as obtained from Eq. (1.13) for different values of λ are plotted against the corresponding values of λ, we get curves as shown in fig. (1.3).

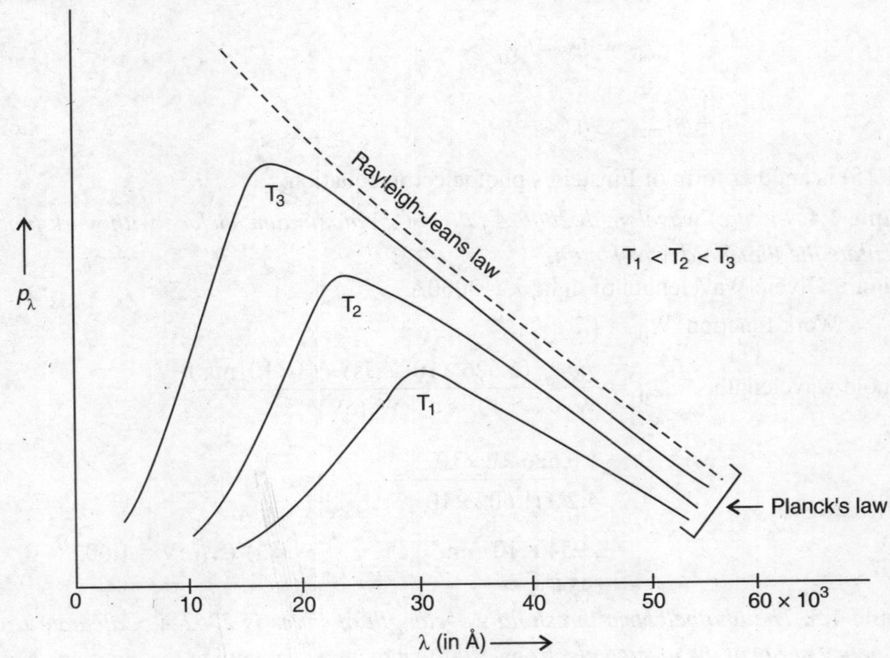

Fig. 1.3

These curves obtained for temperatures T_1, T_2 and T_3 ($T_1 < T_2 < T_3$) agree very well with the experimental results over the whole range of wavelengths.

Further, the Rayleigh-Jeans law, Wien's law and the stefan-Boltzman formula are derived using Planck's law given by Eq. (1.13). Here lies the grand success of quantum theory proposed by Max. Planck.

1.7 PHOTOELECTRIC EFFECT

When an electromagnetic radiation of sufficiently high frequency, as ultraviolet (UV) light and X-rays is incident on a clean metal surface, electrons are emitted out from it. This phenomenon is known as the photoelectric effect and the emitted electron are called as photoelectrons.

The whole range of Electro-magnetic (E-M) spectrum right from Infra-red rays, visible Ultra-violets, *X*-rays, γ-rays produce this effect depending upon the metal. Most metals give this effect when exposed to UV rays or *X*-rays. Sodium is a low *z* material (Here $z = 3$, atomic number) gives this effect when exposed to all radiations of wavelengths smaller than 5455Å. Caesium produces this effect when exposed to all radiations to wavelengths smaller than 6438A°.

Einstein's photoelectric equation. For photoelectric emission of an electron from the metal $h\nu$ must be equal to or greater than W_0 ($h\nu \geq W_0$). Here W_0 is the photoelectric work function of the metal. When an electron at the metal surface absorbs the energy $h\nu$, a certain *mimum* part of this energy is used up by the electron to do work equal to W_0 to overcome the attractive forces of the positive ions of the metal. The remaining *maximum* energy ($h\nu - W_0$) is in the form of maximum kinetic energy $\left(\frac{1}{2} mv_{max}^2\right)$ of the electron emitted from the metal surface.

$$\therefore \qquad \frac{1}{2} mv_{max}^2 = h\nu - W_0 \qquad ...(1.14)$$

This is known as Einstein's photoelectric equation.

Putting $W_0 = h\nu_0$ where ν_0 is called the threshold frequency for a given metal surface. Substituting, we get

$$\frac{1}{2}mv_{max}^2 = h\nu - h\nu_0$$

or $$\frac{1}{2}mv_{max}^2 = h(\nu - \nu_0) \qquad ...(1.15)$$

Eq. (1.15) is another form of Einstein's photoelectric equation.

Example 1.4. *Light of wavelength 2000 Å falls on an aluminium surface with work function 4.2 eV. Calculate the threshold wavelength.*

Solution : Given: Wavelength of light, $\lambda = 2000$Å

Work function, $W_0 = 4.2$ eV

Threshold wavelength, $$\lambda_0 = \frac{hc}{W_0} = \frac{(6.626 \times 10^{-34}\ \text{Js}) \times (3 \times 10^8\ \text{m/s})}{W_0(\text{eV})}$$

$$= \frac{6.626 \times 3 \times 10^{-26}}{4.2 \times (1.602 \times 10^{-19})}$$

$$= 2.954 \times 10^{-7}\ \text{m} \qquad (\because 1\text{eV} = 1.602 \times 10^{-19}\ \text{J})$$

$$= \mathbf{2954\ Å}$$

Example 1.5. *The photoelectric threshold wavelength of silver is 2762 Å. Calculate (a) the maximum kinetic energy of the ejected electrons, (b) the maximum velocity of the electrons ejected, when the silver surface is illuminated with wavelength 2000 Å.*

Solution : Given : $\lambda = 2000 \times 10^{-10}\ \text{m} = 2 \times 10^{-7}$ m.

and $\lambda_0 = 2762 \times 10^{-10}\ \text{m} = 2.762 \times 10^{-7}$ m

(*a*) $$E_{max} = h(\nu - \nu_0)$$

$$= hc\left(\frac{1}{\lambda} - \frac{1}{\lambda_0}\right) = 6.63 \times 10^{-34} \times 3 \times 10^8 \left(\frac{10^7}{2} - \frac{10^7}{2.762}\right)$$

$$= 6.63 \times 3 \times 10^{-19}\left(\frac{1}{2} - \frac{1}{2.762}\right)$$

$$= 6.63 \times 3 \times 10^{-19} \times 0.138$$

$$= \mathbf{2.745 \times 10^{-19}\ J}$$

(*b*) The maximum velocity is given by

$$\frac{1}{2}mv_{max}^2 = E_{max}$$

$$\therefore \quad V_{max} = \left(\frac{2 \times E_{max}}{m}\right)^{\frac{1}{2}}$$

$$= \left(\frac{2 \times 2.745 \times 10^{-19}}{9.11 \times 10^{-31}}\right)^{\frac{1}{2}} = 10^5\left(\frac{2 \times 274.5}{9.11}\right)^{\frac{1}{2}}$$

$$= \mathbf{7.76 \times 10^5\ m/sec.}$$

1.8 COMPTON EFFECT

When a beam of monochromatic* X-rays strikes a targets, the X-rays will be dispersed in all possible directions after interacting with the atoms in the target. This phenomenon is known as

* Monochromatic means single wavelength

scattering and the angle between the directions of incident and scattered rays is called scattered angle. Scattering process differs from reflection, refraction and diffraction in that there is no fixed angular relationship between the incident and scattered rays. According to classical theory, there should be no change in wavelength in scattered rays.

However, American physicist, Arthur H. compton in 1923 showed experimentally that for each angle of incidence, two peaks appear corresponding to scattered X-ray photon with two different wavelengths. The change in wavelength was observed.

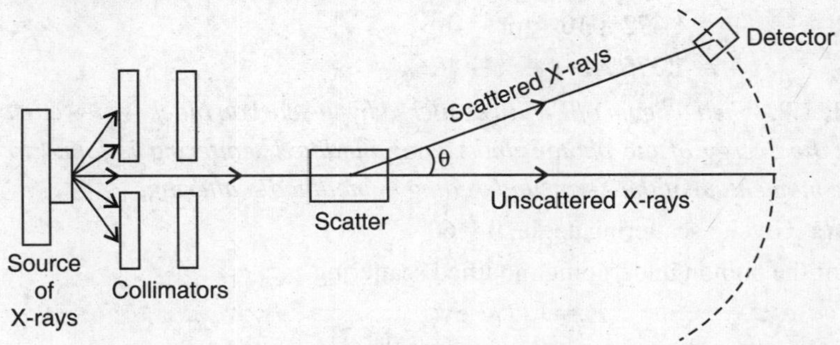

Fig. 1.4. Arrangement for the Study of compton effect.

The wavelength of one peak does not change with the angle of incidence. This is called **primary** or **unmodified component.** We denote it by λ_i. The wavelength of the other peak varies strongly with the angle of incidence and hence it is called **modified component**. It is denoted by λ_i. This effect is known as compton effect. The change in wavelength $\Delta\lambda$ is called the **Compton Shift**. Compton applied Planck-Einstein hypothesis and developed a theory which is consistent with the experimental results. The relation for $\Delta\lambda$, the compton shift is given by

$$\Delta\lambda = \lambda_f - \lambda_i = \frac{h}{m_0 c}(1 - \cos\theta) \quad ...(1.16)$$

Here, m_0 is the rest mass of the stationary electron in the target. The quantity h/m_0c has a value of 0.02426 Å.

$$\frac{h}{m_0 c} = \frac{(6.626 \times 10^{-34}\ \text{Js})}{(9.1 \times 10^{-31}\ \text{kg})\ (3 \times 10^8\ \text{m/s})} = 0.02426\text{Å}$$

Example 1.6. *If the incident radiation is 1.372 Å, find the wavelength of scattered radiation at angle 30°.*

Solution : Given : Wavelength of incident, $\lambda_i = 1.372$ Å.

Angle of scattering, $\theta = 30°$

Wavelength of scattered radiation,

$$\lambda_f = \lambda_i + \frac{h}{m_0 c}(1 - \cos\theta) = 1.372 \times 10^{-10} + \frac{6.626 \times 10^{-34}}{9.1 \times 10^{-31} \times 3 \times 10^8}(1 - \cos 30)$$

$$= 1.372 \times 10^{-10} + 0.02426 \times 10^{-10} \times (1 - 0.866)$$

$$= [1.372 + (0.02426 \times 0.134)] \times 10^{-10}$$

$= 1.375 \times 10^{-10}$ m = **1.375Å.**

Example 1.7. *In Compton scattering, the incident protons have wavelength 3.0×10^{-10} m. Calculate the wavelength of scattered radiation if they are viewed at an angle of 60° to the direction of incidence.*

Solution : Wavelength of scattered radiation, $\lambda_f = \lambda_i + \frac{h}{m_0 c}(1 - \cos\theta)$

$$\therefore \quad \lambda_f = 1.372 \times 10^{-10} + \frac{6.626 \times 10^{-34}}{9.1 \times 10^{-31} \times 3 \times 10^8}(1 - \cos 60°)$$

$$= [1.372 + 0.02426\,(1 - 0.5)] \times 10^{-10} = [1.372 + (0.02426 \times 0.5)] \times 10^{-10}\text{ m}$$

$$= 1.372 \times 10^{-10}\text{ m}$$

$$= \mathbf{1.375\ Å.}$$

Example 1.8. *When a beam of X-rays scattered by the carbon block is viewed at the scattering angle of 60°, the energy of the photon undergoing modified seattering is found to be 12273 eV. Determine the momentum of the X-ray undergoing unmodified scattering.*

Solution : Given : scattering angle, $\theta = 60°$.

Energy of the photon undergoing modified scattering

$$E = 12273\text{ eV}$$

Wavelength of the photon undergoing modified scattering is

$$\lambda_f = \frac{hc}{E_f} = \frac{(6.626 \times 10^{-34}\text{ Js})\,(3 \times 10^8\text{ m/s})}{(12273 \times 1.602 \times 10^{-19})\text{ J}} = \frac{12400}{12273}\text{Å} = 1.01\text{Å}$$

$$\lambda_i = \lambda_f - \frac{h}{m_0 c}(1 - \cos\theta)$$

$$= 1.01 - 0.02426\,(1 - 0.5)\text{ Å}$$

$$= 0.998\text{ Å} = 0.998 \times 10^{-10}\text{ m}$$

Momentum,

$$p_i = \frac{h}{\lambda_i} = \frac{6.626 \times 10^{-34}\text{ Js}}{0.998 \times 10^{-10}\text{ m}}$$

$$= 6.64 \times 10^{-24}\left(\frac{(\text{kg}.m^2/\text{sec}^2)\,.\,\text{sec}}{m}\right) \qquad (\because 1\text{J} = 1\text{kg m}^2/\text{sec}^2)$$

$$= \mathbf{6.64 \times 10^{-24}\ kg.m/sec.}$$

1.9 THE WAVE-PARTICLE DUALITY OF LIGHT

The phenomena of interference, diffraction and polarization can be explained only when light is treated as a continuous wave i.e., wave character. However, the photoelectric effect and the Compton effect, as seen in previous articles, can be explained only by considering the particle nature of light. This suggests that on the one hand light resembles a collection of particles having energy E and momentum p and on the other hand it may be regarded as a continuous electromagnetic wave of frequency ν. In macroscopic world, corpuscular nature and wave nature appear to be mutually exclusive. A particle is precisely localized in space, while a continuous wave cannot be attributed to a particular location in space. However, it is observed that light behaves as a stream of particles in some Phenomena while as a advancing wave in some other phenomena. Thus, light exhibits both the natures. This is known as *wave-particles duality of light*.

The manifestation of wave-particles dualism can be better understood if we consider the entire electromagnetic (E-M) spectrum. At the lower frequency end are radius waves whose wavelengths are too large (few hundred metres). Therefore, the energy available at any point is insignificantly small and the particle nature cannot be observed. On the other hand, the for end of E-M spectrum, consisting of UV rays, X-rays or γ-rays the frequency is too high or wavelength is too short (few nine metres) that the energy is concentrated in a point of very dimension and the wave properties are less noticeable. Thus, at lower frequencies the wave behaviour is prominent and at higher frequencies the particle nature dominates. The visible region represents the transition region where both the aspects can be observed.

1.10 THE DE BROGLIE HYPOTHESIS

In 1924, Louis de-Broglie, a French physicists suggested that the wave-particle dualism need not be a special feature of light alone but the material particles must also exhibit such that dual behaviour. He reasoned out that since nature exhibits a great amount of symmetry, particles as well as waves exhibit properties of both particles and waves. This is known as de-Broglie hypothesis.

According to de-Broglie, a moving particle is associated with a wave which is known as *de-Broglie wave* or *matter wave*.

The relationship between the energy E of a photon and the frequency ν of the associated electromagnetic wave if given by Planck-Einstein equation.

$$E = h\nu \qquad ...(1.17)$$

According to the theory of relativity, the relation between the total relativisible energy E and the relativistic momentum p of a particle of rest mass m_0 is given by

$$E^2 = p^2c^2 + m_0^2c^4$$

Since photons have zero rest mass, the momentum of a photon is obtained by putting $m_0 = 0$ in this equation.

Thus, $$E^2 = p^2c^2 \quad \text{or} \quad E = pc$$

$$\therefore \quad p = \frac{E}{c} \qquad ...(1.18)$$

Substituting Eq. (1.17) and $c = \nu\lambda$ in Eq. (1.18), we get

$$p = \frac{h\nu}{\nu\lambda} = \frac{h}{\lambda}$$

or $$\lambda = \frac{h}{p} \qquad ...(1.19)$$

This equation represents the wave-particle relation for photons. In this expression, λ is the wavelength of the wave associated with photons of momentum p.

In 1924, de-Broglie extended the wave-particle relationship for photons to all particles in motion (*i.e.*, electrons, protons, neutrons, atoms, molecules etc.). Hence a particle of mass 'm' moving with a velocity 'v' and carrying a momentum $p = mv$ must be associated with a wave of wavelength.

$$\lambda = \frac{h}{mv} \qquad ...(1.20)$$

Eq. (1.20) is the de-Broglie equation and wavelength λ is called the *de-Broglie wavelength*.

de-Broglie wavelength in terms of kinetic energy.

Kinetic energy of the particle is

$$E_{\text{k.E.}} = \frac{1}{2}mv^2$$

or $$m \cdot E_{\text{k.E}} = \frac{1}{2}m^2v^2 \quad \text{or} \quad m^2v^2 = 2mE_{\text{k.E.}}$$

or $$p^2 = 2m\,E_{\text{k.E.}}$$

$\therefore$ $$p = \sqrt{2m\,E_{k.E.}} \qquad \text{...(1.21)}$$

Substituting the value of p in Eq. (1.19), we have

$$\lambda = \frac{h}{\sqrt{2mE_{k.E.}}} \qquad \text{...(1.22)}$$

Example 1.9. *Calculate the de-Broglie wavelength of an electron moving with velocity* $\frac{3}{5}c$.

Solution : We have $v = \frac{3}{5}c$ or $\frac{v}{c} = \frac{3}{5}$

Now $$\lambda = \frac{h}{mv}$$

Since v is comparble with the speed of light, m is the relativistic mass given by

$$m = m_0 / \sqrt{1 - v^2/c^2}$$

$\therefore$ $$\lambda = \frac{h}{m_0 v}\sqrt{1 - \frac{v^2}{c^2}}$$

$$= \frac{6.63 \times 10^{-34}}{9.11 \times 10^{-31} \times \frac{3}{5}c}\sqrt{1 - \frac{9}{25}}$$

$$= \frac{6.63 \times 10^{-34} \times 4}{9.11 \times 10^{-31} \times 3 \times 3 \times 10^8} = \frac{6.63 \times 4 \times 10^{-11}}{9.11 \times 3 \times 3}$$

$$= 0.323 \times 10^{-11}\ \text{m}$$

$$= \mathbf{0.0323\ Å}.$$

Example 1.10. *Obtain an expression for the de-Broglie wavelength for a particle of mass m moving with velocity which is comparable wuth the velocity of light.*

Solution : The relativistic kinetic energy K of a particle is related to its total relativistic energy E by the equation:

$$E = K + m_0c^2$$

$$\sqrt{p^2c^2 + m_0^2c^4} = K + m_0c^2$$

Squaring both sides, we get

$$p^2c^2 + m_0{}^2c^4 = K^2 + 2Km_0c^2 + m_0{}^2c^4$$

$$p^2c^2 = K^2 + 2Km_0c^2$$

$\therefore$ $$p = \sqrt{\frac{K\,(K + 2m_0c^2)}{c}}$$

Hence, $$\lambda = \frac{h}{p} = \frac{hc}{\sqrt{K\,(K + 2m_0c^2)}}$$

or $$\lambda = \frac{h}{\sqrt{2m_0K\,(1 + K/2m_0c^2)}}$$

Example 1.11. *The most rapidly moving fast electron in metallic sodium, at the absolutr zero of temperture, has a kinetic energy 3eV. Show that its de-Broglie wavelength is 7Å.*

Solution : $K = 3\text{eV} = 3 \times 1.6 \times 10^{-19}$ J

For the electron,

$$m_0c^2 = 9.11 \times 10^{-31} (3 \times 10^8)^2 = 9.11 \times 9 \times 10^{-15} \text{ J}$$

$$= \frac{9.11 \times 9 \times 10^{-15}}{1.6 \times 10^{-19}} = \frac{9.11 \times 9}{1.6} \times 10^4 \text{ eV}$$

$$= 5.12 \times 10^5 \text{ eV}$$

∴ the kinetic energy of the electron is small as compared with m_0c^2. Hence the de-Broglie wavelength is given by

$$\lambda = \frac{h}{\sqrt{2m_0K}} = \frac{6.63 \times 10^{-34}}{\sqrt{2 \times 9.11 \times 10^{-31} \times 3 \times 1.6 \times 10^{-19}}}$$

$$= 7.1 \times 10^{-10} \text{ m} = \mathbf{7.1 \text{ Å}}$$

Example 1.12. *Calculate the energy of the neutron whose de-Broglie wavelength is 10 Å. (Given : mass of neutron = 1.674 × 10^{-27} kg).*

Solution : We know that

$$\lambda = \frac{h}{p} = \frac{h}{\sqrt{2m\,E}}$$

$$\therefore \quad E = \frac{h^2}{2m\lambda^2}$$

Putting $h = 6.623 \times 10^{-34}$; $m = 1.674 \times 10^{-27}$ kg (mass of neutron); $\lambda = 10$ Å $= 10^{-9}$ m

We get,

$$E = \frac{(6.623 \times 10^{-34})^2}{2 \times 1.674 \times 10^{-27} \times (10^{-9})^2} = \frac{43.864 \times 10^{-68}}{3.348 \times 10^{-45}}$$

$$= \mathbf{13.101 \times 10^{-23} \text{ Joules}}$$

or

$$E = \frac{13.101 \times 10^{-23}}{1.6 \times 10^{-19}} \text{ eV}$$

$$= \mathbf{8.188 \times 10^{-4} \text{ eV}}$$

1.11 WAVE VELOCITY AND GROUP VELOCITY

An electron in motion is associated with a wave whose wavelength λ is given by the de-Broglie relation, and can be considered as a 'wave packet' of small extension formed by the superposition of a large number of waves of wavelengths slightly different from the wavelengths of the associated wave. Such a wave packet possesses both wave and particle properties. The regular seperation λ_{ave} between successive maxima in a wave packet is characteristic of a wave and at the same time it has particle-like localization in space.

Fig. 1.5 shows a pulse consisting of a large number of waves-having slightly different wavelengths produces only one maxima and it is called a *wave packet* or *wave group*.

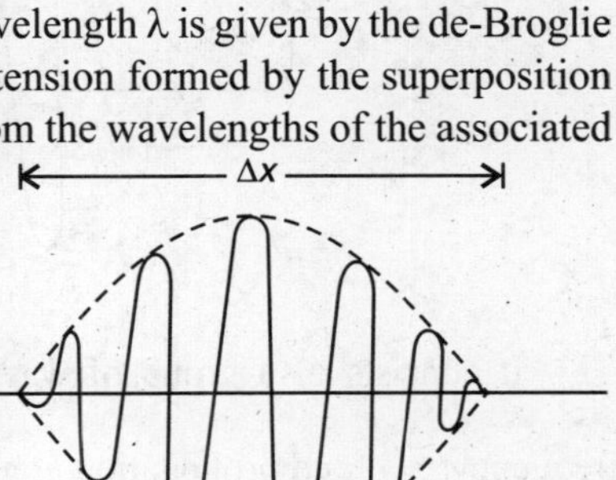

Fig. 1.5. Wave packet

When it advances (travels), the phase velocities of different components are different, the observed velocity is the velocity with which the maximum amplitude of the group advances is called *group velocity (Vg)*.

Simplest way to describe a wave packet is in terms of a superposition of two individual harmonic waves having same amplitude A but slightly different angular frequencies (w_1 and w_2). The expression for waves are:

$$y_1 = A \cos (w_1 t - k_1 x) \quad ...(1.23)$$

and

$$y_2 = A \cos (w_2 t - k_2 x) \quad ...(1.24)$$

Where $\frac{w_1}{k_1}$ and $\frac{w_2}{k_2}$ represent their respective phase (wave) velocities, w being equal to $2\pi/\lambda$ (or $2\pi\nu$) and $k = \frac{2\pi}{\lambda}$, the propogation constant. It is further assumed that $\frac{w_1}{k_1} \neq \frac{w_2}{k_2}$ *i.e.*, it is a *dissipative medium*. The resultant displacement is given by

$$y = y_1 + y_2 = A [\cos (w_1 t - k_1 x) + \cos (w_2 t - k_2 x)]$$

$$= 2A \cos \left[\frac{(w_1 + w_2)}{2} t - \frac{(k_1 + k_2)}{2} x\right] \cos \left[\frac{(w_1 - w_2)}{2} t - \frac{(k_1 - k_2)}{2} x\right]$$

$$= 2A \cos (wt - kx) \cos \left(\frac{\Delta w}{2} t - \frac{\Delta k}{2} x\right) \quad ...(1.25)$$

where

$$w = \frac{w_1 + w_2}{2}; k = \frac{k_1 + k_2}{2}$$

$$\Delta w = w_1 - w_2 \text{ and } \Delta k = k_1 - k_2$$

The resultant wave has two parts:

***(i)* Phase velocity:** A wave of frequency w, propogation constant k and velocity

$$v_p = \frac{w}{k} = \frac{2\pi\nu}{2\pi/\lambda} = \nu\lambda \quad ...(1.26)$$

This is the *phase velocity* or *wave velocity.*

***(ii)* Group velocity:** A second wave of frequency $\frac{\Delta w}{2}$, propogation constant $\frac{\Delta k}{2}$ and velocity

$$v_g = \frac{\Delta w}{\Delta k} \quad ...(1.27)$$

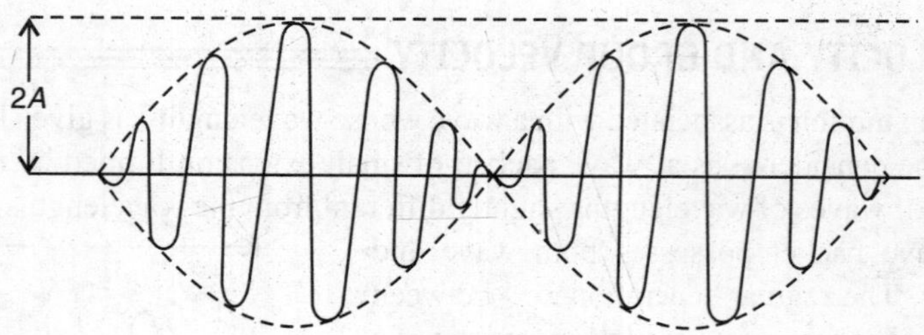

Fig. 1.6

It consists of a group of waves of the first type and is a very slowly moving envelop of frequency $\frac{\Delta w}{2}$ and propogation constant $\frac{\Delta k}{2}$. The modulated pattern moves with a velocity

$$v_g = \frac{\Delta w}{\Delta k}, \text{ known as } \textit{group velocity}.$$

Relation between phase velocity and group velocity. We know that

$$v_p = \frac{w}{k} \qquad \therefore w = kv_p$$

and
$$v_g = \frac{dw}{dk} = \frac{d(kv_p)}{dk} = v_p + \frac{k\, dv_p}{dk} \qquad \text{...(1.28)}$$

But
$$k = \frac{2\pi}{\lambda} \qquad \therefore dk = -\frac{2\pi}{\lambda^2} d\lambda$$

and
$$\frac{k}{dk} = -\frac{\lambda}{d\lambda}$$

Substituting in Eq. (1.28), we get

$$v_g = v_p - \lambda \frac{dv_p}{d\lambda} \qquad \text{...(1.29)}$$

Example 1.13. *A fast moving neutron is found to have an associated de-Broglie wavelength of 2×10^{-12} m. Find the kinetic energy, phase velocity and group velocity of de-Broglie waves. (Given mass of neutron = 1.675×10^{-27} kg)*

Solution : (*i*) Kinetic energy of the fast moving neutron is given by

$$\text{K.E.} = \frac{p^2}{2m}$$

and momentum,
$$p = \frac{h}{\lambda} = \frac{6.62 \times 10^{-34}}{2 \times 10^{-12}} = 3.35 \times 10^{-22} \text{ kg .m. sec}^{-1}.$$

Substituting, we get

$$\text{K.E.} = \frac{(3.35 \times 10^{-22})^2}{2 \times 1.675 \times 10^{-27}}$$

$$= 3.280 \times 10^{-17} \text{ Joules}$$

$$= \frac{3.280 \times 10^{-17}}{1.6 \times 10^{-19}} \text{eV} = \mathbf{2.05 \times 10^2 \text{ eV}}$$

(*ii*) Here, group velocity $v_g = v$, the particle velocity.

$\therefore$ Velocity of moving particle, $v = \frac{h}{m\lambda}$. Now $\lambda = \frac{h}{p} = \frac{h}{mv}$

$$= \frac{6.62 \times 10^{-34}}{1.675 \times 10^{-27} \times 2 \times 10^{-12}}$$

$$= 1.976 \times 10^5 \text{ ms}^{-1}$$

$\therefore$ Group velocity, $v_g = 1.976 \times 10^5 \text{ ms}^{-1}$

(*iii*) Phase velocity, $v_p = \frac{w}{k}$

Using the relation, $E = h\nu = \frac{h}{2\pi} . 2\pi\nu = \hbar w$

and $p = \frac{h}{\lambda} = \frac{h}{2\pi} . \frac{2\pi}{\lambda} = \hbar k$

We have, $$v_p = \frac{w}{k} = \frac{\hbar w}{\hbar k} = \frac{E}{p} \quad ...(i)$$

For a particle of mass m moving with a velocity v

$$E = \frac{1}{2}mv^2 \text{ and } p = mv$$

$$\therefore \quad E = \frac{p^2}{2m}$$

Substituting, $E = \dfrac{p^2}{2m}$ in (i), we get

$$v_p = \frac{p^2}{2mp} = \frac{p}{2m} = \frac{h}{2m\lambda} = \frac{6.62 \times 10^{-34}}{2 \times 1.675 \times 10^{-27} \times 2 \times 10^{-12}}$$

$$= \mathbf{0.988 \times 10^5\ ms^{-1}}$$

Note. From the above calculations, we find that phase velocity v_p is the one half of the group velocity v_g = (*i.e.*, $v_p = v_g/2$) or the particle velocity.

In order that a moving localised particle may be represented as a wave, it is neccessary that the phase (or wave) velocity should be equal to the particle velocity, otherwise the matter wave would not be able to keep pace with the particle whose motion it guides. Thus, a localised particle can not be represented by a single wave. It can be only represented by a group of waves and in such a case group velocity of the wave group is equal to the particle velocity.

Example 1.14. *Find the phase and group velocity of an electron whose de-Broglie wavelength is 1.2Å (neglect relativity effect).*

Solution : Proceeding exactly that of Example 1.13, we have

(*i*) Group velocity, $$v_g = v = \frac{h}{m\lambda} = \frac{6.62 \times 10^{-34}}{9.1 \times 10^{-31} \times 1.2 \times 10^{-10}}$$

$$= 6.06 \times 10^6\ ms^{-1}$$

(*ii*) Phase velocity, $$v_p = \frac{p^2}{2mp} = \frac{h}{2m\lambda} = \frac{6.62 \times 10^{-34}}{2 \times 9.1 \times 10^{-3} \times 1.2 \times 10^{-10}}$$

$$= 3.03 \times 10^6\ ms^{-1}$$

Again, we condude that $v_g = 2v_p$

1.12 HEISENBERG UNCERTAINTY PRINCIPLE

The wave nature of moving particles leads to some inevitable consequences. Classically the position and momentum can be measured to a very high accuracy simultaneously at any instant of its state. However, when a moving particle is conceptualized as a de-Broglie wave packet, such a precision becomes resticted.

The particle is located within the region Δx (refer fig. 1.5), the linear spread of the wave packet. The probability of finding the particle is maximum at the centre of the wave packet and falls to zero at its ends. Therefore, there is an ***uncertainity*** *Δx in the position of the particle.* Further, the wave packet is consitued by number of waves having a range of wavelengths. The spread in wavelengths $\Delta\lambda$ is related to the spread in dimension Δx. As the momentum of the associated particle is related to wavelength through a relation $p = hk = h/\lambda$, there arises an *uncertainty (Δp) in momentum.* The two uncertainties are interrelated by *Heisenberg's uncertainty relation* or *uncertainty prionciple.*

Statement: Heisenberg's uncertainty principle states that *it is impossible to measure both position and momentum of a particle simultaneously to any desired degree of accuracy, and the product of uncertainties is equal to, or greater than Planck's constant h.*

i.e., $$\Delta x \,.\, \Delta p_x \geq h \quad ...(1.30)$$

For motion along X-axis, where Δx is the fundamental error or uncertainty in the measurement of the position and Δp_x the fundamental error or uncertainty in the measurement of momentum (velocity).

Further, $$\Delta x \,.\, \Delta p_x \geq \frac{1}{2}\frac{h}{2\pi}$$

or $$\Delta x \,.\, \Delta p_x \geq \frac{1}{2}\hbar \quad ...(1.31)$$

Example 1.15. *Calculate the uncertainty in measurement of momentum of an electron if the uncertainty in locating it is 1Å.*

Solution : Uncertainty in locating the electron

$$\Delta x = 1\text{Å} = 10^{-10}\,\text{m}$$

∴ Uncertainty in measurement of momentum

$$\Delta p = \frac{\hbar}{2\Delta x}$$

$$= \frac{6.62\times10^{-34}}{2\times2\times3.142\times10^{-10}}$$

$$= \mathbf{5.2 \times 10^{-25}\ kg\ ms^{-1}}$$

Example 1.16. *An electron has a momentum 5.4 × 10^{-26} kgms^{-1} with an accuracy of 0.05%. Find the minimum uncertainty in the location of the electron.*

Solution : Momentum of the electron = 5.4×10^{-26} kg ms^{-1}

∴ Uncertainty in the measurement of momentum

$$\Delta p = \frac{5.4\times10^{-26}\times0.05}{100} = 27\times10^{-30}\,\text{kg ms}^{-1}$$

Hence, uncertainty in the location of electron

$$\Delta x = \frac{\hbar}{2\Delta p} = \frac{6.62\times10^{-34}}{2\times2\times3.142\times27\times10^{-30}}$$

$$= \mathbf{1.95 \times 10^{-6}\ m}$$

Example 1.17. *The position and momentum of 0.5 KeV electron are simultaneously determined. If its position is located within 0.2 mm, what is the percentage uncertainty in its momentum ?*

Solution : Energy of the electron

$$E = 0.5 \times 10^3\ \text{eV} = 0.5 \times 10^3 \times 1.6 \times 10^{-19}\ \text{J}$$
$$= 8 \times 10^{-17}\ \text{J}$$

As it is a non- relativistic case, momentum of the electron

$$p = \sqrt{2mE} = (2\times9\times10^{-31}\times8\times10^{-17})^{\frac{1}{2}}$$
$$= 1.2 \times 10^{-23}\ \text{kg ms}^{-1}$$

Uncertainty in position, $\Delta x = 0.2$ mm = 0.,2 × 10^{-9} m = 2 × 10^{-10} m

Uncertainty in momentum Δp is given by

$$\Delta p \,.\, \Delta x = \frac{\hbar}{2}$$

$$\therefore \qquad \Delta p = \frac{\hbar}{2.\Delta x} = \frac{h}{2 \times 2\pi \times \Delta x} = \frac{7 \times 6.6 \times 10^{-34}}{2 \times 2 \times 22 \times 2 \times 10^{-10}}$$

$$= 2.6 \times 10^{-25}\ \text{kgms}^{-1}$$

$$\therefore \quad \text{percentage uncertainty in momentum} = \frac{\Delta p}{p} \times 100$$

$$= \frac{2.6 \times 10^{-25}}{1.2 \times 10^{-23}} \times 100$$

$$= \mathbf{2.2\%}.$$

Example 1.18. *An electron and a 150 gm baseball are travelling at 220 m/s measured to an accuracy of 0.065%. Calculate and compare uncertainty in position of each of the bodies.*

Solution : Given : (*i*) $V_e = 220$ m/s ; accuracy = 0.065 %

(*ii*) $M = 150$ gm = 0.15 kg

$$V_M = 220\ \text{m/s; accuracy} = 0.065\ \%$$

(*i*) For electron :

$$\Delta V_e = V_e \times 0.065\ \% = 220\ \text{m/s} \times \frac{0.065}{100} = 0.143\ \text{m/s}$$

$$\therefore \qquad \Delta x_e = \frac{\hbar}{m \,.\, \Delta v_e} = \frac{1.05 \times 10^{-34}\ \text{J.s}}{(9.11 \times 10^{-31}\ \text{kg}) \times (0.143\ \text{m/s})}$$

$$= 0.80 \times 10^{-3}\ \text{m}$$

(*ii*) For baseball :

$$\Delta V_m = 0.143\ \text{m/s (same as given in problem)}$$

$$\therefore \qquad \Delta x_m = \frac{\hbar}{M.\Delta V_M} = \frac{1.05 \times 10^{-34}\ \text{J.s}}{(0.15\ \text{kg}) \times (0.143\ \text{m/s})}$$

$$= \mathbf{4.9 \times 10^{-33}\ m}$$

Thus, the uncertainty in the measurement of position of the electron is very large compared to its dimension whereas the uncertainty in position of the baseball is extremely low.

Example 1.19. *The life time of an excited state of nucleus is 10^{-22} s. What is the uncertainty in the energy of γ-rays photon emitted ?*

Solution : According to Heisenberg's uncertainty principle,

$$\Delta E.\Delta t = \frac{\hbar}{2}$$

or

$$\Delta E = \frac{\hbar}{2.\Delta t} = \frac{\frac{h}{2\pi}}{2\,.\,\Delta t}$$

$$= \frac{1.054 \times 10^{-34}\ \text{Js}}{2 \times 10^{-12}\ \text{s}} \qquad \left(\because \frac{h}{2\pi} = 1.054 \times 10^{-34}\ \text{Js}\right)$$

$$= \mathbf{0.527 \times 10^{-22}\ Joules}$$

or

$$= \frac{0.527 \times 10^{-22}}{1.6 \times 10^{-19}} = \mathbf{3.3 \times 10^{-4}\ eV}$$

Example 1.20. *What will be the uncertainty in the velocity of an electron which is contained in a 10 Å box ?*

Solution : According to Heisenberg's uncertainty principle

$$\Delta x \,.\, \Delta p = \frac{\hbar}{2} = \frac{h}{4\pi}$$

Given : $\Delta x = 10 \text{ Å} = 10 \times 10^{-10} \text{ m} = 10^{-9} \text{ m}$

$$\therefore \quad \Delta x \,.\, \Delta (\text{m.v}) = \frac{h}{4\pi}$$

$$\Delta V = \frac{h}{4\pi m \,.\, \Delta x} = \frac{6.63 \times 10^{-34}}{4 \times 3.14 \times 9.1 \times 10^{-31} \times 10^{-9}}$$

$$= \mathbf{5.75 \times 10^4 \ ms^{-1}}$$

Example 1.21. *Calculate the smallest possible uncertainty in position of an electron moving with velocity 3×10^7 ms^{-1}.*

Solution : Let Δx_{min} be the minimum uncertainty in position of the electron and Δp_{max}, the maximum uncertainty of electron momentum, then

$$\Delta x_{min} \,.\, \Delta p_{max} \geq \frac{\hbar}{2}$$

But $$p = mv = \frac{m_0 v}{\sqrt{1 - \frac{v^2}{c^2}}}$$

$$\Delta x_{min} \geq \frac{\hbar \sqrt{1 - \frac{v^2}{c^2}}}{2 \, m_0 v}$$

$$\geq \frac{h \sqrt{1 - \frac{v^2}{c^2}}}{4 \, \pi \, m_0 v}$$

$$\geq \frac{6.62 \times 10^{-34} \sqrt{1 - \frac{(3 \times 10^7)^2}{(3 \times 10^8)^2}}}{4 \times 3.14 \times 9.1 \times 10^{-31} \times 3 \times 10^7}$$

$$\geq 1.9 \times 10^{-12} \text{ m}$$

Thus, $$\Delta x_{min} \cong 0.019 \times 10^{-10} \text{ m} = \mathbf{0.019 Å}$$

1.13 WAVE FUNCTION (Ψ)

Waves, as we know, represent the propogation of a disturbance in a medium. Light waves, sound waves, water waves, waves on a string are examples of waves. All of them are characterized by same quantity that varies with position and time. Light waves are represented by electromagnetic field variations while sound waves are represented by pressure variations in terms of compressions and rarefactions. In case of de-Broglie wave, we can not specify in a similar manner. Electrons sometimes behave like classical particles and sometime like classical waves, depending upon the experimental arrangement. Since electron have wave properties, it may be assumed that a quantity

ψ (read as psi) represents a de-Broglie wave just like electric vector $\vec{E}$ in light waves. The quantity ψ is called the *wave function.*

Mathematically, wave function ψ describes the motion of an electron. Max Born (1882 – 1970), a German physicist, showed in 1926 that the square of the absolute value of wave function, $|\psi|^2$ is proportional to the probability of finding a particle in unit volume in space, centred at the point where ψ is evaluated, at the coordinate (x, y, z, t). Thus, the probability of finding a particle between x and $x + dx$, y and $y + dy$, z and $z + dz$ is

$$P \propto \psi \psi^* \, dx \, dy \, dz$$

where ψ^* is complex conjugate of ψ. The complex conjugate of a function is obtained by reversing the sign of the imaginary is, that is in the function, the quantity $\psi\psi^*$ is real and also written as $|\psi|^2$. Therefore, probability can be written as

$$P \propto |\psi|^2 \, dx \, dy \, dz \quad \text{or} \quad P \propto |\psi|^2 \, dV \qquad ...(1.32)$$

Conditions for well-behaved function.

Thus, the wave function ψ gives the information about the particle in terms of (x, y, z, t) within the wave packet. The conditions for a well behaved functions are:

(*i*) It should be normalized wave function.

(*ii*) It should be single valued function of space and time

$$\psi = f(x, y, z, t)$$

(*iii*) It must be finite.

(*iv*) ψ and its derivatives $\frac{\partial \psi}{\partial x}, \frac{\partial \psi}{\partial y}, \frac{\partial \psi}{\partial z}$ must be continuous everywhere in the region where ψ is defined.

Normalization condition

The probability of finding a particle if at all it exists, somewhere in the universe must be unity. Since the probability of its being located in an elemental volume (dV) is proportional to $|\psi|^2 dx.dy.dz$, it is convenient to choose the constant of proportionality such that the sum of probabilities over all values of x, y, z must be unity. In other words,

$$\iiint_{-\infty}^{+\infty} \psi \, \psi^* \, dx \, dy \, dz = 1$$

or

$$\iiint_{-\infty}^{+\infty} |\psi|^2 \, dx \, dy \, dz = 1 \qquad ...(1.33)$$

This condition is known as the *normalization condition.* This wave function that satisfies the above condition is said to be *normalized.* Whenever, wave functions are normalized, $|\psi|^2 \, dV$ equals the probability that the particle will be located in an elemental volume dV. Thus,

$$\text{Probability, } P = |\psi(x, y, z)|^2 \, dV \qquad ...(1.34)$$

The value of probability P lies between 0 and 1. If it has a value of 0.3 or 0.74, it means that there is a 30% or 74% of chance of finding the particle within the selected volume. Thus, it is not possible to specify the exact position of the particle with complete certainty. It is possible to specify only the probability of locality it. Max Born was awarded Nobel Prize in physics in 1954; for this inovative idea.

Example 1.22. *The wave function of electron moving on x-axis like a wave is given by $\psi(x) = 2 \sin 2\pi x$. Calculate the probability of forming out an electron from 25 cm to 50 cm length as the x-axis.*

Solution : The probability of locating an electron is given by

$$P = \int_{x_1}^{x_2} |\psi|^2 \, dx$$

$$= \int_{0.25}^{0.50} (2 \sin 2\pi x)^2 . dx = 4 \int_{0.25}^{0.50} \sin^2 2\pi x \, dx$$

$$= 4.\frac{1}{2}\left[x - \frac{\sin 4\pi x}{4\pi}\right]_{0.25}^{0.50} \qquad \left(\because \text{Using } \sin^2\theta = \frac{1-\cos 2\theta}{2}\right)$$

$$= 2\left[0.50 - \frac{\sin 2\pi}{4\pi} - 0.25 + \frac{\sin \pi}{4\pi}\right]$$

$$= 2\,[0.25 - 0 - 0] = 2 \times 0.25$$

$$= \mathbf{0.5}$$

or 50% probability of electron to be located there.

Example 1.23. *Using uncertainty principle, prove non-existence of electrons and existence of protons and neutrons in nucleus.*

Solution : The size of the nucleus is about 10^{-14} m. Let us suppose that electron is confined inside the nucleus. Then uncertainty in the position $\Delta x = 10^{-14}$ m.

Using Heisenberg's uncertainty principle, we write

$$\Delta p.\,\Delta x \geq \frac{h}{4\pi}$$

or

$$\Delta p \geq \frac{h}{4\pi\,.\,\Delta x}$$

$$\geq \frac{6.63 \times 10^{-34}\ \text{J.s}}{4 \times 3.14 \times 10^{-14}\ \text{m}}$$

$$\geq 0.527 \times 10^{-20}\ \text{Ns} \qquad ...(i)$$

This means the magnitude of the momentum of the particles in the nucleus must be of this order *i.e.*

$$|\vec{p}| = 0.527 \times 10^{-20}\ \text{Ns.}$$

Since the mass of electron = 9.1×10^{-31} kg. The magnitude of momentum is relativistic. The relativistic formula for the energy E of the electron is

$$E^2 = p^2c^2 + m_0^2c^4 \qquad ...(ii)$$

The energy associated with the electron when it is at rest *i.e.* m_0c^2 is of the order of 0.511 MeV., which is much smaller than the first term (p^2c^2) and hence it can be neglected.

∴ Eq. (*iii*) now becomes,

$$E^2 = p^2c^2 \qquad \text{or } E = pc$$

∴

$$E = (0.527 \times 10^{-20}) \times (3 \times 10^8)\ \text{J}$$

$$= \frac{0.527 \times 3 \times 10^{-12}}{1.6 \times 10^{-19}} \cong 10\ \text{MeV}$$

Thus, if all electrons are present in the nucleus, than on being emitted, then it should have kinetic energy of the order of 10 MeV. However, the particle emitted from the nucleus have only 2 to 4 MeV energy. For example: for protons and neutrons, $M_0 = 1.67 \times 10^{-27}$ kg. This is non-relativistic case since its velocity is smaller than velocity of light as calculated under :

$\therefore$ Velocity $$|\vec{v}| = \frac{|\vec{p}|}{m_0} \cong 3 \times 10^6 \text{ ms}^{-1}$$

In this case, K.E., $$E = \frac{p^2}{2m_0} = \frac{(0.527 \times 10^{-20})^2}{2 \times 1.67 \times 10^{-27}} \text{ Joule}$$

$$\cong \frac{(0.527 \times 10^{-20})^2}{2 \times 1.67 \times 10^{-27} \times 1.6 \times 10^{-19}} \text{eV}$$

$$\cong 52 \text{ KeV.} \quad ...(iii)$$

Since, this energy is smaller than the energy emitted by electron, both the particle protons and neutrons can exist insides the nucleus.

1.14 SCHRODINGER WAVE EQUATION

Erwin Schrodinger (1887–1961), an Austrian theoretical physicist, formulates the wave function,which plays an important role in quantum mechanics. This equation can not be derived from any fundamental laws. It plays the same role in quantum mechanics as Newton's second law of motion in classical mechanics.

Let us consider a micro-particle. We associate a wave function φ with the motion of this micro-particle. φ function represents the wave field of the particle.

Wave function for a particle in motion

Let us consider a particle of mass m moving with a velocity v in positive x-direction. Let p_x be its accurately known momentum and E its total energy. Then position of the particle would be completely uncertain. The wave associated with such a particle should be a plane continuous simple harmonic wave travelling in the positive. x-direction and is given by

$$\psi(x, t) = Ae^{-iw\left(t - \frac{x}{v}\right)} \quad ...(1.35)$$

where $i = \sqrt{-1}$, A is the amplitude and w is the angular frequency of the wave. Eq. (i) can be expressed in terms of p_x and E.

The wavelengths λ of the associated wave is given by de-Broglie hypothesis:

$$\lambda = \frac{h}{p_x}$$

$$\therefore \quad p_x = \frac{h}{\lambda} = \frac{h}{2\pi} \cdot \frac{2\pi}{\lambda} = \hbar \cdot \frac{2\pi}{\upsilon / \nu}, \text{ where } \nu \text{ is the frequency}$$

$$= \hbar \cdot \frac{2\pi\nu}{v} = \hbar \frac{w}{\upsilon}$$

$$\therefore \quad \frac{w}{\upsilon} = \frac{p_x}{\hbar} \quad ...(i)$$

The frequency ν of the associated wave is given by Planck's hypothesis:

$$E = h\nu = \frac{h}{2\pi} \cdot 2\pi\nu = \hbar w \quad ...(ii)$$

$$\therefore \quad w = \frac{E}{\hbar}$$

Now from Eq. (1.35),

$$\psi(x, t) = Ae^{-i\left(wt - \frac{wx}{\upsilon}\right)}$$

Now substituting for w and w/v in this equation, we get

$$\psi(x, t) = Ae^{-i\left(\frac{Et}{\hbar} - \frac{xp_x}{\hbar}\right)}$$

$$= Ae^{+\frac{i}{\hbar}(p_x.x - Et)} \qquad ...(1.36)$$

This is the expression for the wave function $\psi(x, t)$ of a particle in motion in one dimension (along +ve x-direction) with momentum p_x and total energy E.

Time-dependent Schrodinger's wave equation.

If the speed of the particle is small as compared to that of light, the total energy E is the sum of its kinetic energy $p^2{}_x/2m$ and its potential energy V.

Then,
$$E = \frac{p_x^2}{2m} + V \qquad ...(1.37)$$

If the motion of the particle is due to the action of force $F(x)$ which is a function of x, then V will be a function only of x. Multiplying both sides of Eq. (1.37) by the wave function $\psi(x, t)$, we get

$$\frac{p_x^2\psi}{2m} + V\psi = E\psi \qquad ...(1.38)$$

We know that

$$\psi = Ae^{\frac{i}{\hbar}(p_x x - Et)} \quad (\because \text{ Eq. } 1.36) \qquad ...(1.39)$$

Differentiating Eq. (1.39) w.r.t. to x,

$$\frac{\partial\psi}{\partial x} = A\left(\frac{i}{\hbar}\right)p_x\, e^{\frac{i}{\hbar}(p_x.x - Et)}$$

$$= \frac{i}{\hbar}p_x\, A\, e^{\frac{i}{\hbar}(p_x.x - Et)}$$

$$= \frac{i}{\hbar}p_x\,\psi \qquad ...(1.40)$$

Again differentiating, we get

$$\frac{\partial^2\psi}{\partial x^2} = \left(\frac{i}{\hbar}\right)p_x\frac{\partial\psi}{\partial x} = \left(\frac{i}{\hbar}\right)p_x\left(\frac{i}{\hbar}\right)p_x\psi$$

$$= -\frac{p_x^2}{\hbar^2}\psi$$

$$\therefore \quad p_x{}^2\psi = -\hbar^2\frac{\partial^2\psi}{\partial x^2} \qquad ...(1.41)$$

Differentiating Eq. (1.39) w.r.t to t,

$$\frac{\partial\psi}{\partial t} = A\left(\frac{i}{\hbar}\right)(-E)\, e^{\frac{i}{\hbar}(p_x\, x - Et)}$$

$$= -\left(\frac{i}{\hbar}\right) E\, A\, e^{\frac{i}{\hbar}(p_x \cdot x - Et)}$$

$$= -\frac{i}{\hbar} E\psi$$

$$\therefore \qquad E\psi = -\frac{\hbar}{i} \frac{\partial \psi}{\partial t} \qquad ...(1.42)$$

Substituting these expressions for $p_x^2\psi$ and $E\psi$ in Eq. (1.38), we get

$$-\frac{\hbar^2}{2m}\frac{\partial^2 \psi}{\partial x^2} + V(\psi) = -\frac{\hbar}{i}\frac{\partial \psi}{\partial t}$$

or

$$-\frac{\hbar^2}{2m}\frac{\partial^2 \psi}{\partial x^2} + V(\psi) = i\hbar\frac{\partial \psi}{\partial t} \qquad ...(1.43)$$

This expression is known as schrodinger's time-dependent wave equation in one dimension. In three dimensions, the time-depended schrodinger's wave equation is

$$-\frac{\hbar^2}{2m}\left(\frac{\partial^2 \psi}{\partial x^2} + \frac{\partial^2 \psi}{\partial y^2} + \frac{\partial^2 \psi}{\partial z^2}\right) + V\psi = i\hbar\frac{\partial \psi}{\partial t}$$

This equation is written as,

$$-\frac{\hbar^2}{2m}\nabla^2\psi + V\psi = i\hbar\frac{\partial \psi}{\partial t} \qquad ...(1.44)$$

$$\left[-\frac{\hbar^2}{2m}\nabla^2 + V\right]\psi = i\hbar\frac{\partial \psi}{\partial t} \qquad ...(1.45)$$

where ∇^2 is the Laplacian operator defined by

$$\nabla^2 = \frac{\partial^2}{\partial x^2} + \frac{\partial^2}{\partial y^2} + \frac{\partial^2}{\partial z^2}$$

Time-independent schrodinger's wave equation.

The three dimensional time-dependent schrodinger's wave equation is

$$-\frac{\hbar^2}{2m}\nabla^2\psi + V\psi = i\hbar\frac{\partial \psi}{\partial t} \qquad ...(i)$$

and explicitly in terms of $\vec{r}$ and t, we have

$$-\frac{\hbar^2}{2m}\nabla^2\psi(\vec{r}, t) + V\,\psi(\vec{r}, t) = i\hbar\frac{\partial \psi(\vec{r}, t)}{\partial t} \qquad ...(ii)$$

The one dimensional Schrodinger's wave equation is

$$-\frac{\hbar^2}{2m}\frac{\partial^2 \psi}{\partial x^2} + V\psi = i\hbar\frac{\partial \psi}{\partial t} \qquad ...(iii)$$

and explicitly in terms of x and t, we have

$$-\frac{\hbar^2}{2m}\frac{\partial^2 \psi(x, t)}{\partial x^2} + V\psi(x, t) = i\hbar\frac{\partial \psi(x, t)}{\partial t} \qquad ...(iv)$$

When the potential energy V does not depend explicitly on time and is a function of x only and the total energy E is constant, the wave function $\psi(x, t)$ can be written as the product of two separate functions $u(x)$ as a function only of x and $f(t)$ a function only of t.

$$\therefore \qquad \psi(x, t) = u(x) \cdot f(t) \qquad ...(v)$$

The above relation is briefly written as $\psi = uf$

Hence, $$\frac{\partial \psi}{\partial x} = f\frac{\partial u}{\partial x} \text{ and } \frac{\partial^2 \psi}{\partial x^2} = f\frac{\partial^2 u}{\partial x^2}$$

and $$\frac{\partial \psi}{\partial t} = u\frac{\partial f}{\partial t} \text{ and } \frac{\partial^2 \psi}{\partial t^2} = u\frac{\partial^2 f}{\partial t^2}$$

Now $$\psi = A\, e^{\frac{i}{\hbar}(x \cdot p_x - Et)}$$

$$\therefore \qquad \frac{\partial \psi}{\partial t} = -\frac{i}{\hbar} E\, A\, e^{\frac{i}{\hbar}(xp_x - Et)} = -\frac{i}{\hbar} E\psi = \frac{-i}{\hbar} Euf \qquad ...(vi)$$

Substituting in Eq. (*iii*), we have

$$-\frac{\hbar^2}{2m} \cdot f\frac{\partial^2 u}{\partial x^2} + Vfu = i\hbar\left(-\frac{i}{h}\right) Euf \qquad ...(viii)$$

Dividing both sides by *f*, we have

$$-\frac{\hbar^2}{2m}\frac{\partial^2 u}{\partial x^2} + Vu = Eu$$

or, $$\frac{\partial^2 u}{\partial x^2} + \frac{2m}{\hbar^2}(E - V)\, u = 0 \qquad ...(1.39)$$

This is one dimensional time-independent (*i.e.*, the steady state) Schrodinger's wave equation. In three dimensional, the steady state Schrodinger's wave equation is

$$\nabla u^2 + \frac{2m}{\hbar^2}(E - V)\, u = 0 \qquad ...(1.40)$$

where $u(x, y, z)$ is the position-dependent wave function.

Example 1.24. *Normalize the one dimensional wave function given by*

$$\psi(x) = A \sin\left(\frac{\pi x}{a}\right) \qquad 0 < x < a$$

$$\psi(x) = 0 \qquad \textit{outside}$$

Solution : The wave function $\psi(x)$ is said to be normalized if it satisfies the condition

$$\int_{-\infty}^{+\infty} \psi\, \psi^*\, dx = 1$$

Applying, we get

$$\int_0^a \psi\psi^*\, dx = 1$$

or $$\int_0^a A \sin\left(\frac{\pi x}{a}\right) A^* \sin\left(\frac{\pi x}{a}\right) dx = 1$$

or $$AA^* \int_0^a \sin^2\left(\frac{\pi x}{a}\right) dx = 1$$

Now $$\cos 2\theta = 1 - 2\sin^2\theta \qquad \therefore \qquad \sin^2\theta = \frac{1 - \cos 2\theta}{2}$$

$$\therefore \quad AA^* \int_0^a \frac{1-\cos\left(2\frac{\pi x}{a}\right)}{2} dx = 1$$

$$AA^* \left[\int_0^a dx - \int_0^a \cos\left(\frac{2\pi x}{a}\right) dx \right] = 2$$

$$AA^* \left[x - \frac{a}{2\pi}\left(\frac{\sin 2\pi x}{a}\right) \right]_0^a = 2$$

$$AA^*[a - \theta + 0] = 2 \quad \text{or} \quad AA^* = \frac{2}{a}$$

But $AA^* = |A|^2$ Hence $|A|^2 = \frac{2}{a}$,

Assuming A to be real, we get $A = \sqrt{2/a}$

Therefore, normalized wave function is

$$\psi(x) = \sqrt{\frac{2}{a}} \sin\left(\frac{\pi x}{a}\right)$$

1.15 ENERGY LEVELS

In the previous section, we learnt that light energy cannot take arbitrary values but must be multiples of photon energy $h\nu$. It was similarly established that the electrons in an atom cannot have arbitrary amounts of energy but they take only discrete energies. This is a consequence of the Bohr's postulate of permitted orbits. In each of the permitted (stationary) orbits, the electron has a specific amount of energy and cannot possess any other energy value. The energy of electron at a very large distance from the nucleus is purely kinetic energy and it is assigned a positive value. As the distance between nucleus and electron decreases, the kinetic energy decreases. When the electron is just about to come under the influence of nucleus, its energy becomes zero. When the electron becomes bound to the nucleus and orbits round it, electron energy is potential in nature and treated as a negative quantity. Among the infinite number of mechanically possible orbits, only a certain set of orbits are actually permissible for electron due to quantum restrictions. These permitted orbits are characterized by allowed energy values. We therefore say that electron in an atom can occupy only certain energy levels. The energy levels are schematically represented by horizontal lines drawn to an energy scale. For example, electron orbits and the corresponding energy levels of the hydrogen atom are shown in Fig.1.7. The atom is in its lowest possible energy level when the electron is in

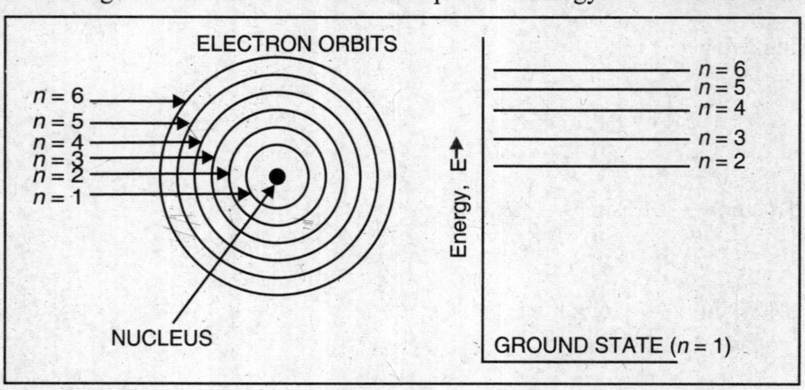

Fig. 1.7. Electron orbits and the corresponding energy levels of the hydrogen atom

the innermost orbit, closest to the nucleus. Then, the atom is said to be in the *ground state.* An electron in this ground state is stable and moves in this orbit continuously without emitting energy. If electron in ground state absorbs energy in some way, it will go to a higher energy level and the atom is said to be *excited.* The passing of an electron from one energy level to another level within the atom occurs in a jump which is called a *quantum transition.* The electron transitions may be induced by a variety of ways. Interaction with light photons is one of the means of supplying energy to orbital electrons which causes upward electron transitions and sends the atom into its excited state.

1.16 POPULATION

The atoms of each chemical element have their own characteristic system of energy levels. The energy difference between the successive energy levels of an atom is of the order of 1 eV to 5 eV. The energy levels are common to all the atoms in a system which is composed of identical atoms. We can therefore say that a certain number of atoms occupy a certain energy state. The number of atoms per unit volume that occupy a given energy state is called the *population* of that energy state. The population N of an energy level E depends on the temperature T. Thus,

$$\boxed{N = e^{-E/kT}} \quad \text{Boltzmann's equation} \qquad ...(1.41)$$

where k is known as the Boltzmann constant.

In a material, atoms are distributed differently in different energy states. The atoms normally tend to be at their lowest possible energy level which need not be the ground state. At temperatures above 0 K, the atoms always have some thermal energy and therefore, they are distributed among the available energy levels according to their energy.

1.17 THERMAL EQUILIBRIUM

At thermal equilibrium, the number of atoms (population) at each energy level decreases with the increase of energy level, as shown in Fig. 1.8.

If we, in particular, consider two energy levels E_1 and E_2 as in Fig. 1.9, their population can be computed with the help of Boltzmann equation (1.41). Thus,

and
$$N_1 = e^{-E_1/kT} \qquad ...(1.42)$$

$$N_2 = e^{-E_2/kT} \qquad ...(1.43)$$

The ratio of the population in these two states, N_2/N_1, is called the *relative population*. It is given by

$$\frac{N_2}{N_1} = \frac{e^{-E_2/kT}}{e^{-E_1/kT}}$$

or
$$\boxed{\frac{N_2}{N_1} = e^{-(E_2-E_1)/kT}} \qquad ...(1.44)$$

It follows from the above that

$$\boxed{N_2 = N_1 e^{-\Delta E/kT}} \qquad ...(1.45)$$

where
$$\Delta E = (E_2 - E_1)$$

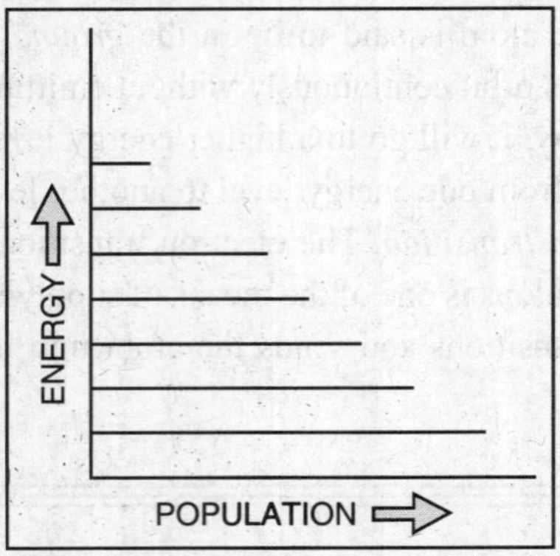

Fig. 1.8. Relative populations of energy levels as a function of energy above the ground state at thermal equilibrium.

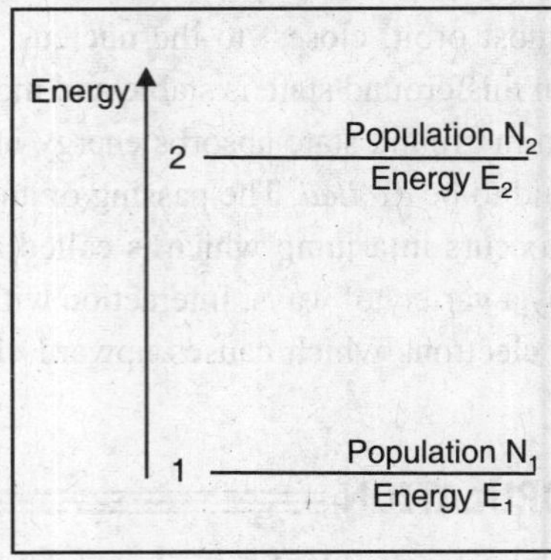

Fig. 1.9. Two energy level system.

The plot of the equation (1.45) is known as *Boltzmann distribution.*

Eq. (1.44) suggests that the relative population N_2/N_1 is dependent on two factors: (*i*) the energy difference $(E_2 - E_1)$ and (*ii*) the temperature T. For the sake of estimation, let us assume hydrogen gas to be a monoatomic gas and find out the atomic population at room temperature at the first excited level. Here $E_1 = -13.6$ eV, $E_2 = -3.39$ eV and $T = 300$ K. Thus,

$$+\Delta E = +(E_2 - E_1) = +[(-3.39) - (-13.6)] \text{ eV} = +10.21 \text{ eV}$$

$$kT = (8.6 \times 10^{-5} \text{ eV/K})(300 \text{ K}) = 0.025 \text{ eV}$$

$$\frac{N_2}{N_1} = e^{\frac{-10.21eV}{0.025eV}} = e^{-408.4} \approx 0$$

It means that at room temperature all atoms are in the ground state. If temperature is raised, say to 6000K

$$\frac{N_2}{N_1} = e^{\frac{-10.21}{0.516}} = e^{-19.79} = 2.5 \times 10^{-9}$$

We thus find that in a material at thermal equilibrium, more atoms are in the lower energy state than in the higher energy state. We call such a distribution of atoms as *normal distribution.* Thermal collisions, on one hand, make some atoms jump into higher energy state but, on the other hand, bring some atoms down to lower energy state. On the whole, the Boltzmann distribution will be maintained.

There is an important implication of Eq. (1.44) that we should understand. The fraction of excited atoms would be large, if the energy levels are close or the temperature is very high. In the limiting case $E_2 - E_1 = 0$, we find that

$$\frac{N_2}{N_1} = e^0 = 1$$

$$\therefore \quad N_2 = N_1$$

Similarly, when $T \to \infty$,

$$\frac{N_2}{N_1} = e^0 = 1$$

$$\therefore \quad N_2 = N_1$$

Therefore, *as long as the material is in thermal equilibrium, the population of the higher state cannot exceed the population of the lower state.* This is a very important conclusion to be borne in mind.

Example 1.25. *Find the relative population of the two states in a ruby laser that produces a light beam of wavelength 6943 Å at 300 K and 500 K.*

Solution : The population ratio is given by

$$\frac{N_2}{N_1} = e^{-(E_2 - E_1)/kT}$$

$$E_2 - E_1 = h\nu = \frac{hc}{\lambda} = \frac{12400}{\lambda} \text{ eV}$$

$$\therefore \quad E_2 - E_1 = \frac{12400}{\lambda} \text{ eV} = 1.79 \text{ eV}$$

$$(a) \quad \therefore \quad \frac{N_2}{N_1} = \exp\left[\frac{-1.79 \text{ eV}}{(8.61\times10^{-5} \text{ eV/K})\times 300 \text{ K}}\right] = e^{-69.3}$$

$$= \mathbf{8 \times 10^{-31}}$$

$$(b) \quad \left[\frac{N_2}{N_1}\right]_{500\text{K}} = \exp\left[\frac{-1.79 \text{ eV}}{(8.61\times10^{-5} \text{ eV/K})\times 300 \text{ K}}\right] = e^{-69.3}$$

$$= e^{-41.58}$$

$$= \mathbf{8.7 \times 10^{-19}}$$

Example 1.26. *Find the ratio of population of the two states in a He-Ne laser that produces light of wavelength 6328 Å at 27° C.*

Solution : $$\frac{N_2}{N_1} = e^{-(E_2 - E_1)/kT}$$

$$\therefore \quad E_2 - E_1 = \frac{12400}{6328} \text{ eV} = 1.96 \text{ eV}$$

$$\therefore \quad \frac{N_2}{N_1} = \exp\left[\frac{-1.96 \text{ eV}}{(8.61\times10^{-5} \text{ eV/K})\times 300 \text{ K}}\right] = e^{-75.88}$$

$$= \mathbf{1.1 \times 10^{-33}}$$

Example 1.27. *Find the ratio of population of two energy states of the Ruby laser, the transition between which is responsible for the emission of photon of wavelength λ = 6928Å. Assume the ambient temperature is 18 K.* ***(GBTU, 2010)***

Solution : Given $\lambda = 6928 \text{ Å} = 6928 \times 10^{-10}$ m

$T = 18$ K

$$E_2 - E_1 = h\nu = \frac{hc}{\lambda}$$

$$= 6.6 \times 10^{-34} \times 3 \times 10^8 / 6928 \times 10^{-10}$$

$$= \mathbf{0.0028 \times 10^{-16} \text{ J.}}$$

The ratio of population of two energy states is given by

$$\frac{N_2}{N_1} = e^{-\left(\frac{h\nu}{kT}\right)}$$

$$= e^{-\left(\frac{0.0028\times10^{-16}}{1.38\times10^{-23}\times 18}\right)} = e^{-(0.0001127\times10^7)} = e^{-1127} \cong 0 \; (approx.)$$

Example 1.28. *The ratio of population of two energy levels out of which upper one corresponds to a metastable states is 1.059 × 10^{-30}. Find the wavelength of light emitted at temperature T = 330 K.* ***(GBTU, 2010)***

Solution : Given :

$$\frac{N_2}{N_1} = e^{\left(-\frac{h\nu}{kT}\right)} = 1.059\times10^{-30}$$

$$T = 330 \text{ K}$$

$$h = 6.6\times10^{-34} \text{ J sec.}$$

$$k = 1.38\times10^{-23} \text{ J/K}$$

∴ Frequency,

$$\nu = -\log_e\left[\frac{N_2}{N_1}\right]\frac{kT}{h} = -\log_e(1.059\times10^{-30})\times\frac{kT}{h}$$

$$\lambda = \frac{c}{\nu} = 3\times10^8/\nu$$

∴

$$\lambda = \frac{-3\times10^8\times1.38\times10^{-23}\times6.6\times10^{-34}}{\log_e(1.059\times10^{-30})\times1.38\times10^{-23}\times330}$$

$$= \mathbf{1.1\times10{-}7\ m}$$

1.18 ABSORPTION AND EMISSION OF LIGHT

In an atom, an electron in the ground state is stable and moves continuously in that orbit without radiating energy. When the electron receives an amount of energy equal to the difference of energy of the ground state and one of the excited states (*i.e.* outer orbits), it absorbs energy and jumps to the excited state. There are a variety of ways in which the energy may be supplied to the electron. One way is to illuminate the material with light of appropriate frequency $\nu = (E_2 - E_1)/h$. The photons of energy $h\nu = (E_2 - E_1)$ induce electron transitions from the energy level E_1 to the level E_2, as shown in Fig.1.10.

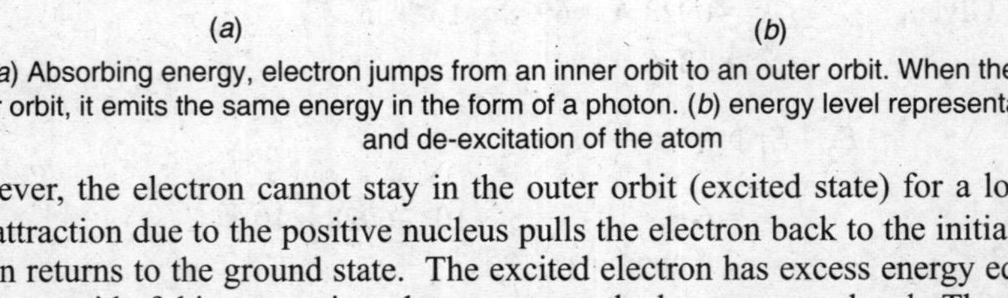

Fig. 1.10 (*a*) Absorbing energy, electron jumps from an inner orbit to an outer orbit. When the electron returns to the inner orbit, it emits the same energy in the form of a photon. (*b*) energy level representation of excitation and de-excitation of the atom

However, the electron cannot stay in the outer orbit (excited state) for a longer time. The Coulomb attraction due to the positive nucleus pulls the electron back to the initial inner orbit and the electron returns to the ground state. The excited electron has excess energy equal to $(E_2 - E_1)$ and it has to get rid of this energy in order to come to the lower energy level. The only mechanism through which the electron can lose its excess energy is through the emission of a photon. Therefore, the excited electron emits a photon of energy $h\nu = (E_2 - E_1)$ and returns to the ground state. This is the visualization of emission of light according to Bohr's quantum theory.

When we see light from any source, we actually "see" electrons jumping from excited states to lower states. This type of emission of light which occurs on its own is known as *spontaneous emission* and is responsible for the light coming from candles, electric bulbs, fire, stars, sun etc. conventional sources of light.

1.19 EINSTEIN'S PREDICTION

Einstein predicted in 1917 that there must be a second emission process to establish thermodynamic equilibrium. For example, if we illuminate a material with light of suitable frequency, the atoms in it absorb light and go to higher energy state.

The excited atoms tend to return randomly to the lower energy state. As the ground state population is very large, more and more atoms are excited under the action of incident light and it is likely that a stage may be reached where all atoms are excited. This violates thermal equilibrium condition. Therefore, Einstein suggested that there could be an additional emission mechanism, by which the excited atoms can make downward transitions. He predicted that the photons in the light field induce the excited atoms to fall to lower energy state and give up their excess energy in the form of photons. He called this second type of emission as *stimulated emission.*

1.20 THE THREE PROCESSES

When radiation is incident on a medium, three quantum mechanical process may occur:

(*a*) The photons of the incident radiation are absorbed by the atoms in the medium and the atoms go into excited states. This process is called absorption.

(*b*) The excited atoms return to the ground state on their own accord and emit the energy difference in the form of photons. This process is called spontaneous emission.

(*c*) The excited atoms get de-excited by the incident photons and they fall to the ground state emitting the extra energy in the form of photons. This process is called stimulated emission.

The process of absorption causes loss of photons. In the process of spontaneous emission light is emitted in all directions. This process dominates in all conventional and natural light sources.

In the process of simulated emission, photons are multiplied and their characteristics are related to each other. Hence, amplification of light takes place. This process is made to dominate in the laser light source.

Let us consider a medium consisting of identical atoms capable of being excited from the energy level 1 to the energy level 2 by absorption of photons. Let the levels be denoted by E_1 and E_2 and their populations be N_1 and N_2 respectively. Let the atoms be in thermal equilibrium. In the equilibrium condition, the number of atomic transitions upward must be equal to the number of downward transitions. Thus no net photons are generated or lost. However, when the atoms are subjected to an external light of frequency ν, the following three processes occur in the medium.

(*a*) Absorption

An atom residing in the lower energy level E_1 may absorb the incident photon and jump to the excited state E_2 as depicted in Fig.1.11. This transition is known as *induced* or *stimulated absorption* or simply as *absorption.* Corresponding to each absorption transition, one photon disappears from the incident light field and one atom adds to the population at the excited energy level E_2. This process may be represented as

$$A + h\nu \rightarrow A^* \qquad ...(1.46)$$

where A denotes an atom in the lower state and A^* an excited atom.

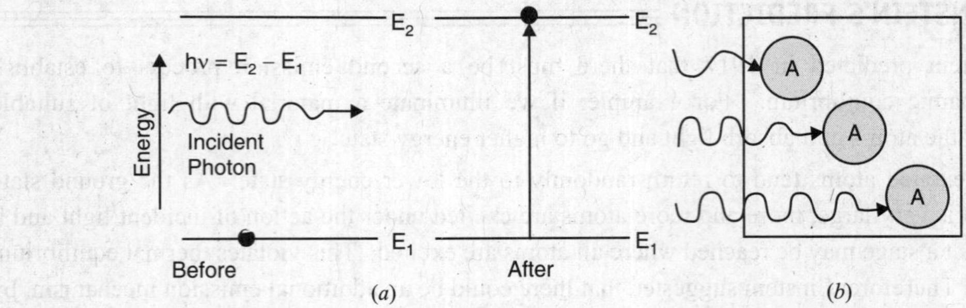

Fig. 1.11 Absorption process (*a*) Induced absorption (*b*) Material absorbs photons.

The number of atoms per unit volume that makes upward transitions from the lower level to the upper level per second is called the *rate of absorption transition*. It is represented by

$$R_{abs} = \frac{-dN_1}{dt}$$

where – (dN_1/dt) stands for the rate of decrease of population at the lower level E_1. The rate of absorption transition can also be represented by the rate of the increase of population at the upper level E_2. Thus,

$$R_{abs} = \frac{dN_2}{dt}$$

$$\therefore \quad R_{abs} = \frac{-dN_1}{dt} = \frac{dN_2}{dt} \qquad ...(1.47)$$

The number of absorption transitions occurring in the material at any instant will be proportional to the population in the lower level and the number of photons per unit volume in the incident beam. The rate of absorption may be expressed conveniently as

$$\boxed{R_{abs} = B_{12}\,\rho(\nu)\,N_1} \qquad ...(1.48)$$

where B_{12} is a constant of proportionality, and $\rho(\nu)$ is the energy density of incident light. B_{12} is known as the *Einstein coefficient for induced absorption* and it indicates the probability of an induced transition from level $1 \rightarrow 2$.

At thermal equilibrium, the population in the lower energy state is far larger than that in the higher energy state. Therefore, as light propagates through the medium, it gets absorbed.

(*b*) Spontaneous Emission

An excited atom can stay at the excited level for an average lifetime τ_{sp}. If it is not stimulated by any other agent during its short lifetime, the excited atom undergoes a transition to the lower energy level on its own. During the transition, it gives up the excess energy in the form of a photon, as shown in Fig. 1.12. This process in which an excited atom emits a photon all by itself and without any external impetus is known as *spontaneous emission*.

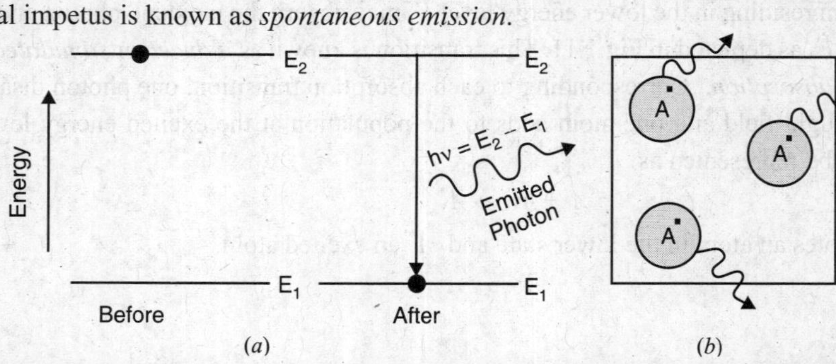

Fig. 1.12. Spontaneous emission process (*a*) emission (*b*) Material emits photons haphazardly.

The process is represented as,

$$A^* \rightarrow A + h\nu \qquad ...(1.49)$$

The rate of spontaneous transitions, R_{sp}, is given by,

$$R_{sp} = \frac{-dN_2}{dt} = \frac{N_2}{\tau_{sp}} \qquad ...(1.50)$$

The number of photons generated will be proportional to the population of the excited level only and may be expressed as follows :

$$\boxed{R_{sp} = A_{21}N_2} \qquad ...(1.51)$$

where A_{21} is known as the *Einstein coefficient for spontaneous emission* and is a function of frequency and properties of the material. It indicates the probability of a spontaneous transition from level $2 \rightarrow 1$. It is to be noted from eq.(1.51) that the process of spontaneous emission is independent of the light energy.

A comparison of equations (1.50) and (1.51) establishes that

$$\boxed{A_{21} = 1/\tau_{sp}} \qquad ...(1.52)$$

Thus, the reciprocal of the coefficient A_{21} is the time for the spontaneous transition τ_{sp} and represents the *lifetime of the spontaneous emission* also. It is the average time for which the excited atom would remain in the upper level before undergoing spontaneous transition.

It follows from quantum mechanical considerations that spontaneous transitions take place from a given state only to states lying lower in energy. Therefore, spontaneous transitions cannot occur from level E_1 to level E_2. The rate of spontaneous transitions from E_1 to E_2 is zero.

$$\therefore \qquad \boxed{A_{12} = 0} \qquad ...(1.53)$$

Important Features :

1. The process of spontaneous emission is not amenable for control from outside.
2. It is essentially probabilistic in nature.

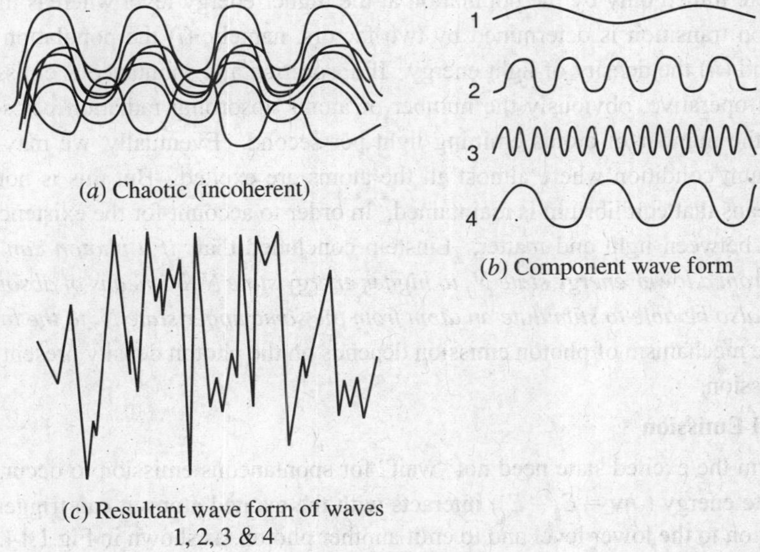

(*a*) Chaotic (incoherent)

(*b*) Component wave form

(*c*) Resultant wave form of waves 1, 2, 3 & 4

Fig. 1.13. Incoherent radiation (*a*) Incoherent waves, (*b*) component wave form (*c*) Resultant wave

3. The instant of transition, direction of propagation, the initial phase and the plane of polarization of each photon are all random. They are different for different photons emitted by various atoms. Thus, an atom can radiate into any of the 4π steridians with any sense of polarization.
4. The light is not monochromatic because of various line broadening processes.
5. Because of lack of directionality, the light spreads in all directions around the source. The light intensity goes on decreasing rapidly with distance from the source.
6. The light is incoherent. It results from a superposition of wavetrains of random phases as depicted in Fig. 1.13. The net intensity due to superposition of incoherent waves is proportional to the number of radiating atoms. Thus,

$$I_{Total} = NI \qquad ...(1.54)$$

where I is the intensity of light emitted by one atom.

The process of spontaneous emission dominates in conventional light sources.

If only the two processes, namely absorption and spontaneous emission exist, then under the equilibrium condition,

$$\frac{dN_1}{dt} = \frac{dN_2}{dt}$$

$$B_{12}\,\rho(\nu)\,N_1 = A_{21}\,N_2$$

or

$$\rho(\nu) = \frac{A_{21}}{B_{12}}\frac{N_2}{N_1}$$

$$\therefore \qquad \rho(\nu) = \frac{A_{21}}{B_{12}}\,e^{-h\nu/kT}$$

The above equation does not correspond to Planck's equation for radiation in thermal equilibrium. Planck's equation was confirmed experimentally and derived theoretically.

From the equations (1.51) and (1.48), we understand that the probability of a spontaneous transition is determined only by the population at the higher energy level whereas the probability of an absorption transition is determined by two factors, namely, (*i*) the population at the lower energy level and (*ii*) the density of light energy. If absorption and spontaneous emission were the only processes operative, obviously the number of atoms absorbing radiation per second would be more than the number of atoms emitting light per second. Eventually, we may end up with a non-equilibrium condition where almost all the atoms are excited. But this is not observed in practice. It means that equilibrium is maintained. In order to account for the existence of the state of equilibrium between light and matter, Einstein concluded that *if a photon can stimulate an atom to move from a lower energy state N_1 to higher energy state N_2 by means of absorption, then a photon should also be able to stimulate an atom from the same upper state N_2 to the lower state N_1.* This alternative mechanism of photon emission depends on the photon density present and is called stimulated emission.

(*c*) Stimulated Emission

An atom in the excited state need not "wait" for spontaneous emission to occur. If a photon with appropriate energy ($h\nu = E_2 - E_1$) interacts with the excited atom, it can trigger the atom to undergo transition to the lower level and to emit another photon, as shown in Fig.1.14.

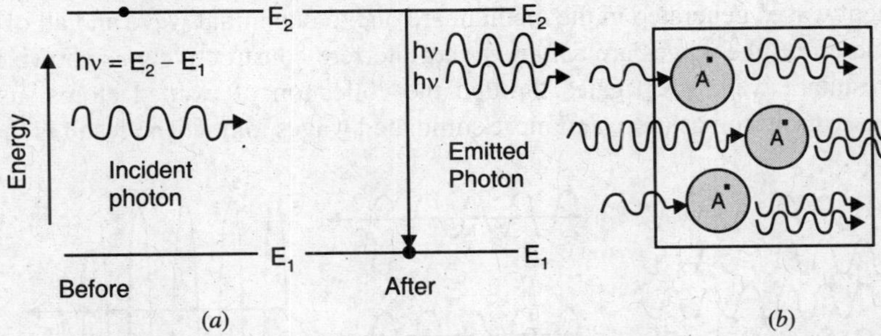

Fig. 1.14. Stimulated emission process (*a*) emission, (*b*) Material emits photons in a coordinate manner.

The process of emission of photons by an excited atom through a forced transition occurring under the influence of an external agent is called *induced or stimulated emission.* The process may be represented as

$$A^* + h\nu \rightarrow A + 2h\nu \qquad ...(1.55)$$

The rate of stimulated emission of photons is given by

$$R_{st} = B_{21}\, \rho(\nu)\, N_2 \qquad ...(1.56)$$

where B_{21} is the *Einstein coefficient for stimulated emission.*

Important Features :

1. The process of stimulated emission is controllable from outside.
2. The photon emitted in this process propagates in the same direction as that of the stimulating photon.
3. The emitted photon has exactly the same frequency, phase and plane of polarization as those of the incident photon.
4. The light produced in this process is directional, coherent and monochromatic.
5. **Light Amplification :** The outstanding feature of this process is the multiplication of photons. For one photon hitting an excited atom, there are two photons emerging. The two photons are in phase and travel along the same direction. These two photons stimulate two excited atoms in their path and produce a total four photons which are in phase and travel along the same direction. These four photons can in turn stimulate four excited atoms and generate eight photons and so on. The number of photons builds up in an avalanche like manner, as illustrated in Fig.1.15.

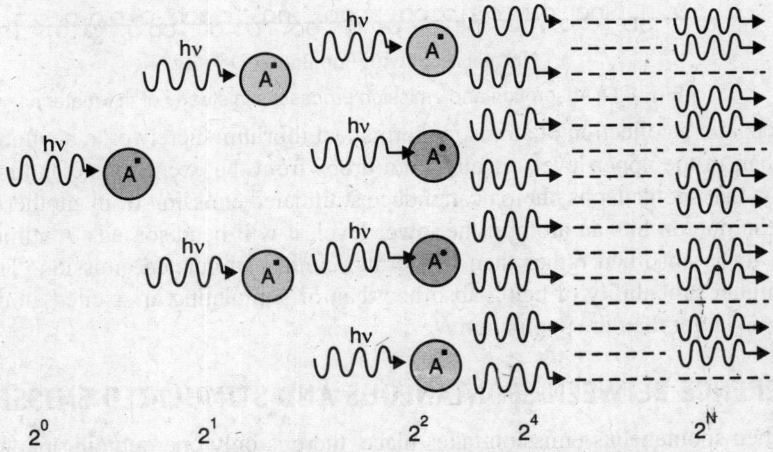

Fig. 1.15 Multiplication of simulated photons into an avalanche

All the light waves generated in the medium are due to one initial wave and all of them are in phase. Therefore, the waves are *coherent* and interfere constructively as shown in Fig.1.16. As the resultant wave propagates through the collection of excited atoms, its amplitude continuously grows on as more and more stimulated waves join it. The result is that light gets amplified.

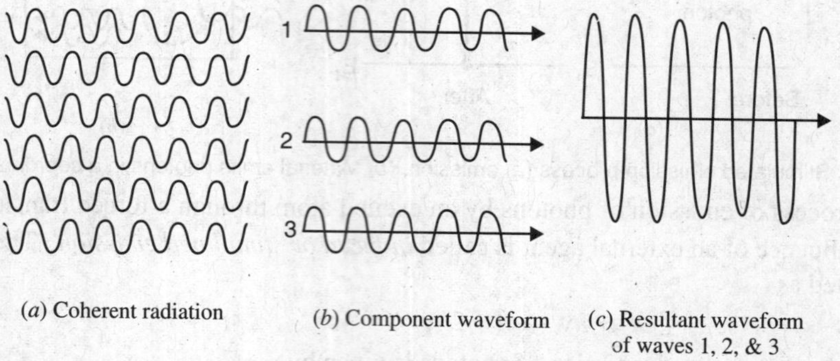

Fig. 1.16 Coherent radiation (*a*) coherent wave form (*c*) Resultant wave

6. **High Intensity :** Because of constructive interference of the waves, the net intensity of the resultant light will be proportional to the square of the number of atoms emitting light. Thus,

$$I_{Total} = N^2 I \qquad ...(1.57)$$

Hence, the light produced due to stimulated emissions is of very high intensity than the light generated through spontaneous emissions.

The above three processes occur together and under steady state condition, there would be a balance achieved between the absorption and emission processes (Fig. 1.17).

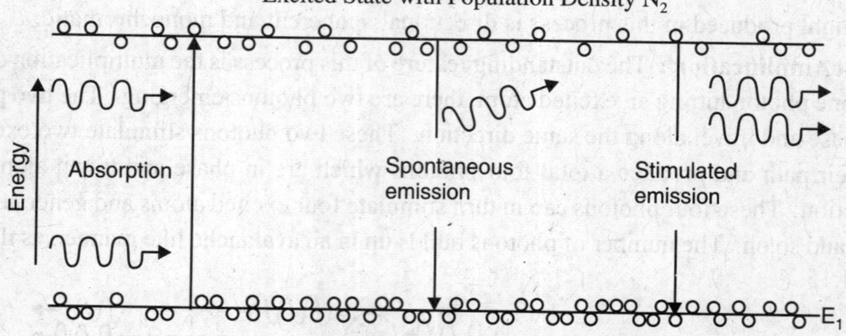

Fig. 1.17 Absorption and emission processes in steady of the material.

If we consider a collection of atoms in thermal equilibrium, there would be more atoms in the lower level than in the upper level. If one atom drops from the excited state and emits a photon spontaneously, that spontaneous photon can induce stimulated emission from another excited atom. However, if the photon hits an atom in the lower level, it will be absorbed. A stimulated photon is also likely to be absorbed rather than triggering another stimulated emission. Thus, a photon has a much higher probability of being absorbed than of stimulating an excited atom because the population N_1 is larger than the population N_2.

1.21 DIFFERENCE BETWEEN SPONTANEOUS AND STIMULATED EMISSION

1. When spontaneous emission takes place, there is only one outgoing photon in random direction having energy $h\nu$ while in stimulated emission, There are two outgoing photons moving in the same direction.

2. In spontaneous emission, there is no phase relation between the photons emitted while in stimulated emission, photons have same frequency and are *in phase* within the incident photons.
3. In spontaneous emission *incoherent* radiation is achieved while stimulated emission the radiations are unidirectional and coherent.
4. In spontaneous emission, the probable rate of transition from excited level E_2 to lower level E_1 is proportional to the number of excited electrons remaining in E_2 level, while in case of stimulated emission, the rate of emission is proportional to the number of atoms in the excited state and the energy density of the incident radiation.

1.22 EINSTEIN RELATIONS

Under thermal equilibrium, the mean population N_1 and N_2 in the lower and upper energy levels respectively must remain constant. This condition requires that the number of transitions from E_2 to E_1 must be equal to the number of transitions from E_1 to E_2. Thus,

$$\left.\begin{array}{l}\text{The number of atoms absorbing}\\ \text{photons per second per unit volume}\end{array}\right\} = \left\{\begin{array}{l}\text{The number of atoms emitting photons}\\ \text{per second per unit volume}\end{array}\right.$$

$$\left.\begin{array}{l}\text{The number of atoms absorbing}\\ \text{photons per second per unit volume}\end{array}\right\} = B_{12}\,\rho(\nu)\,N_1$$

$$\left.\begin{array}{l}\text{The number of atoms emitting}\\ \text{photons per second per unit volume}\end{array}\right\} = A_{21}\,N_2 + B_{21}\,\rho(\nu)\,N_2$$

In equilibrium condition, the number of transitions from E_2 to E_1 must be equal to the number of transitions from E_1 to E_2. Thus,

$$B_{12}\,\rho(\nu)\,N_1 = A_{21}N_2 + B_{21}\,\rho(\nu)\,N_2 \qquad \text{...(1.58)}$$

$$\rho(\nu)\,(B_{12}\,N_1 - B_{21}N_2) = A_{21}N_2$$

$$\therefore \quad \rho(\nu) = A_{21}\,N_2/B_{12}\,N_1 - B_{21}\,N_2$$

On dividing both the numerator and denominator on the right hand side of the above equation with $B_{12}\,N_2$, we get

$$\boxed{\rho(\nu) = \frac{A_{21}/B_{12}}{N_1/N_2 - B_{21}/B_{12}}} \qquad \text{...(1.59)}$$

It follows from eq. (1.44) that

$$N_1/N_2 = e^{(E_2-E_1)/kT}$$

As

$$(E_2 - E_1) = h\nu$$

$$N_1/N_2 = e^{h\nu/hT}$$

$$\therefore \quad \rho(\nu) = \frac{A_{21}}{B_{12}}\left[\frac{1}{e^{h\nu/kT} - B_{21}/B_{12}}\right] \qquad \text{...(1.60)}$$

To maintain thermal equilibrium, the system must release energy in the form of electromagnetic radiation. It is required that the radiation be identical with black body radiation and be consistent with Planck's radiation law for any value of T. According to Planck's law

$$\rho(\nu) = (8\pi\,h\nu^3\,\mu^3/c^3)\,\frac{1}{e^{h\nu/kT} - 1} \qquad \text{...(1.61)}$$

where μ is the refractive index of the medium and c is the velocity of light in free space.

Energy density $\rho(\nu)$ given by Eq. (1.60) will be consistent with Planck's law (Eq. 1.61) only if

$$\boxed{B_{21} = B_{12}} \qquad ...(1.62)$$

and
$$A_{21}/B_{12} = (8\pi h\nu^3 \mu^3/c^3) \qquad ...(1.63)$$

Therefore,
$$\boxed{B_{12} = B_{21} = \frac{c^3}{8\pi h\nu^3\mu^3} A_{21}} \qquad ...(1.64)$$

Equations (1.62) and (1.63) are known as the *Einstein relations.* Equation (1.64) gives the relationship between the A and B coefficients.

The first relation (1.62) shows that the coefficients for both absorption and stimulated emission are numerically equal. The equality implies the following. When an atom with two energy levels is placed in the radiation field, the probability for an upward (absorption) transition is equal to the probability for a downward (stimulated emission) transition.

The second relation (1.63) shows that the ratio of coefficients of spontaneous versus stimulated emission is proportional to the third power of frequency of the radiation. This is why it is difficult to achieve laser action in higher frequency ranges such as x-rays.

Example 1.29. *The wavelength of emission is 6000 Å and the lifetime* τ_{sp} *is* 10^{-6} *s. Determinie the coefficient for the stimulated emission.*

Solution : The coefficient for stimulated emission is given by

$$B_{21} = \frac{c^2 A_{21}}{8\pi \mu^3 h\nu^3} \qquad \text{Eq. (1.64)}$$

But
$$A_{21} = \frac{1}{\tau_{sp}} \text{ and } \frac{c^3}{\nu^3} = \lambda^3. \text{ Taking } \mu = 1, \text{ we get}$$

$$B_{21} = \frac{\lambda^3}{8\pi h\tau_{sp}} = \frac{\left(6000 \times 10^{-10}\right)^3 \text{ m}^3}{8 \times \pi \times 6.626 \times 10^{-34} \text{ Js} \times 10^{-6}\text{ s}}$$

$$= \frac{216 \times 10^{-21}}{166.6 \times 10^{-40}} \cdot \frac{\text{m}^3}{\text{J.s}^2} = \mathbf{1.3 \times 10^{19} \text{ m / kg.}}$$

1.23 CONDITIONS FOR LARGE STIMULATED EMISSIONS

The key to laser action is the existence of stimulated emission. In practice, the absorption and spontaneous emissions always occur together with stimulated emission. Let us now study the conditions under which the number of stimulated emissions can be made larger than those of the other two processes.

(a) From equations (1.51) and (1.56), we can write for the ratio of the stimulated transitions to spontaneous transitions as

$$R_1 = \frac{\text{Stimulated transitions}}{\text{Spontaneous transitions}}$$

$$= B_{21}\,\rho(\nu)\,N_2/A_{21}\,N_2$$

$$= (B_{21}/A_{21})\,\rho(\nu) \qquad ...(1.65)$$

Using eq.(1.61) for $\rho(\nu)$, we get

$$R_1 = (B_{21}/A_{21}) \left[\frac{8\pi h\nu^3\mu^3}{c^3} \frac{1}{e^{h\nu/kT} - 1}\right] \qquad ...(1.66)$$

From eq.(1.62, 1.63) and (1.64), we can write

$$\frac{B_{21}}{A_{21}} = \frac{B_{12}}{A_{21}} = \frac{c^3}{8\pi h\nu^3\mu^3} \quad ...(1.67)$$

Using the eq.(1.67) into eq(1.66), we obtain

$$R_1 = (c^3/8\,\pi\,h\nu^3\,\mu^3)\left[(8\pi\,h\nu^3\mu^3/c^3)\frac{1}{e^{h\nu/kT}-1}\right]$$

or

$$\boxed{R_1 = \frac{1}{(e^{h\nu/kT}-1)}} \quad ...(1.68)$$

If we assume $\nu = 5 \times 10^{14}$ Hz (yellow light) and $T = 2000$ K, the value of $h\nu / kT$ is 11.99.

$$\therefore \quad R_1 = \frac{1}{e^{11.99}-1} = 6\times10^{-6}$$

The above result shows that in the optical region spontaneous emissions dominate over the stimulated emissions.

The equation (1.65) suggests that the light field density $\rho(\nu)$ present within the material is required to be enhanced if we want large number of stimulated emissions.

(***b***) The ratio of stimulated transitions to absorption transitions is given by

$$R_2 = \frac{\text{Stimulated transitions}}{\text{Absorption transitions}}$$

$$= \frac{B_{21}\,\rho(\nu)\,N_2}{B_{12}\,\rho(\nu)\,N_1} \quad ...(1.69)$$

As

$$B_{21} = B_{12}$$

$$\boxed{R_2 = N_2/N_1} \quad ...(1.70)$$

At thermodynamic equilibrium,

$$N_2/N_1 << 1$$

Therefore at equilibrium, absorption transitions overwhelm stimulated transitions. A photon of the light field may hit an excited atom leading to stimulated emission, or be absorbed on hitting an atom in the ground state. As $N_2 << N_1$ at thermodynamic equilibrium, a photon has a much higher probability of being absorbed than of stimulating an excited atom. As a result, the absorption process dominates stimulated emission and the medium will absorb the incident light. If, on the other hand, more atoms are in the excited state, *i.e.* $N_2 > N_1$, photons are more likely to cause stimulated emission than absorption. Therefore, in order to achieve more stimulated emissions, the population N_2 of the excited state should be made larger than the population N_1 of the lower energy state.

To sum up, three conditions are to be satisfied to make stimulated transitions overwhelm the other transitions: (*i*) the population at excited level should be greater than that that at the lower energy level, (*ii*) the ratio B_{21}/A_{21} should be large and (*iii*) a very high density of radiation density should be present in the medium. A medium amplifies light only when these three conditions are fulfilled.

1.24 SPONTANEOUS AND STIMULATION EMISSION IN OPTICAL REGION

Rate of spontaneous emission,

$$R_{sp} = A_{21}\,N_2 \qquad (\because \text{Eq. 1.51}) \quad ...(i)$$

and rate of stimulated emission

$$R_{st} = B_{21}\,\rho(\nu)\,N_2 \qquad (\because \text{Eq. 1.56}) \qquad ...(ii)$$

Dividing Eq. (*ii*) by Eq. (*i*), we get

$$R_1 = \frac{\text{Stimulated transitions}}{\text{Spontaneous transitions}} = \frac{B_{21}\,\rho(\nu)\,N_2}{A_{21}\,N_2}$$

$$= \frac{B_{21}}{A_{21}}\rho(\nu) \qquad ...(iii)$$

Applying Planck's radiation law,

$$\rho(\nu) = \frac{8\pi h\nu^3}{c^3\,(e^{h\nu/kT}-1)} \qquad ...(iv)$$

From Eq. (*iii*), we conclude that stimulated transitions will dominate spontaneous transitions, if the radiation density $\rho(\nu)$ is very large and B_{21}/A_{21} is also very large. Using Eq. (*iv*) in Eq. (*iii*), we get

$$R_1 = \left(\frac{B_{21}}{A_{21}}\right)\left[\frac{8\pi h\nu^3}{c^3}\cdot\frac{1}{e^{h\nu/kT}-1}\right]$$

But $$\frac{B_{21}}{A_{21}} = \frac{c^3}{8\pi h\nu}$$

i.e., $$R_1 = \frac{1}{e^{h\nu/kT}-1}$$

Considering $\nu = 5 \times 10^{14}$ Hertz and T = 300*k*, then

$$R_1 = 10^{-58}$$

Thus, in optical region, stimulated emission is negligible compared to spontaneous emission.

Example 1.30. *(a) At what temperature are the rates of spontaneous and stimulated emission equal ? Assume $\lambda = 5000$ Å.*

(b) At what wavelength are they equal at 300 K ?

Solution : If the rates of spontaneous and stimulated emission are equal, then

$$R = \left[e^{h\nu/kT}-1\right]^{-1}$$

$$\lambda = 5000 \text{ Å},\ \nu = \frac{c}{\lambda} = \frac{3\times10^8 \text{ m/s}}{5000\times10^{-10}\text{ m}}$$

$$= 6 \times 10^{14} \text{ Hz}$$

$$\therefore \quad \frac{h\nu}{kT} = \frac{(6.626\times10^{-34}\text{ J.s})\,(6\times10^{14}\text{ s}^{-1})}{(1.38\times10^{-23}\text{ J/K})\text{T}}$$

$$= \frac{28.81\times10^3}{T}\text{K}$$

(*a*) $$\therefore\ \exp\left[\frac{28.81\times10^3}{T}\text{K}\right] = 2$$

or $$\frac{28.81\times10^3}{T}K = ln2 = 0.693$$

$$\therefore \qquad T = \frac{28.81 \times 10^3}{T} K = \mathbf{41{,}573\ K}$$

$$(b) \quad \therefore \qquad \frac{h\nu}{kT} = \frac{(6.626 \times 10^{-34}\,\text{J.s})\,\nu}{(1.38 \times 10^{-23}\,\text{J/K})\,(300\,\text{K})}$$

$$= (1.6 \times 10^{-13}\,\text{s})\nu$$

$$\therefore \qquad \exp\,[(1.6 \times 10^{-13}\,\text{s})\nu] = 2$$

oı

$$(1.6 \times 10^{-13}\,\text{s})\nu] = 0.693$$

$$\therefore \qquad \nu = \frac{0.693}{1.6 \times 10^{-13}}\,\text{Hz} = 4.3 \times 10^{12}\ \text{Hz}$$

$$\therefore \qquad \lambda = \frac{c}{\nu} = \frac{3 \times 10^8\,\text{m/s}}{43 \times 10^{12}\ \text{s}^{-1}} = \mathbf{69.8\ \mu m}$$

Example 1.31. *Find the ratio of spontaneous emission to stimulated emission for a cavity of temperature 50K and wavelength 10^{-5} m.*

Solution :

$$\frac{\text{Rate of spontaneous emission}}{\text{Rate of Stimulated emission}} = \frac{A_{21}}{B_{21}} = e^{\left(\frac{h\nu}{kT}\right)} - 1$$

$$= e^{\left(\frac{hc}{\lambda kT}\right)} - 1$$

$$= e^{\left[\frac{6.625 \times 10^{-34} \times 3 \times 10^8}{10^{-5} \times 1.38 \times 10^{-23} \times 50}\right]} - 1$$

$$= e^{(28.8)} - 1$$

$$= e^{28.8} \qquad (\because e^{28.8} >> 1)$$

1.25 CONDITION FOR LIGHT AMPLIFICATION

Let us consider a beam of light propagating through a material medium. If the photons strike lower state (unexcited) atoms, they may be absorbed and removed from the stream of photons which therefore loses energy. However, if the photons strike excited atoms, more photons can be produced which are added to the light beam and increase its energy. Since the probabilities of absorption and stimulated emission are the same, both attenuation and amplification of the light beam occur simultaneously. We will now show that amplification can predominate only if there are more atoms in the higher level than in the lower level.

Let there be n photons per unit volume in the light beam. As the beam travels through the medium, some photons are absorbed due to absorption transitions and some photons are generated due to emission transitions. We will not take into account the photons generated by spontaneous emission as these photons go in random directions and do not contribute to the light beam propagating through the medium. Thus, we consider only the photons generated by stimulated emissions.

Let $-(dn/dt)$ be the net rate of loss of photons from the beam as it travels through an elemental volume of the medium having a thickness Δx and an area of unity.

The net rate of loss of photons from the light beam must be equal to the difference between the net rates of absorption and stimulated emission transitions. Thus,

$$-\frac{dn}{dt} = B_{12}\,\rho(\nu)\,N_1 - B_{21}\,\rho(\nu)\,N_2$$

or

$$-\frac{dn}{dt} = (N_1 - N_2)\,\rho(\nu)\,B_{12} \qquad ...(1.71)$$

If the energy density of the light field in the medium is $\rho(\nu)$, then the intensity I is

$$I = \rho(\nu)\, \upsilon \qquad ...(1.72)$$

where υ $(= c/\mu)$ is the velocity of light in the medium.

As $\rho(\nu) = nh\nu$,

$$I = nh\, \nu\upsilon \qquad ...(1.73)$$

The loss of photons, $-dn$ in a small thickness dx of medium may be written as,

$$-dn = \frac{dI\,(x)}{dx} \cdot \frac{dx}{h\,\nu\upsilon}$$

The net rate of loss during a time interval dt is given by,

$$-\frac{dn}{dt} = \frac{dI\,(x)}{dx} \cdot \frac{1}{h\nu} \qquad ...(1.74)$$

$dx/dt = \upsilon$ is used in obtaining the above equation.

Using eq.(1.2) into eq.(1.74), we get

$$-\frac{dn}{dt} = +\,\alpha I\,(x) \cdot \frac{1}{h\nu}$$

Using Eq. (1.40) into the above, we get

$$\therefore \qquad -\frac{dn}{dt} = \alpha\rho\,(\nu)\,\upsilon \cdot \frac{1}{h\nu} \qquad ...(1.75)$$

Comparing equations (1.71) and (1.75), we obtain

$$\alpha\rho\,(\nu)\,\upsilon\frac{1}{h\nu} = (N_1 - N_2)\,\rho(\nu)\,B_{12}$$

$$\therefore \qquad \alpha = (N_1 - N_2)\,\frac{B_{12}\,h\nu}{\upsilon} \qquad ...(1.76)$$

Eq. (1.76) relates the absorption coefficient α to the difference in populations $(N_1 - N_2)$ of the two energy levels. For a material in thermal equilibrium, $N_1 > N_2$ and α is positive.

If N_2 is somehow made greater than N_1, then α becomes a negative quantity and the relation (1.4) takes the following form.

$$I = I_0 e^{(-\alpha)x}$$

$$\text{or} \qquad I = I_0 e^{\gamma x} \qquad ...(1.77)$$

where γ $(= -\,\alpha)$ is referred to as the *gain coefficient* per unit length. As the gain coefficient γ is a positive quantity, the equation (1.77) implies that the intensity of light grows exponentially as the light beam travels through the medium. This is clearly amplification of light. Incorporating γ, we rewrite eq.(1.76) as,

$$\boxed{\gamma = (N_2 - N_1)\,\frac{B_{12}\,h\nu}{\upsilon} \quad \textit{Condition for amplification}} \qquad ...(1.78)$$

γ will be positive if $(N_2 - N_1) > 0$, that is,

$$\boxed{N_2 > N_1} \qquad ...(1.79)$$

The condition (1.79) is known as *population inversion*, because it is the inverse of the normal situation. Eq. (1.78) thus indicates that population inversion is a necessary condition to be satisfied for causing the amplification of incident light.

Using the relations (1.64) for B_{12}, (1.52) for A_{21} and $c = \mu\,\upsilon$, we can rewrite equation (1.78) as

$$\gamma = (N_2 - N_1).\,\frac{\nu^2}{8\,\pi\,\nu^2\tau_{sp}} \qquad ...(1.80)$$

1.26 LINESHAPE FUNCTION

For the sake of simplicity, we have assumed till now that the emission of light by all atoms in the specimen is strictly monochromatic and the frequency is given by

$$\nu_0 = \frac{E_2 - E_1}{h}$$

However, this is not true. The light emitted by an atom is in reality a burst of decaying exponential, as shown in Fig.1.18. The actual frequency of the light, although centered at ν_o, has a spread over a certain range $\Delta\nu$, as depicted in Fig. 1.19. Fourier analysis of the decaying light shows that

$$\Delta\nu \approx \frac{1}{\tau_{sp}} \quad ...(1.81)$$

$\Delta\nu$ is known as the *linewidth* and is defined as the frequency difference corresponding to the intensity $\frac{1}{2}$ *I*.

Similarly, atoms are capable of absorbing light not just at a single frequency but over a band of frequencies. It implies that atoms interact with light over a range of frequencies. If we plot the intensity of radiation as a function of frequency, we would again obtain the bell-shaped curve shown in Fig. 1.18.

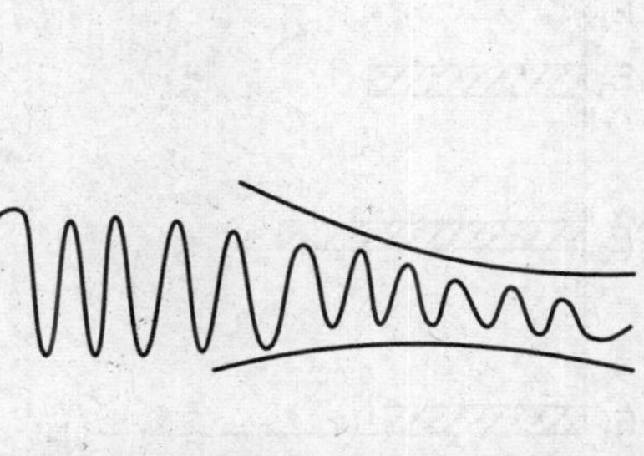

Fig. 1.18 A decaying electric field of light burst.

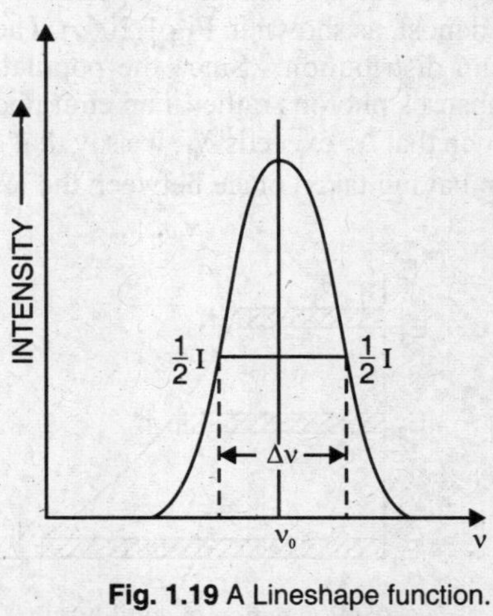

Fig. 1.19 A Lineshape function.

The shape of the curves is described by the *lineshape function* $g(\nu)$. The function $g(\nu)$ is usually normalized according to

$$\int_{-\infty}^{+\infty} g(\nu)\, d\nu = 1 \quad ...(1.82)$$

We may consequently view $g(\nu)\, d\nu$ as the probability that a given transition between the energy levels E_2 and E_1 will result in the emission (or absorption) of a photon whose frequency lies between ν and $\nu + d\nu$. Hence a photon of energy $h\nu$ may not necessarily stimulate another photon of energy $h\nu$, but it, may stimulate a photon which has an energy between $h\nu$ and $h(\nu + d\nu)$.

We now assume that radiation density $\rho(\nu)\, d\nu$ incident on the collection of atoms has frequencies between ν and $\nu + d\nu$. Further, out of the total N_1 and N_2 atoms per unit volume only $N_1\, g(\nu)\, d\nu$ and $N_2\, g(\nu)\, d\nu$ atoms per unit volume are capable of interacting with radiation of frequencies lying between ν and $\nu + d\nu$. Incorporating these ideas, Eq. (1.78) takes the form

$$\gamma = (N_2 - N_1)\,\frac{B_{21} h\nu_0}{\upsilon}\,g(\nu) = (N_2 - N_1)\,\frac{\upsilon^2\,g(\nu)}{8\pi\nu^2\tau_{sp}} \qquad ...(1.83)$$

where ν_o is the central frequency.

1.27 POPULATION INVERSION

When an atomic system is in thermal equilibrium, photon absorption and emission processes take place side by side, but because $N_1 > N_2$, absorption dominates. However, laser operation requires obtaining stimulated emission exclusively. To achieve a high percentage of stimulated emission, a majority of atoms should be at the higher energy level than at the lower level. *The non-equilibrium state in which the population N_2 of the upper energy level exceeds to a large extent the population N_1 of the lower energy level is known as the state of population inversion.*

Extending the Boltzmann distribution, Eq. (1.44), to this non-equilibrium state of population inversion, it is seen that N_2 can exceed N_1 only if the temperature *were* negative. In view of this, the state of population inversion is sometimes referred to as a *negative temperature state*. It does not mean that we can attain temperatures below absolute zero. The terminology underlines the fact that the state of population inversion is a non-equilibrium state. It should be borne in mind that the population inversion state is attained at normal temperatures.

Let us consider for the moment a system that has three energy states E_1, E_2 and E_3. With the system in equilibrium, the uppermost level E_3 is populated least and the lowest level E_1 is populated most, as shown in Fig.1.20(*a*). The dotted curve shown in Fig. 1.20(*a*) represents a normal Boltzmann distribution. Since the population in the three states is such that $N_3 < N_2 < N_1$, the system absorbs photons rather than emit photons. However, if the system is supplied with external energy such that N_2 exceeds N_1, we say that the system reached population inversion, the population inversion having taken place between the levels E_2 and E_1, as shown in Fig. 1.20(*b*).

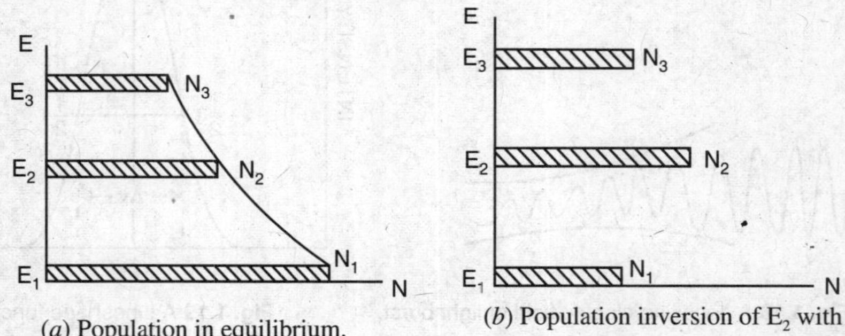

(*a*) Population in equilibrium.

(*b*) Population inversion of E_2 with respect to E_1.

Fig. 1.20 Three level system.

Under the population inversion condition, the stimulated emission can produce a cascade of light. The first few randomly emitted spontaneous photons trigger stimulated emission of more photons and those stimulated photons induce still more stimulated emissions and so on. As long as the excited state population is more than the lower level population, stimulated emissions are more likely than absorption ; and consequently, light gets amplified as shown in Fig. 1.21. As soon as the population at lower level becomes equal to or larger than that at the excited state, population inversion ends, stimulated emissions diminish and amplification of light ceases.

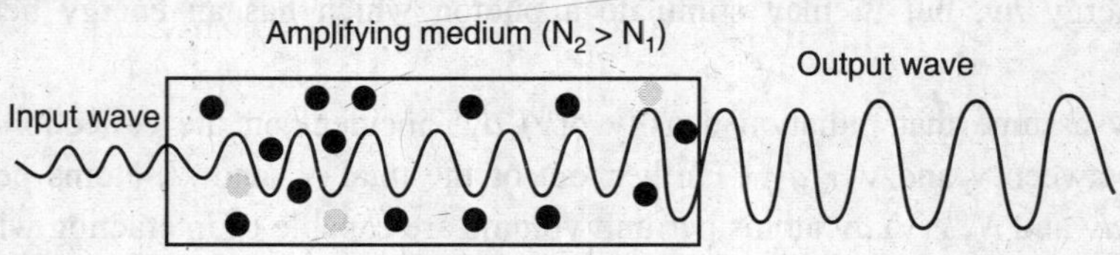

Fig. 1.21 Amplification of a light wave in a medium with population inversion.

1.28 PUMPING

In order to realize and maintain the state of population inversion, it is necessary that atoms must be continuously promoted from the lower level to the excited level. Energy is to be supplied somehow to the laser medium to raise atoms from the lower level to the excited level and for maintaining population at the excited level at a value greater than that of the lower energy. The process by which atoms are raised from the lower level to the upper level is called *pumping.*

The usual method of exciting atoms to higher energy levels is to heat the material. From Eq. (1.17) it is seen that as long as $E_2 > E_1$, $N_2 < N_1$. Heating the material only increases the average energy of atoms but does not make N_2 greater than N_1. Therefore, population inversion cannot be achieved by heating the material.

1.29 PUMPING METHODS

In order to create the state of population inversion, we have to selectively excite the atoms in the material to particular energy levels. There are a variety of methods employed for this purpose. The most common methods of pumping make use of light and electrons. We confine our study here only to these techniques. These are:

(*a*) Optical pumping
(*b*) Electrical pumping
(*c*) X-ray pumping
(*d*) Chemical pumping

(*a*) Optical Pumping :

Optical pumping is the use of photons to excite the atoms. A light source such as a flash discharge tube is used to illuminate the laser medium and the photons of appropriate frequency excite the atoms to an uppermost level. From there, they drop to the metastable upper laser level to create the state of population inversion. Optical pump sources are flash discharge tubes, continuously operating lamps, spark gaps or an auxiliary laser is sometimes used as the pump source.

The pump photon must have higher frequency than the emitted photon. This is because the atoms are to be raised above the upper laser level from a lower level which is below or at the lower laser level. This is one of the factors which reduces the laser efficiency.

The pumping level of the atom must not be a narrow level. It should be sufficiently wide, spanning a range of energies. If it is narrow, one can use a pump photon of only a specific frequency. Such a situation severely restricts the choice of sources and a large portion of the source power would go wasted. Fortunately, in a majority of cases the upper levels are wide bands and atoms can be excited to many of the upper levels. Therefore, light sources emitting a broad range of wavelengths like a flash lamp can be used to excite atoms.

Optical pumping is suitable for any laser medium which is transparent to pump light. Optical pumping is used for solid state crystalline lasers and liquid tunable dye lasers.

(*b*) Electrical Pumping :

Electrical pumping can be used only in case of laser materials that can conduct electricity without destroying lasing activity. This method is limited to gases. In case of a gas laser, a high voltage pulse initially ionizes the gas so that it conducts electricity. An electric current flowing through the gas excites atoms to the excited level from where they drop to the metastable upper laser level leading to population inversion.

In semiconductor lasers also electrical pumping is used, but here it is not the atoms that are excited. It is the current carriers namely electrons and holes which are excited and a population inversion is achieved in the junction region. The electrons recombine with holes in the junction regions producing laser light. Thus, in semiconductor lasers, a direct conversion of electrical energy into light energy takes place.

(*c*) X-ray Pumping :

In some cases atoms of solids can be excited by X-ray beam and pumping can be achieved.

(*d*) Chemical Pumping :

When excitation is produced by exothermic chemical reactions, it requires highly reactive and often explosive gaseous mixture.

1.30 ACTIVE MEDIUM

Atoms in general are characterized by a large number of energy levels. However, all types of atoms are not suitable for laser operation. Even in a medium consisting of different species of atoms, only a small fraction of atoms of a particular species are suitable for stimulated emission and laser action. Those atoms which cause light amplification are called *active centers*. The rest of the medium acts as host and supports active centers. The medium hosting the active centers is called an *active medium. An active medium is thus a medium which, when excited, reaches the state of population inversion, and eventually causes light amplification*. The active medium may be a solid, a liquid or a gas.

1.31 METASTABLE STATES

An atom can be excited to a higher level by supplying energy to it. Normally, excited states have short lifetimes and release their excess energy in a matter of nanoseconds (10^{-9} sec) by spontaneous emission. Atoms do not stay at such excited states long enough to be stimulated to emit their energy. Though, the pumping agent continuously raises the atoms to the excited level, many of them rapidly undergo spontaneous transitions to the lower energy level. Population inversion cannot be therefore established. For establishing population inversion, the excited atoms are required to " wait" at the upper lasing level till a large number of atoms accumulate at that level. Thus, what is needed is an excited state with a longer lifetime. Such longer-lived upper levels from where an excited atom does not return to lower level at once, but remains excited for an appreciable time, are known as *metastable states.* Phosphors are an example of materials having metastable states. They emit persistent light called *phosphorescence* because of metastable states existent in them.

Atoms stay in metastable states for about 10^{-6} to 10^{-3} s. This is 10^3 to 10^6 times longer than the time of stay of atom at excited levels. Therefore , it is possible for a large number of atoms to accumulate at a metastable level. The metastable state population can exceed the population of a lower level and lead to the state of population inversion.

If the metastable states do not exist, there could be no population inversion, no stimulated emission and hence no laser operation.

Thus, the foundation to the laser operation is the existence of metastable states.

1.32 PUMPING SCHEMES

Atoms in general are characterized by a large number of energy levels. Among these energy levels, two, three or four levels will be pertinent to the pumping process. Accordingly, pumping schemes are classified as two-level, three-level and four-level schemes. Among them, the two-level scheme will not lead to laser action. The three-level and four-level schemes are important and are widely employed.

(*a*) Two Level Pumping Scheme :

It appears that the most simple and straight-forward method to establish population inversion is to pump an excess of atoms into the higher energy state by applying intense radiation. But basically, a two-level pumping scheme, shown in Fig. 1.9, is not suitable for attaining population inversion. We can establish it through the following considerations.

Population inversion requires a build-up of population in the upper laser level. It is possible only if the upper level is populated faster than it decays, so that the population of the upper level exceeds that of the lower level. It, in turn, is possible only if the lifetime of spontaneous emission is longer. Thus, it is required that the lifetime Δt at upper level E_2 must be longer. According to Heisenberg uncertainty principle, the linewidth ΔE_2 of level E_2 and the lifetime Δt are related by.

$$\Delta E_2 \,.\, \Delta t \geq \hbar \qquad ...(1.84)$$

Δt will be longer if ΔE_2 is smaller which implies that the upper energy level E_2 must be narrow. However, if E_2 is narrow, we have to use only a specific frequency photon $(\nu = E_2 - E_1 / \hbar)$ to pump atoms. It means that the pump source should be highly monochromatic. In practice, monochromatic source of required frequency may not exist. Even if it exists, the pumping efficiency would be very low. The result is that enough population cannot be excited to level E_2. Further, pumping radiation on one hand excites the ground state atoms and on the other hand induces transitions from the upper level to the lower level. It means that pumping operation simultaneously populates and depopulates the upper level. Hence, population inversion cannot be attained in a two-level scheme. All that we may achieve at best is a system of equally populated levels.

One must therefore look for materials with either three or four energy level systems, which operate with *different frequencies* of pumping and lasing transitions.

The transition between the two levels that generate stimulated emission is called a *lasing transition*. The terminal level is called the *lower lasing level* and the upper level as *upper lasing level*. The upper lasing level should be a metastable level. The uppermost level to which atoms are excited is known as the *pumping level*. The transition between the ground level and the pump level is called the *pump transition*.

(*b*) Three Level Pumping Scheme :

A model of a three level pumping scheme is illustrated in Fig. 1.22. A three level scheme is one in which the lower laser level is either the ground state or a level whose separation from the ground state is small compared to kT.

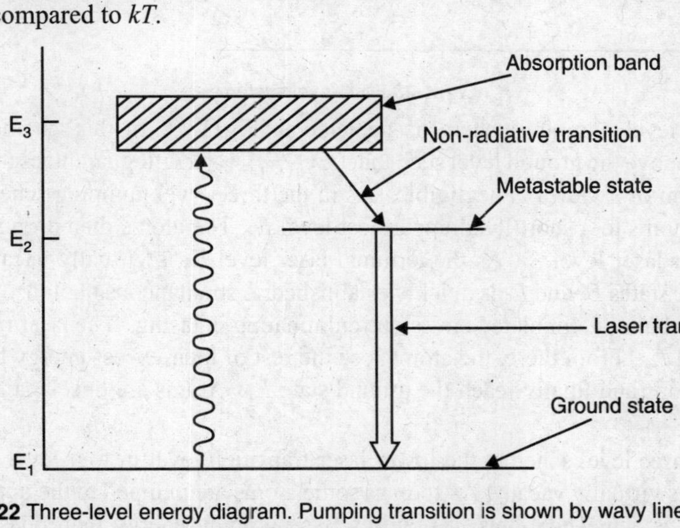

Fig. 1.22 Three-level energy diagram. Pumping transition is shown by wavy line, nonradiative transition by simple arrow, laser transition by hollow arrow.

Initially, the population distribution among the three levels obeys the Boltzmann law. When the atoms are subjected to an intense radiation of pumping frequency $\nu_p = (E_3 - E_1 / h)$, the atoms are pumped to the higher level E_3. Some of the excited atoms make spontaneous transitions to the ground state but many of them undergo spontaneous non-radiative transitions to the metastable level E_2. As spontaneous transitions from E_2 to E_1 do not occur often, the atoms accumulate at the metastable level E_2. The build-up of atoms at E_2 continues because of pumping process. Eventually, the population N_2 at E_2 exceeds the population N_1 at E_1 and population inversion is attained. Now, a photon of $h\nu$ (= $E_2 - E_1$) can induce stimulated emission and laser action.

For better pumping efficiency, the level E_2 should be a band of energy levels instead of being a single narrow line. It allows use of a pumping radiation of wider bandwidth to excite more atoms. However, the major disadvantage of a three level scheme is that it requires very high pump powers. In this scheme, the terminal level of the laser transition is simultaneously the ground state. Therefore, inversion of population requires more than half of the ground state atoms to be lifted to the higher energy level. As the ground state is heavily populated, large pumping power is to be used to depopulate the ground level to the required extent.

The three level scheme can produce light only in pulses. Once stimulated emission commences, the metastable state E_2 gets depopulated very rapidly and the population of the ground state increases quickly. As a result the population inversion ends. One has to wait till the population inversion is again established. Thus, three level lasers operate in *pulsed mode.*

(*c*) Four Level Pumping Scheme

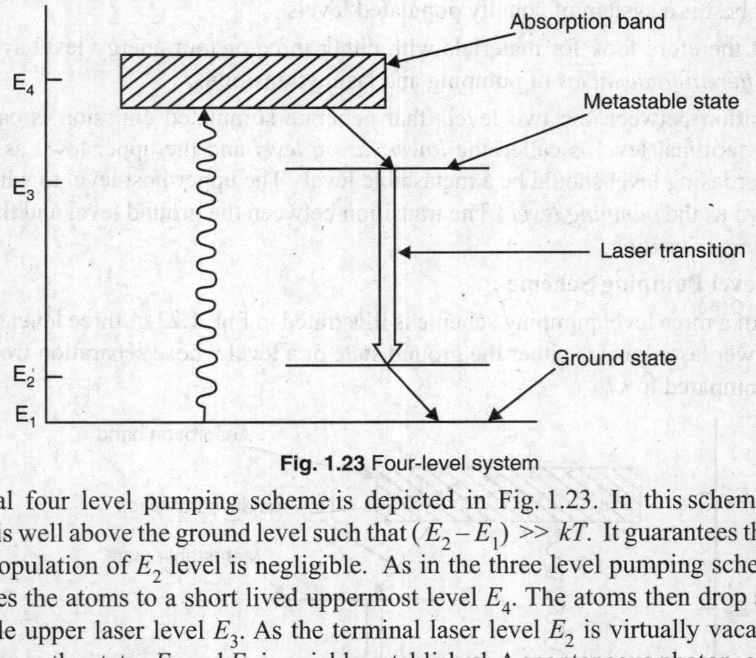

Fig. 1.23 Four-level system

A typical four level pumping scheme is depicted in Fig. 1.23. In this scheme, the terminal laser level E_2 is well above the ground level such that $(E_2 - E_1) >> kT$. It guarantees that the thermal equilibrium population of E_2 level is negligible. As in the three level pumping scheme, the pump energy elevates the atoms to a short lived uppermost level E_4. The atoms then drop spontaneously to a metastable upper laser level E_3. As the terminal laser level E_2 is virtually vacant, population inversion between the states E_3 and E_2 is quickly established. A spontaneous photon of energy $h\nu = E_3 - E_2$ can initiate a chain of stimulated emissions culminating in lasing. The laser transition takes the atoms to the level E_2. From there, the atoms lose the rest of their excess energy by radiative or non-radiative transitions and finally reach the ground state E_1. Atoms are once again available for excitation.

In contrast to three level scheme, the lower laser transition level in four level scheme is not the ground state and is virtually vacant. As soon as some atoms are pumped to the upper laser level, population inversion is achieved. Thus, it requires less pumping energy than does a three level laser. This is the major advantage of this scheme. Further, the lifetime of the lower laser transition level E_2 is much shorter as it is not a metastable state. Hence atoms in level E_2 quickly drop to the ground state. This steady depletion of E_2 level helps sustain the population inversion by avoiding an accumulation of atoms in the lower lasing level. Therefore, four level lasers can operate in a *continuous wave (cw) mode.*

1.33 AMPLIFICATION AND GAIN

When an active medium is in the inverted state, a photon of appropriate energy can stimulate the emission of a cascade of photons. This is the process of amplification. The initial photon may be looked upon as the input signal, the active medium as the quantum optical amplifier and the emerging light as the amplified output. The degree of amplification is measured as gain which is the increase in intensity when a light beam passes through an active medium. The gain may be expressed as

$$G = \frac{1}{I} \cdot \frac{dI}{dx} \qquad ...(1.85)$$

The *gain* may be otherwise defined as the amount of stimulated emission which a photon can generate as it travels a given distance. For example, if $G = 4$ per cm, it means that one photon produces four photons per each centimeter it travels in the medium. Unfortunately, laser materials have a very low gain, of the order of 0.0001/cm to 0.01/cm It means that the photon has to travel a long length of the laser material for producing large amplification. As an example, if we have a material whose gain is 0.001/cm and if we wish to achieve a light amplification of 1000 times, it is calculated that the light has to travel about 69 meters in the medium. Such long distances are obviously not practical. However, the important point to note here is that the amount of amplification increases rapidly with the distance.

In practice, laser materials are not used to amplify light from some outside source. Laser, despite its name, is basically a generator of light.

One of the ways of making light to pass through a long length of laser medium is by keeping mirrors on both sides of a short laser rod or tube. The light bounces back and forth between the mirrors and makes many passes through the medium increasing the effective distance of travel by many times. Such an arrangement of mirrors transforms the simple amplifying medium into a source of light.

Although population inversion is necessary for light amplification, it alone is not sufficient to make the stimulated emissions dominate other processes. According to the relation (1.65), large radiation energy density $\rho(\nu)$ is required to be present in the active medium for this purpose. The pair of mirrors help to maintain a large radiation density in the medium.

1.34 AMPLIFIER

A rod shaped laser medium without end mirrors acts as an amplifier. A light beam of appropriate frequency incident on one end triggers stimulated emission and an amplified output will emerge from the other end. Such amplifiers are used to increase laser pulse powers to higher levels and to amplify weak signals in fiber optic communication systems.

1.35 OPTICAL RESONATOR

Basically a laser is very much similar to an electronic oscillator. In an electronic oscillator, an amplifier tuned to a specific frequency is provided with a positive feedback, as shown in Fig. 1.24. When the oscillator is switched on, any noise signal of the appropriate frequency appearing at the input will be amplified. The amplified output is fed back to the input and is amplified again, and so on. A stable output is quickly reached and the oscillator acts as a source of the particular frequency. In case of low frequency electronic oscillators, the feedback may be given using electric wires. However at high frequencies, waves can not be confined in wires and hence different kinds of feedback mechanisms are to be used. At microwave frequencies, for example, cavities are used.

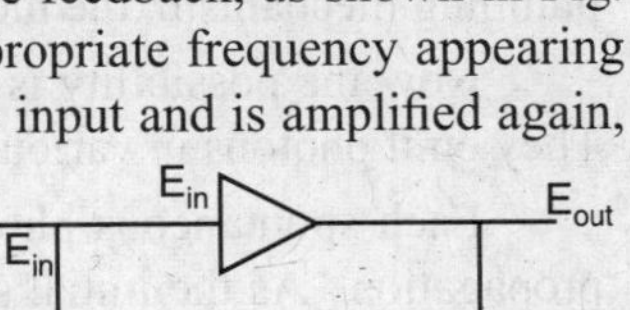

Fig. 1.24 Feedback Oscillator

The cavities are completely enclosed metal boxes, such as in Fig. 1.25(*a*), in which a standing wave is produced. If we have to use such a closed cavity at light frequency, its dimensions should be of the order of a wavelength, *i.e.* of the size of 0.5 mm. It is not possible to fabricate such microcavities. Townes and and Schawlow showed that there is no necessity for a closed cavity at light frequencies and an open cavity in the form of two parallel mirrors will serve the purpose.

Therefore, in case of lasers, feedback is obtained by placing the active medium between a pair of mirrors which are facing each other. The role of the input signal is played by a chance photon spontaneously emitted along the optic axis of the laser rod or tube. The photon triggers many stimulated emissions in its wake and the photons at the end are reflected back into the medium by the mirror. The photons reverse their direction and travel through the medium till they reach the mirror at the other end. Substantial amplification occurs as the light beam is reflected several times at the mirrors and travel through the active medium gaining strength in each passage.

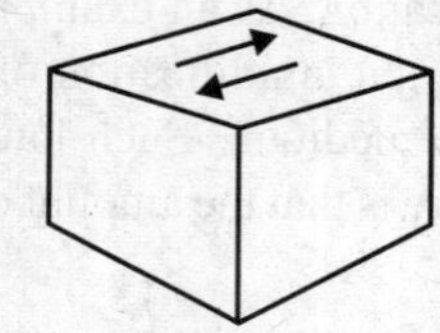

(*a*)

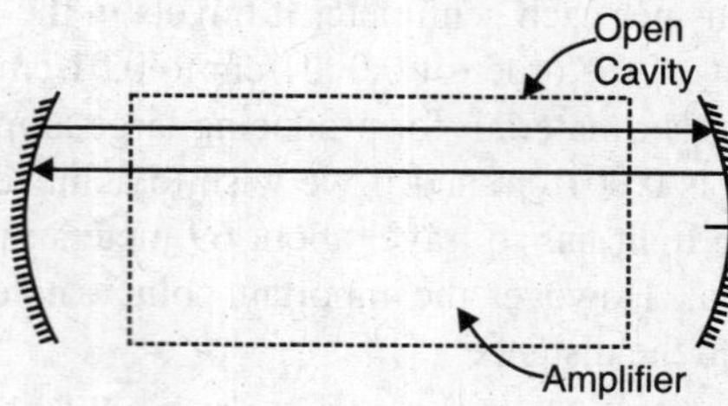

(*b*)

Fig. 1.25 (a) Closed cavity used at microwave frequencies (b) Open cavity, Fabry-Perot resonator, used at optical frequencies.

In electronic oscillators, it is the tank circuit or resonating circuit that builds up large output with moderate input. In case of lasers, the pair of mirrors constitute an *optical resonator*. This structure is actually an open resonator as the sides are open. It is known as a **Fabry – Perot resonator**. The mirrors could be either plain or curved and are designed such that one of them reflects all the light that reaches it while the other one reflects most of the light incident on it. A small portion of light is transmitted through this mirror as the laser output beam.

The basic components of a laser may be represented in a block diagram, Fig. 1.26. A laser consists of an active medium (*AM*) placed between two mirrors (*M*, *OM*) and is pumped by a pumping agent (*P*). An additional element (*IE*) may be sometimes located within the cavity to achieve certain desired characteristics in the laser output.

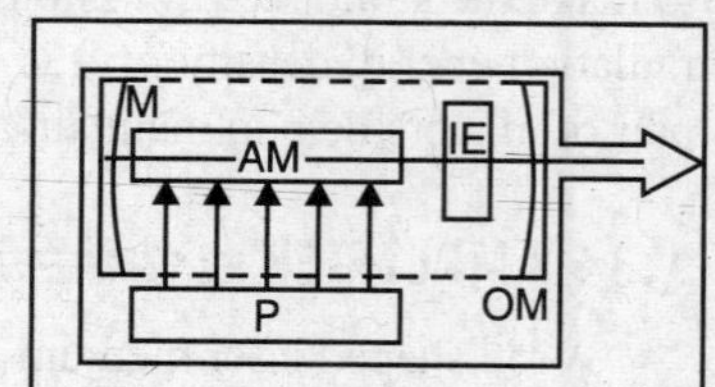

Fig. 1.26. Block diagram of a laser.

1.36 ACTION OF OPTICAL RESONATOR

The successive stages of the action of an optical resonator are depicted in Fig. 1.27. To start with, the active centres in the laser medium are in the ground state (Fig. 1.27*a*). Through a suitable pumping mechanism, the medium is taken into the state of population inversion (Fig. 1.27*b*).

Now the possibility is that some of the large number of excited atoms decay spontaneously. They emit photons in various directions.

Each spontaneous photon can trigger many stimulated transitions along the direction of its propagation. As the initial spontaneous photons are emitted in different directions, the stimulated photons will also travel in different directions. Many of them leave the medium without reinforcing their strength. In the absence of end mirrors, the net effect would have been that incoherent light is produced without a specific direction. On the other hand, in the presence of end mirrors, a specific direction is imposed on photons. Photons propagating along the optic axis of the pair of mirrors are augmented whereas photons emitted in any other direction will pass out of the sides of the resonator and are lost, as shown in Fig.1.27(*c*).

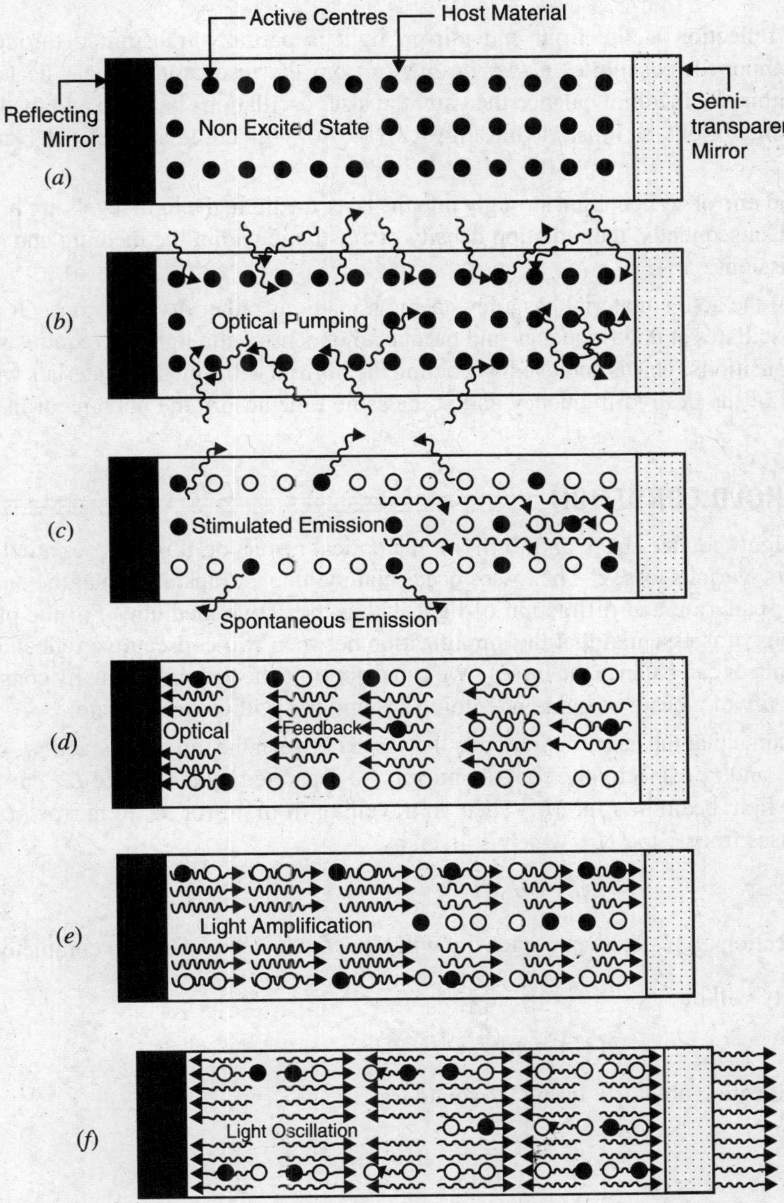

Fig. 1.27 Light amplification and oscillation due to the action of optical resonator.

A majority of photons travelling along the axis are reflected back on reaching the end mirror. They propagate towards the opposite mirror and on their way they stimulate more and more atoms and build up their strength, as shown in Fig. 1.27(*d*). Thus, the mirrors provide positive feedback of light into the medium so that stimulated emission acts are sustained and the laser operates as an oscillator.

The photons that strike the opposite mirror are reflected once more into the medium Fig. 1.27(*e*). The photons will travel along the axis generating more photons and cause more and more amplification. As the photons are thus reflected back and forth between the two mirrors, stimulated emissions sharply increase. Thus by allowing the light beam to traverse the same active medium several times, the effective length of the medium is increased and the amplification of light is augmented.

At each reflection at the front end mirror, light is partially transmitted through it. The transmitted component constitutes a *loss of energy* from the resonator. When the losses at the mirrors and within the medium balance the gain, the laser oscillations build up and the laser output beam will become bright, as illustrated in Fig. 1.27(*f*). A highly collimated intense beam emerges from the laser.

As the end mirrors reflect light strongly into the laser medium, the light levels are higher within the resonator. Consequently, the radiation density $\rho(\nu)$ is large within the medium and sustains the stimulated emissions.

In general the active material has a few lasing transitions rather than only one. It is desirable that the laser oscillates at one frequency and produce monochromatic light. It requires suppression of undesired transitions. This is achieved by coating the mirrors with suitable materials which highly reflect photons of the desired frequency and at the same time absorb the photons of the undesired frequencies.

1.37 THRESHOLD CONDITION

As the light bounces back and forth in the optical resonator, it undergoes amplification as well as it suffers various losses. The losses occur mainly due to transmission at the output mirror and due to the scattering and diffraction of light within the active medium. For the proper build up of oscillations, it is essential that the amplification between two consecutive reflections of light from rear end mirror can balance the losses. We can determine the threshold gain by considering the change in intensity of a beam of light undergoing a round trip within the resonator.

Let us assume that the laser medium fills the space between the mirrors M_1 and M_2 which have reflectivities r_1 and r_2 respectively. Let the mirrors be separated by a distance L. Further, let the intensity of the light beam be I_0 at M_1. Then, in travelling from mirror M_1 to mirror M_2, the beam intensity increases from I_0 to $I(L)$, which is given by,

$$I(L) = I_0 e^{(\gamma-\alpha_s)L} \quad \text{...(1.86)}$$

After reflection at M_2, the beam intensity will be $r_2 I_0 e^{(\gamma-\alpha_s)L}$ and after a complete round trip the final intensity will be

$$I(2L) = r_1 r_2 I_0 e^{(\gamma-\alpha_s)2L} \quad \text{...(1.87)}$$

The amplification obtained during the round trip is,

$$G = \frac{I(2L)}{I_0} = r_1 r_2 e^{(\gamma-\alpha_s)2L}$$

The product $r_1 r_2$ represents the losses at the mirrors whereas α_s includes all the distributed losses such as scattering, diffraction and absorption occurring in the medium. The losses are balanced by gain, when $G \geq 1$ or $I(2L) = I_0$. It leads to the condition that,

$$r_1 r_2 e^{2(\gamma-\alpha_s)L} \geq 1$$

$$e^{2(\gamma-\alpha_s)L} \geq \frac{1}{r_1 r_2}$$

Taking logarithms on both the sides , we get

$$2L(\gamma - \alpha_s) \geq -\ln r_1 r_2$$

$$\gamma - \alpha_s \geq -\frac{1}{2L} \ln r_1 r_2$$

$$\therefore \quad \gamma \geq \alpha_s - \frac{1}{2L} \ln r_1 r_2$$

or

$$\boxed{\gamma \geq \alpha_s + \frac{1}{2L} \ln \frac{1}{r_1 r_2}} \qquad ...(1.88)$$

Eq. (1.88) is known as the *condition for lasing.* It shows that the initial gain must exceed the sum of the losses in the cavity. This condition is used to determine the threshold value of pumping energy necessary for lasing action.

γ, the amplification of the laser will be dependent on how hard the laser medium is pumped. As the pump power is slowly increased, a value of γ_{th} called *threshold value* will be reached and the laser starts oscillating. The threshold value γ_{th} is given by

$$\therefore \quad \boxed{\gamma_{th} = \alpha_s + \frac{1}{2L} \ln \frac{1}{r_1 r_2}} \qquad ...(1.89)$$

Therefore, for the laser to oscillate, $\gamma \geq \gamma_{th}$. Equation (1.89) is known as the *threshold condition for lasing* and states the criterion when the net gain would be able to counteract the effect of losses in the cavity. The value of γ must be at least γ_{th} for laser oscillations to commence. If $\gamma > \gamma_{th}$, the waves grow and the amplifier reaches saturation. It lowers the value of γ in turn and eventually an equilibrium value is attained at γ_{th}.

Example 1.32. *The length of a laser tube is 150 mm and the gain factor of the laser material is 0.0005/cm. If one of the cavity mirrors reflects 100% light that is incident on it, what is the required reflectance of the other cavity mirror ?*

Solution :

$$\gamma \geq \frac{1}{2L} \ln \frac{1}{r_1 r_2}$$

$$\therefore \quad r_2 = \frac{1}{r_1 e^{2L\gamma}} = \frac{1}{1 \times e^{2 \times 15 \times 0.0005}}$$

$$= \mathbf{0.985}$$

It means the second mirror should have a reflectance of 98.5%.

1.38 CRITICAL POPULATION INVERSION

We define a quantity N_{th} as

$$N_{th} = (N_2 - N_1)_{th}$$

N_{th} is called the *critical population inversion* or *threshold population inversion density.* It denotes the minimum population inversion density required to start lasing action and then to sustain it. Using Eq. (1.83), we obtain

$$N_{th} = \frac{8\pi\nu^2 \tau_{sp} \gamma_{th}}{\upsilon^2 g(\nu)} = \frac{8\pi\nu_0^2 \tau_{sp} \gamma_{th}\, \Delta\nu}{\upsilon^2} \qquad ...(1.90)$$

We replaced g(ν) by 1/Δν in the above equation using the relation (1.109). We note that the lasing threshold will be achieved readily at the central frequency ν_o. Thus, we write

$$\boxed{N_{th} = \frac{8\pi\nu_0{}^2\ \tau_{sp}\ \Delta\nu}{\upsilon^2} \left(\alpha_s - \frac{1}{2L} \ln r_1 r_2\right)} \qquad ...(1.91)$$

Let us consider the case in which mirror losses and scattering losses are all small and the laser medium is not being pumped. A light pulse starting with an original field strength E_O bounces back and forth between the mirrors. It makes a round trip in a time T given by

$$T = 2L/v$$

The electric field after one round trip will be

$$E(T) = E_0 e^{-2\alpha L} r_1 r_2 \qquad \text{...(1.92)}$$

Obviously, the electric field decays with time. The decaying electric field may be represented by

$$E(T) = E_0 e^{-t/t_c} \qquad \text{...(1.93)}$$

where t_c represents the *cavity decay constant.*

The value of the electric field at $t = T$, may be written as

$$E(T) = E_0 e^{T/t_c} \qquad \text{...(1.94)}$$

Equating Eq. (1.192) with Eq. (1.94), we get

$$e^{-T/t_c} = e^{-2\alpha L} r_1 r_2$$

$$= e^{-(2\alpha L - \ln r_1 r_2)}$$

$$\therefore \qquad T/t_c = 2\alpha L - \ln r_1 r_2$$

Using $T = 2L/v$ in the above equation, we obtain

$$t_c = \frac{1}{\upsilon}\left[\alpha - \frac{1}{2L}\ln r_1 r_2\right]^{-1} \qquad \text{...(1.95)}$$

The fractional loss l_c per round trip is defined as

$$l_c = \frac{E_0 - E(T)}{E_0}$$

When the losses are small, we can write

$$E(T) = E_0\,[1 - 2L\,(\alpha - \ln r_1 r_2)]$$

$$\therefore \qquad l_c = 2L\left[\alpha - \frac{1}{2L}\ln r_1 r_2\right] \qquad \text{...(1.96)}$$

We find that t_c and l_c are related as

$$t_c = \frac{2L}{\upsilon \ell_c} \qquad \text{...(1.97)}$$

Using the expression (1.95) for t_c into Eq. (1.90), we can rewrite

$$N_{th} = \frac{8\pi \nu_0^{\,2}\,\tau_{sp}\,\Delta\nu}{\upsilon^3 \tau_c} \qquad \text{...(1.98)}$$

Further, using Eq. (1.97) into Eq. (1.98), we get

$$\boxed{N_{th} = \frac{4\pi \nu_0^{\,2}\,\tau_{sp}\Delta\nu}{\upsilon^2}\left(\frac{\ell c}{L}\right)} \qquad \text{...(1.99)}$$

Eq. (1.99) shows that lasing is achieved more easily when the r.h.s. of the equation is smaller. In order to achieve critical inversion with the lowest possible pumping power, the atoms must have as narrow a line width, $\Delta\nu$, as possible. Thus the Eq. (1.99) sets the condition on the atoms to be used.

It is also seen from Eq. (1.99) that the laser condition becomes more difficult to be satisfied as the laser frequency increases. It will require that the lifetime t_c of the photons is made as large as possible.

1.39 CONDITION FOR STEADY STATE OSCILLATION

According to wave picture of light, light amplification implies a continuous and marked increase in the amplitude of the light wave. For this to occur, it is necessary that a wave making a complete round trip inside the resonator has to fulfil a phase condition in addition to the amplitude condition given by eq. (1.84). It is necessary that the wave returning to some point in the medium must have the same phase as that of the original wave with any number of reflections from the mirrors. It means that the phase delay between the waves must be some multiple of 2π. Thus, if light starts out at a wave peak when it is reflected from the output mirror, it should be again at the wave peak after one round trip or any number of round trips. This imposes a certain constraint on the relationship between the wavelength λ and the length of the laser rod, L. It is required that the optical path length travelled by a wave between two consecutive reflections at the same end mirror should be an integral multiple of the wavelength. For example, if light starts at a wave peak when it is reflected from the output mirror, it will travel an integral number of wavelengths before it reaches the output mirror again, where it will be again at peak. The waves stimulated by that wave will be at peak at the output mirror as shown in Fig. 1.28. Therefore, they all add up in amplitude by constructive interference and add up in amplitude. Thus, the condition for amplification is

$$\boxed{2\mu L = m\lambda} \qquad (m = 1, 2, 3, ...) \qquad ...(1.100)$$

where μ is the refractive index of active medium and μL is the optical path. When the condition (1.100) is not satisfied, each wave will reach the output mirror in a different phase, and undergo destructive interference and cancel out each other. The cavity length L thus imposes a severe restriction that only those light waves which can fit an integral number of wavelengths within twice the cavity length are amplified strongly. Waves of all other wavelengths are attenuated.

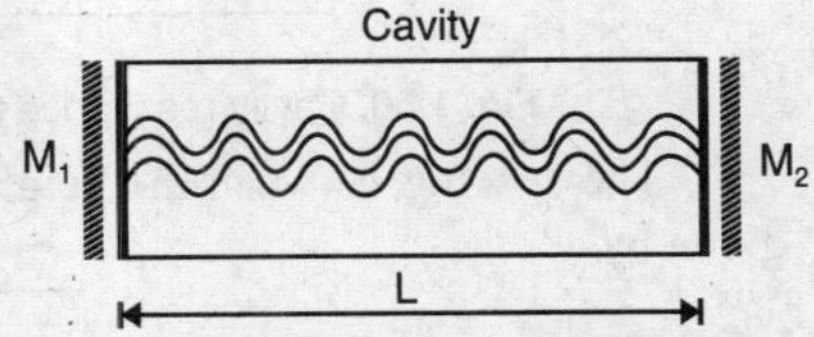

Fig. 1. 28. Condition of resonance

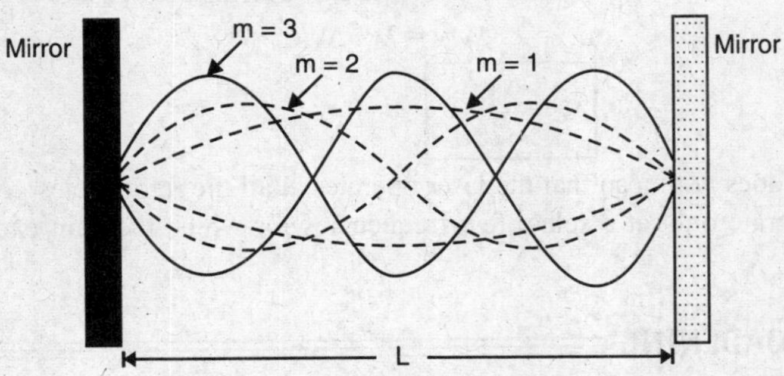

Fig. 1.29. Standing wave pattern and axial modes in optical resonator.

If the space bounded by the two end mirrors in a laser is regarded as a mirror cavity resonator, then the length L of the resonator should accommodate an integral number of standing half-waves as shown in Fig. 1.29. Thus, Eq. (1.100) which may be rewritten as

$$L = \frac{m\lambda}{2\mu} \qquad ...(1.101)$$

expresses the *condition of resonance* between the mirror cavity and the light waves.

1.40 CAVITY RESONANCE FREQUENCIES

It is seen in the previous section that the cavity will be resonant for those waves which fit an integral number of half-wavelength between the mirrors. The wavelengths of such waves are given by Eq.(1.101) as

$$\lambda_m = \frac{2\mu L}{m} \quad ...(1.102)$$

We can express Eq. (1.102) in terms of frequency as

$$\nu_m = \frac{mc}{2\mu L} \quad ...(1.103)$$

Theoretically, the cavity can resonate at a very large number of frequencies that satisfy the condition (1.103). For example, if we take $L = 0.5$ m, $\lambda = 5000$Å and $\mu = 1.5$, we obtain $m = 3 \times 10^6$. It means that the resonator supports 3×10^6 frequencies.

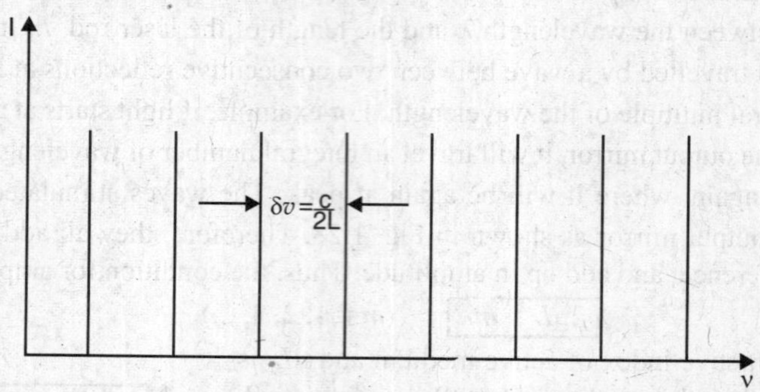

Fig. 1.30. Cavity resonance frequencies representing the possible longitudinal modes.

The spacing between the neighbouring frequencies is constant as shown in Fig. 1.30 and is given by,

$$\Delta\nu = \nu_{m+1} - \nu_m = \frac{c}{2\mu L} \quad ...(1.104)$$

By using $\Delta\nu/\nu = \Delta\lambda/\lambda$, Eq. (1.104) can be written in terms of $\Delta\lambda$ as

$$\Delta\lambda = \lambda.\ \Delta\nu/\nu = \lambda^2\ .\ \Delta\nu/c$$

$$\therefore \quad \Delta\lambda = \frac{\lambda^2}{2\mu L} \quad ...(1.104)$$

However, it does not mean that the laser operates at all the frequencies given by Eq. (1.103). The laser will operate only at a select few frequencies for which the gain exceeds all the cavity losses.

1.41 LINE BROADENING

Whenever an atom absorbs or emits light it is generally mentioned that the frequency at which absorption or emission takes place is given by

$$\nu_0 = \frac{E_2 - E_1}{h} \quad ...(1.105)$$

The above equation is based on the tacit assumption that the discrete energy levels E_2 and E_1 are infinitely narrow and the atomic transition from level E_1 to level E_2 (or from level E_2 to E_1) occurs at a single definite frequency ν_0. However, in actual practice, we find that the light emitted by atoms is not strictly monochromatic. It becomes therefore necessary that we have to modify our visualisation of energy levels. In order to understand the experimental facts, we assume that the energy levels E_2 and E_1 are not sharp but are spread over a range of energies dE_2 and dE_1 respectively, as shown in Fig. 1.31(*a*). Since transitions can occur between dE_2 and dE_1, the light emitted by the atoms will have a frequency spread dv, with the most intense part appearing at the central frequency ν_0. The emitted light is in the form of wave trains with an exponential damping of their amplitudes. If the intensity of light is plotted against the frequency, we obtain a bell shaped curve as shown in Fig. 1.31(*b*). We have already noted, in an earlier section, that the *line broadening* shown in Fig. 1.31(*b*) can be described by the *lineshape function* $g(\nu)$.

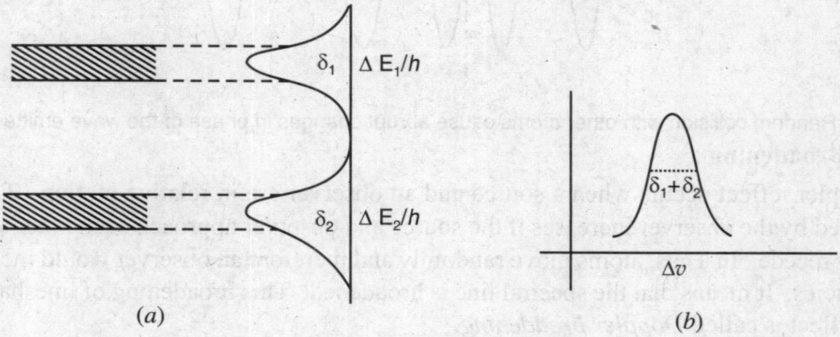

Fig. 1.31

There are a number of mechanisms which lead to line broadening. They are broadly classified into two categories : (*i*) homogeneous and (*ii*) inhomogeneous broadening. If the broadening mechanism effects each individual atom in the sample to the same extent, then the broadening is said to be *homogeneous.* In such a case all of the atoms in the sample will have the same centre frequency ν_0 and the same lineshape. Natural broadening and collision broadening belong to this category. On the other hand, if different atoms in the sample have slightly different frequencies for the same transition, the overall response of the sample broadens out and the broadening is said to be *inhomogeneous*. Doppler broadening and broadening due to crystal defects belong to this category.

Natural Broadening :

The energy levels in an atom are in reality not sharp. If the energy levels are sharp ($dE = 0$), the Heisenberg uncertainty principle

$$\Delta E.\, \Delta t \sim \hbar$$

leads to the impossible conclusion that the atom can stay in the excited state for an infinite length of time. ($\Delta t \to \infty$, if $\Delta E \to 0$). Hence, the uncertainty principle requires that each energy level has a spread, in terms of frequency,

$$\Delta\nu = \frac{1}{2\pi\tau} \qquad \text{...(1.106)}$$

where $\tau = \Delta t = \hbar/\Delta E$ is known as the *mean lifetime* of the atom in the energy state E.

The upper state have a mean lifetime of the order of 10^{-8} s. A spectral line arising from a transition from an excited state to the ground state will have a frequency spread of the order of

$$\Delta\nu = \frac{1}{2 \times 3.14 \times 10^{-8}\, s} \cong 16\ MHz$$

The natural broadening is relatively small in magnitude and is often masked by other mechanisms.

Collision Broadening:

In a gas random collisions occur between the atoms. If an atom which is emitting a wave train undergoes a collision, then the phase of the wave train is suddenly altered. Thus each collision leads to random phase changes and the effect of collisions may be viewed as shortening of the wave train as shown in Fig. 1.32. It is equivalent to broadening of the spectral line. This broadening is known as *collision broadening*.

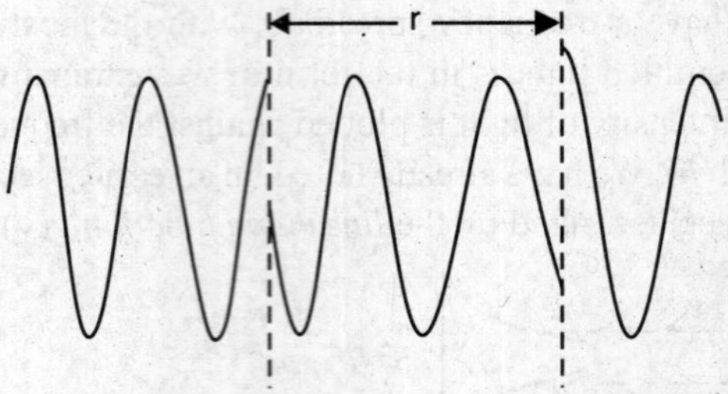

Fig. 1.32. Random collision with other atoms cause abrupt changes in phase of the wave emitted by an atom.

Doppler Broadening :

Doppler effect occurs when a source and an observer are in relative motion. The frequency as measured by the observer increases if the source and observer approach each other and decreases when they recede. In a gas, atoms move randomly and therefore an observer would measure a range of frequencies. It means that the spectral line is broadened. This broadening of lineshape caused by Doppler effect is called *Doppler broadening.*

In actual practice, all the broadening mechanisms will be present simultaneously and are to be taken into consideration.

Homogeneous broadening mechanisms will result in a *Lorentzian lineshape*. In that case the lineshape function is given by

$$g_L(\nu) = \frac{\Delta\nu}{2\pi\left[(\nu-\nu_0)^2 + (\Delta\nu/2)^2\right]} \qquad ...(1.107)$$

where $\Delta\nu$ is the line width. *Line width* is measured as the separation between the two points on the frequency curve (Fig. 1.19) where the value of $g(\nu)$ falls to half of its peak value.

Inhomogeneous broadening mechanisms will result in a *Gaussian lineshape* in cases of gases. The lineshape function is given by

$$g_G(\nu) = \frac{2}{\Delta\nu}\left(\frac{\ell n2}{\pi}\right)^{1/2} \exp\left[-\ell n2\left(\frac{\nu-\nu_0}{\Delta\nu/2}\right)^2\right] \qquad ...(1.108)$$

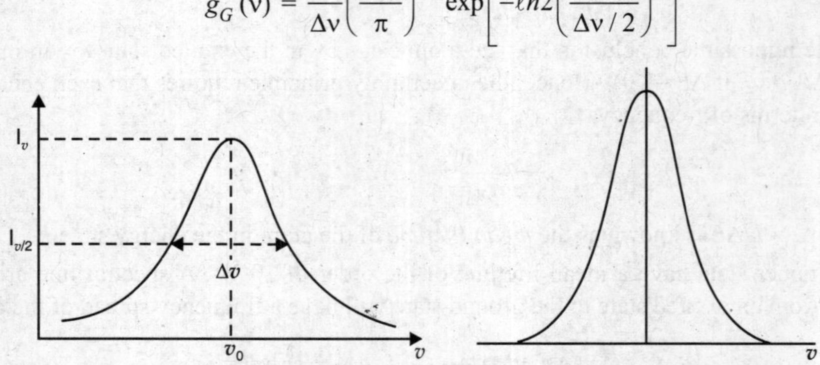

(*a*) Lorentzian Lineshape (*b*) Gaussian Lineshape

Fig. 1.33

Putting $\nu = \nu_0$ in Eq. (1.107), we get

$$g_L(\nu_0) = \frac{2}{\pi\Delta\nu}$$

and from Eq. (1.108), we obtain

$$g_G(v_0) \approx \frac{2}{\Delta\nu}\left(\frac{ln 2}{\pi}\right)^{1/2}$$

The above results may be approximated in general to

$$\boxed{g(\nu_0) \approx \frac{1}{\Delta\nu}} \qquad ...(1.109)$$

1.42 GAIN SATURATION

Initially, population inversion condition is created in the lasing medium by the pumping agent by pumping at a fixed average rate. Now that the population of upper lasing level is more than the lower level, and the rates of induced transitions B_{21} and B_{12} are equal, light of suitable frequency induces more transitions from level E_2 to level E_1 than in the opposite direction. To begin with, the gain of the medium may be well above the threshold value and amplification takes place. Lasing will begin and the strength of the light field within the active medium increases exponentially. The rate at which stimulated emissions take place is proportional to the strength of the light field present. Therefore as the intensity of the light builds up in the medium, the rate at which atoms are removed from the excited state increases. With each stimulated emission of a photon, the population inversion decreases by 2. As the intensity of light due to stimulated emission increases, the degree of population inversion decreases. Consequently, the gain will decrease. The gain ultimately settles down at a value where the rate of production of the excess inverted population is balanced by the rate of decrease through stimulated emission. It happens when the gain just balances the losses in the medium. In terms of population inversion, there is a threshold value N_{th} corresponding to this situation. In the steady state condition $(N_2 - N_1)$ remains equal to N_{th} even though the pumping rate is greater than the threshold pumping rate. To sum up, light amplification in a laser medium cannot increase without limit. As the amplification increases, there is a companion decrease in the population at the upper level. As a result, the population inversion is reduced, the number of stimulated emmision events decrease and the amplification goes down. The reduction in the population inversion and consequent self-adjustment of gain caused by the presence of light field is called *gain saturation*. The gain saturation is the mechanism which adjusts the gain to a value where it just balances the losses in the cavity so that steady oscillations can result.

1.43 GAIN BANDWIDTH

Ideally, a group of identical atoms radiate at the same frequency. However, in practice, because of the various broadening mechanisms there will be a small spread of frequencies about the central value. As a result we have to consider a certain frequency interval called bandwidth corresponding to a (stimulated) transition.

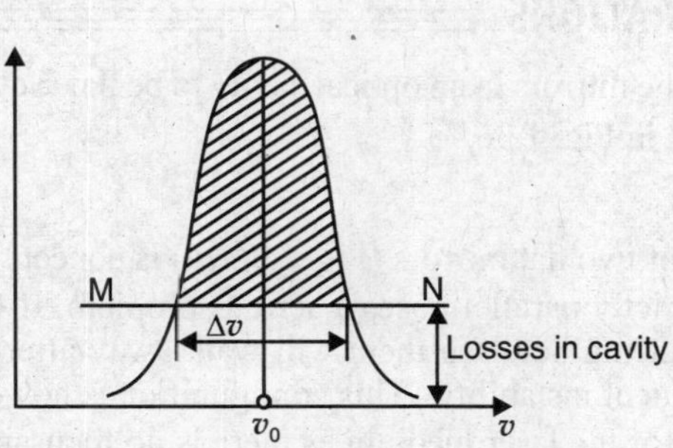

Fig. 1.30. Gain profile with loss level superposed on it.

The limited range of frequencies over which stimulated emission can provide sufficient gain is called the *emission linewidth* or *gain bandwidth.* It is also referred to as *gain profile*. Fig. 1.34 shows a typical gain profile. The line *MN* in Fig. 1.34 indicates the cavity loss level. The curve above the line includes the area of the net gain for the laser. The gain bandwidth Δv is measured at the cavity loss level (line *MN*). Thus Δv constitutes the range of frequencies for which the gain exceeds the cavity losses.

1.44 LASER OPERATING FREQUENCIES

The frequencies at which laser will operate are determined together by the resonant frequencies of the optical cavity resonator and by the laser emission linewidth. If an output is to exist at a particular frequency, the cavity must be resonant at that specific frequency and there must be sufficient gain at that frequency. Laser oscillation can take place only when the gain is large enough to maintain resonance. The actual profile of frequencies radiated by a laser is the product of the envelope of resonant modes and the gain profile. As a result, the laser emission line (Fig. 1.34) transforms into a series of narrower spectral lines corresponding to cavity modes, as shown in Fig. 1.35. Thus, several frequencies lie within the width of a single emission line. Therefore, the output of a laser consists of a number of closely spaced frequencies.

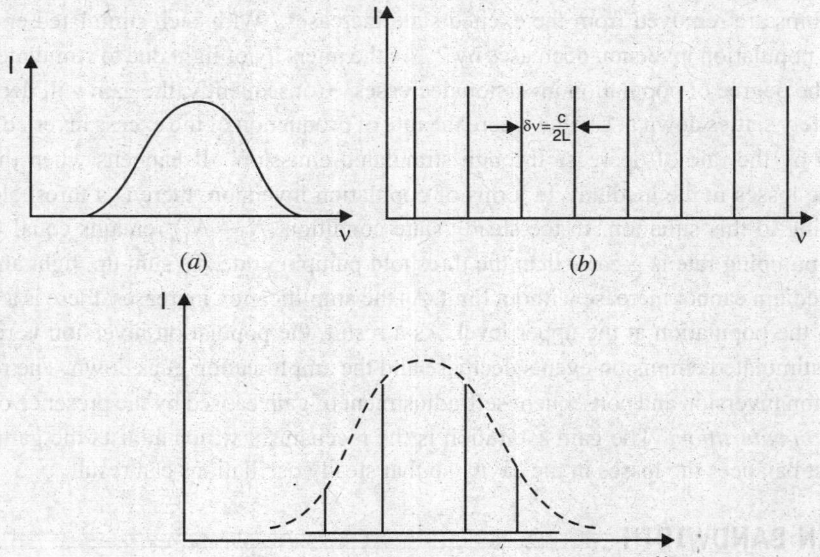

(*a*) (*b*) (*c*)

Fig. 1.35

1.45 CAVITY CONFIGURATIONS

It is not necessary for the mirrors in an optical cavity to be flat. Several configurations are used. Some of them are illustrated in Fig. 1.36.

(*a*) Plane Parallel Cavity :

The cavity consisting of two flat mirrors (Fig. 1.36(*a*)) is conceptually simple, but is unstable. Aligning the two mirrors strictly parallel poses a serious problem. If the mirrors are not precisely parallel, the light beam bouncing between them will walk away after a few reflections, and leads to a loss of light and consequent instability. This configuration is however very efficient. The light beam makes maximum use of the laser medium as there is no focussing of light within the cavity. This configuration is rarely used now a days because of alignment problem and low stability.

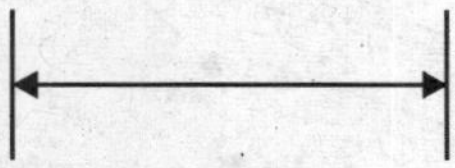

(*a*) Plane parallel resonator.

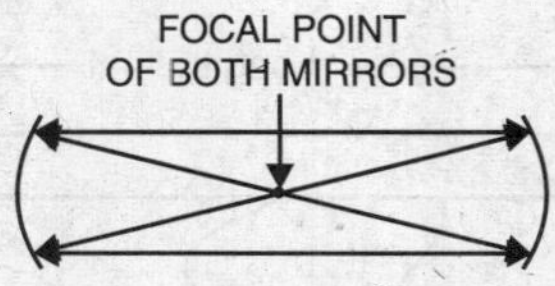

(*b*) Confocal resonator.

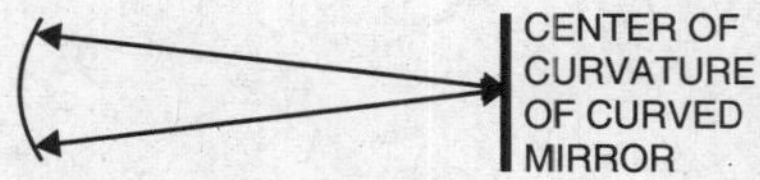

(*c*) Hemispherical resonator.

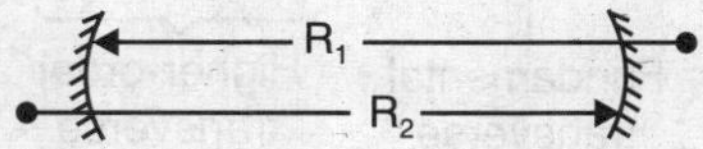

(*d*) Long radius resonator.

Fig. 1.36. Some of the laser resonator configurations

(*b*) Confocal Cavity :

A confocal cavity (Fig. 1.36(*b*)) has two concave mirrors of the same radius of curvature *R*. They are separated by a distance L equal to *R*, *i.e.* $L = R$. The focal length of the mirrors is $L/2$ and the focal points coincide in the centre. Hence the configuration is called "confocal". A confocal cavity is much easier to align. However, the filling is poor and only a small fraction of the medium is utilized.

(*c*) Hemispherical Cavity :

A hemispherical cavity (Fig. 1.36(*c*)) is a hybrid between the plane parallel and the confocal type. It has a concave mirror at one end and a plane mirror at the other. The plane mirror is placed at the centre of curvature of the spherical mirror, *i.e.* $L = R$. Even if the light is not emitted precisely along the axis of the system, the curved mirror can often reflect it back into the cavity. However, in this case also the filling is poor and the output is low.

(*d*) Long Radius Cavity :

A long radius cavity (Fig. 1.36(*d*))has two concave mirrors. Their radii of curvature are significantly longer than the distance between them. That is $R_1 = R_2 > L$. This is a good compromise between the plane parallel and the confocal configurations. Slight misalignment of the cavity mirrors would not cause severe problems because their curvature would automatically focus light back toward the other mirror. This configuration is a stable resonant cavity because the light rays reflected from one mirror to the other will keep bouncing back and forth indefinitely. It is this type of cavity that is used in commercial lasers now a days.

1.46 Modes

We have seen in one of the previous sections that a wave of frequency ν that travels along the axis of the cavity forms a series of standing waves within the cavity. They are discrete resonant conditions determined by the physical dimensions of the cavity.

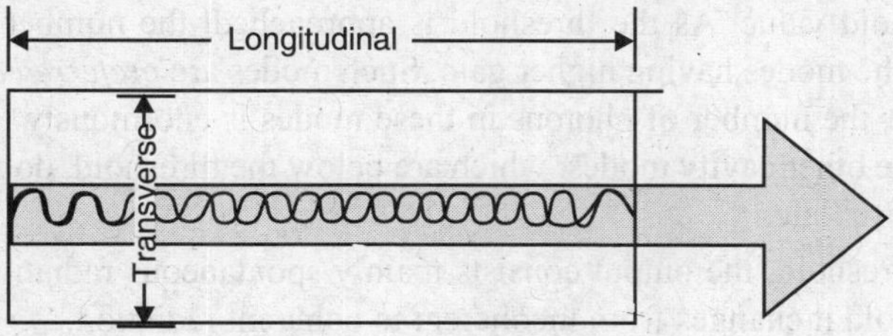

(*a*) Longitudinal travelling waves.

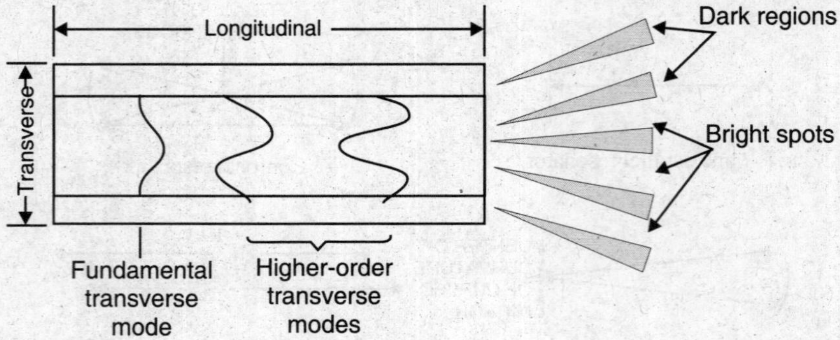

(*b*) Transverse modes

Fig. 1.37

The laser modes governed by the axial dimension of the resonant cavity are called the *longitudinal modes* or *axial modes*. The laser modes determined by the cross-sectional dimension of the optical cavity are called the *transverse modes.*

1.46.1 Longitudinal Modes

If we consider a cavity flanked by two plane parallel mirrors, the standing waves in the cavity satisfy the condition (1.110).

$$\nu_m = \frac{mc}{2\mu L} \quad \text{...(1.110)}$$

m is the number of half – wavelengths or axial modes that fit into the cavity. Each value of m defines an axial mode of the cavity. Axial modes are thus formed by plane waves travelling along the laser cavity on a line joining the centres of the mirrors and consist of a large number of frequencies.

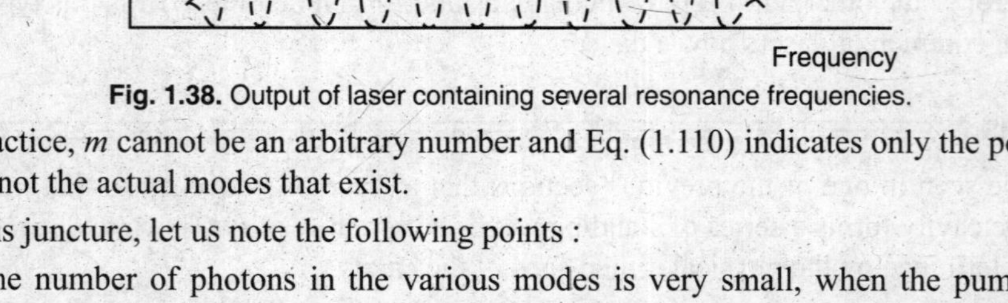

Fig. 1.38. Output of laser containing several resonance frequencies.

In practice, m cannot be an arbitrary number and Eq. (1.110) indicates only the possible axial modes but not the actual modes that exist.

At this juncture, let us note the following points :

(*i*) The number of photons in the various modes is very small, when the pumping rate is below threshold value. As the threshold is approached, the number of photons rapidly increases in the modes having higher gain. Such modes are *preferred cavity modes.* Above the threshold, the number of photons in these modes is enormously large. The number of photons in the other cavity modes, which are below the threshold, do not increase. Hence, they cannot oscillate.

(*ii*) Below the threshold, the output consists mainly spontaneous radiation and is incoherent. At the threshold it changes from incoherent to coherent radiation.

The frequency separation $\Delta\nu$ between adjacent modes is given by Eq. (1.11).

$$\Delta\nu = \frac{c}{2\mu L} \qquad ...(1.111)$$

As $\Delta\nu$ is independent of m, the frequency separation of adjacent modes is the same irrespective of their actual frequencies.

All the axial modes contribute to a single spot of light in the laser output.

Many lasers tend to have naturally many modes. The co-existence of multiple modes reduces the monochromaticity of the laser light, and therefore leads to poor coherence.

1.46.2 Transverse Modes

In addition to the axial modes, a laser output is characterized by *T*ransverse *E*lectro *M*agnetic (TEM) modes. The TEM modes are generally few in number and they are easy to see. If the laser beam is spread out by a negative lens, and focussed on to a screen, several bright patches will be seen on the screen. The patches are separated by intervals called *nodal lines*. Within each patch, the phase of light is the same, but, between patches the phase is reversed (Fig. 1.39).

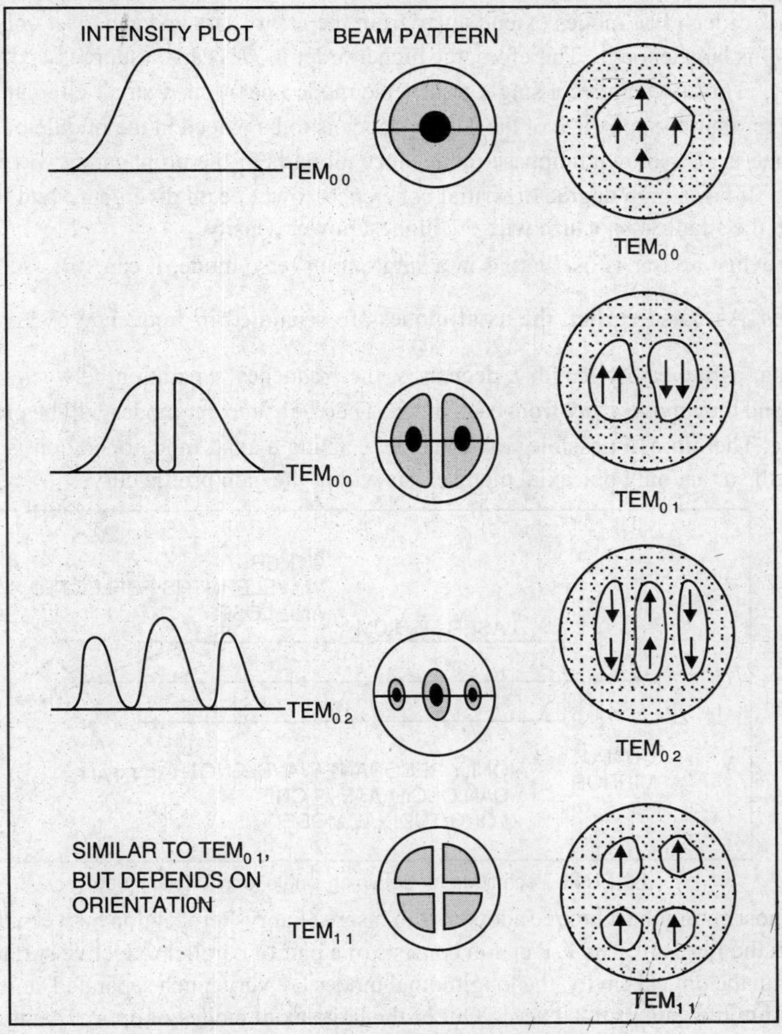

Fig. 1.39. Intensity distribution of transverse modes.

The transverse modes characterize the intensity distribution across the cross-section of the laser beam. In general, the allowed modes are designated as TEM_{mn}, where m and n are integers. The integers m and n represent the number of intensity minima in two orthogonal directions of the laser beam.

The lowest order transverse mode is TEM_{00}. It is the simplest mode and has a smooth cross-section profile with a peak in the middle. TEM_{01} beam has a single minimum dividing the beam into two bright spots. A TEM_{11} beam has two perpendicular minima dividing the beam into four quadrants. If the values of m and n are large, the laser beam contains more bright spots.

Operation of a laser in multimode form provides considerably more power than in single mode operation.

1.47 SINGLE MODE OPERATION

A laser approximates to a monochromatic source when it is operated in a single longitudinal and a single transverse mode. By placing appropriate devices in the laser cavity, one can modify the laser output to contain a single mode.

Higher order TEM modes extend away from the cavity axis and can exist only if the aperture of the cavity is large enough. Therefore, the higher order modes can be suppressed by narrowing the laser cavity. Thus, to achieve a single transverse mode operation, a small circular aperture that is slightly larger than the spot size of the TEM_{00} mode is to be placed in the middle of the cavity. The TEM_{00} mode is also known uniphase mode since all parts of the propagating wavefront are in the same phase. It has a high degree of spatial coherence, lower beam divergence and the beam can be focussed to the smallest spot size with the highest power density.

Even when a laser is oscillating in a single transverse mode, it can still oscillate on several axial modes. As noted earlier, the axial modes are separated in frequency by $\Delta\nu = \frac{c}{2\mu L}$. It is apparent that as the cavity length L decreases, the frequency separation between the axial modes increases, and they move away from each other. Then only a fewer modes will be present within the gain profile. Therefore, a possible way to achieve a single axial mode operation is to decrease the cavity length so that only one axial mode exists within the gain profile curve.

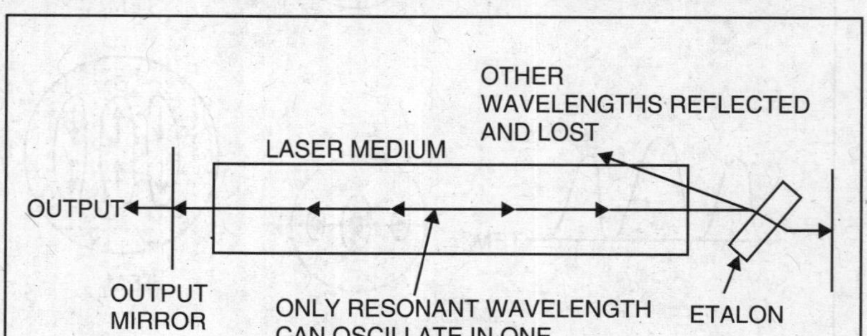

Fig. 1.40. Technique for obtaining single longitudinal mode.

The most common technique adopted is to insert a Fabry-Perot etalon between the laser mirrors as shown in the Fig.1.40. The F-P etalon consists of a pair of parallel reflective surfaces. When it is placed within the optical cavity, the longitudinal modes are very much separated, and only one mode will fall within the gain profile curve. Out of the large axial modes of optical cavity, the mode that coincides with the etalon mode is sustained. All other modes cancel out by destructive interference when they pass through the F-P etalon.

Example 1.33. *The half-width of the gain profile of a He-Ne laser material is about* 2×10^{-3} *nm. What should be the maximum length of the cavity in order to have a single longitudinal mode oscillation ?*

Solution : The separation between successive lines is given by

$$\Delta\lambda = \frac{\lambda^2}{2\mu L}$$

In order to obtain a single mode of oscillation,

$$\Delta\lambda = \text{Half-width of the gain profile}$$

$$\therefore \quad L = \frac{\lambda^2}{2\mu\Delta\lambda} = \frac{\left(6328 \times 10^{-10}\right)^2 m^2}{2 \times 1 \times 2 \times 10^{-12}\, m} \quad (\mu = 1)$$

$$= \frac{4 \times 10^{-13}}{4 \times 10^{-12}} m = \mathbf{10cm}$$

Example 1.34. *The half-width of the gain profile of a He-Ni laser material is about* 2×10^{-3} *nm. If the length of the cavity is 30 cm, how many longitudinal modes can be excited ? The emission wavelength of He-Ne laser is 6328 Å.*

Solution : The distance of separation between successive spectral lines is given by

$$\Delta\lambda = \frac{\lambda^2}{2\mu L} = \frac{(6328 \times 10^{-10}\, m)^2}{2 \times 1 \times 30 \times 10^{-2}\, m} = \frac{4 \times 10^{-13}}{6 \times 10^{-1} m}$$

$$= 0.66 \times 10^{-3} \text{ nm.}$$

μ is taken as 1.

$$\text{Number of modes} = \frac{\text{Half-width of the gain profile}}{\Delta\lambda}$$

$$= \frac{2 \times 10^{-3}\ nm}{0.66 \times 10^{-3}\ nm} = \mathbf{3}$$

1.48 LASER RATE EQUATIONS

We learnt in the previous sections that the populations of energy levels of the lasing medium change under the action of radiation. These changes can be described conveniently by means of rate equations. Rate equations help us in determining the steady state population difference and the threshold pumping rate required to maintain a steady state population inversion. Such a study shows that population inversion cannot take place in a two level system while a minimum pump power is required to obtain inversion in a three level system. On the other hand the attainment of population inversion in a four level system is not dependent on the pump power. We also draw information regarding the optimum power that could be extracted from the laser.

1.48.1 Two Level System

Let us consider a two level system consisting of energy levels E_1 and E_2 which are populated by N_1 and N_2 atoms per unit volume respectively. The total number of atoms which participate in the lasing action is given by

$$N_0 = N_1 + N_2 \qquad \text{...(1.112)}$$

Lasing begins when the population of the level E_2 exceeds that of E_1. The threshold population inversion density is given by

$$N_{th} = N_2 - N_1 \quad ...(1.113)$$

Combining eqs. (1.112) and (1.113), we get

$$N_2 = \frac{N_0}{2} + \frac{N_{th}}{2}$$

That is, $N_2 > N_0/2$

The above condition implies that lasing can begin in a 2-level laser only when more than half of the total population is pumped up to the upper energy level.

We will now examine whether the population inversion state can be reached in a two-level system. Let $\rho(\nu)$ be the energy density of the light of frequency ν incident on the system. The number of atoms per unit volume per unit time which are excited to the upper level are

$$N_{ab} = B_{12}\,\rho(\nu)g(\nu)\,N_1 = W_{12}\,N_1 \quad ...(1.114)$$

where $W_{12} = B_{12}\,\rho(\nu)g(\nu)$

The number of atoms per unit volume per unit time undergoing stimulated emissions from E_2 to E_1 are

$$N_{st} = W_{21}\,N_2$$

Since the stimulated emission probability is equal to the absorption probability,

$$W_{12} = W_{21} \quad ...(1.115)$$

While some of the excited atoms at level E_2 undergo stimulated emissions, some of the others undergo spontaneous emission transitions, which consist of both radiative and non-radiative types. The number of atoms undergoing spontaneous transitions form E_2 to E_1 will be

$$N_{sp} = (A_{21} + S_{21})\,N_2 = T_{21}N_2 \quad ...(1.116)$$

The rate of change of population of E_2 level is given by

$$\frac{dN_2}{dt} = W_{12}N_1 - W_{21}N_2 - T_{21}N_2$$

As $W_{12} = W_{21}$,

$$\frac{dN_2}{dt} = W_{12}(N_1 - N_2) - T_{21}N_2 \quad ...(1.117)$$

Similarly, the rate of change of population at E_1 level is given by

$$\frac{dN_1}{dt} = -\,W_{12}\,(N_1 - N_2) + T_{21}N_2 \quad ...(1.118)$$

In the steady state condition,

$$\frac{dN_2}{dt} = 0 \text{ and } \frac{dN_1}{dt} = 0$$

$$\therefore \quad W_{12}(N_1 - N_2) - T_{21}N_2 = 0$$

and $$-\,W_{12}(N_1 - N_2) + T_{21}N_2 = 0$$

$$\therefore \quad W_{12}(N_1 - N_2) = T_{21}N_2$$

or $$N_1 - N_2 = \frac{T_{21}}{W_{12}}N_2$$

$$\therefore \quad N_1 = N_2\left(1+\frac{T_{21}}{W_{12}}\right) = N_2\left(\frac{W_{12}+T_{21}}{W_{12}}\right)$$

or

$$\boxed{\frac{N_2}{N_1} = \frac{W_{12}}{W_{12}+T_{21}}} \qquad ...(1.119)$$

$$N_2 < N_1 \text{ as } W_{12}+T_{21} > W_{12}$$

It implies that we can never attain a steady state population inversion by *optical pumping* in a two level system.

However, a pn-junction semiconductor laser is a two-level laser where pumping is done by direct electrical pumping.

1.48.2 Three Level Laser

Let us now consider a three level system consisting of energy levels E_1, E_2 and E_3 which are populated by N_1, N_2 and N_3 atoms per unit volume respectively. Let N_0 be the total number of active atoms per unit volume.

$$N_0 = N_1 + N_2 + N_3 \qquad ...(1.120)$$

The rate of change of atomic density N_3 has the following components :

(i) The pump transition to E_3 which raises atoms from the ground level E_1 given by $W_p(N_1 - N_3)$,

(ii) The non-radiative spontaneous transition to the level E_2 given by $S_{32}N_3$, and

(iii) The spontaneous transitions to the level E_1 given by $A_{31}N_3$.

Therefore, the rate equation for N_3 can be written as

$$\frac{dN_3}{dt} = W_p(N_1 - N_3) - A_{31}N_3 - S_{32}N_3 \qquad ...(1.121)$$

The rate of change of atomic density N_2 has the following components :

(i) The stimulated emissions to E_1 which produce laser light given by $W_{21}(N_2 - N_1)$,

(ii) The spontaneous emission to the level E_1 given by $A_{21}N_2$, and

(iii) The spontaneous transition from the level E_3 given by $S_{32}N_3$.

Therefore, the rate equation for N_2 can be written as

$$\frac{dN_2}{dt} = -W_{21}(N_2 - N_1) + S_{32}N_3 - A_{21}N_2 \qquad ...(1.122)$$

The rate of change of atomic density N_1 has the following components:

(i) The pump transition transfers atoms to the level E_3 given by $W_p(N_1 - N_3)$.

(ii) The stimulated emission to the level E_1 given by $W_{21}(N_2 - N_1)$, and

(iii) The spontaneous transitions to the level E_1 given by $A_{21}N_2$.

The rate equation for N_1 can be written as

$$\frac{dN_1}{dt} = -W_p(N_1 - N_3) + W_{21}(N_2 - N_1) + A_{12}N_2 \qquad ...(1.123)$$

Under steady state condition, we must have

$$\frac{dN_3}{dt} = 0, \frac{dN_2}{dt} = 0 \text{ and } \frac{dN_1}{dt} = 0$$

We write from Eq. (1.121), that

$$W_pN_1 = (W_p + A_{31} + S_{32})N_3$$

$$N_3 = \frac{W_p}{W_p + A_{31} + S_{32}} N_1 \qquad ...(1.124)$$

As the probability of spontaneous transition from level 3 to level 2 is much higher than the probability of spontaneous transition from level 3 to level 1.

$$A_{31} << S_{32}.$$

$$\therefore \qquad N_3 = \frac{W_p}{W_p + S_{32}} N_1 \qquad ...(1.125)$$

From Eq. (1.122), we obtain

$$W_{21}N_1 + S_{32}N_3 = (W_{21} + A_{21})N_2$$

Using the value of N_3 from Eq. (1.125) into the above equation, we get

$$\left[W_{21} + \frac{W_p S_{32}}{W_p + S_{32}}\right] N_1 = (W_{21} + A_{21})\, N_2$$

$$\therefore \qquad \frac{N_2}{N_1} = \frac{W_{21}(W_p + S_{32}) + W_p S_{32}}{(W_p + S_{32})(W_{21} + A_{21})}$$

$$\therefore \qquad \frac{N_2 - N_1}{N_1} = \frac{W_p(S_{32} - A_{21}) - S_{32}A_{21}}{(W_p + S_{32})(W_{21} + A_{21})} \qquad ...(1.126)$$

The minimum pumping rate required to achieve population inversion may be given by

$$W_{pt} = \frac{S_{32}A_{21}}{S_{32} - A_{21}} \qquad ...(1.127)$$

As

$$S_{32} << \mathrm{A}_{21}$$

$$W_{pt} \approx A_{21} \qquad ...(1.128)$$

Similarly

$$\frac{N_2 + N_1}{N_1} = \frac{2W_{21}(W_p + S_{32}) + W_p S_{32} + W_p A_{21} + S_{32}A_{21}}{(W_p + S_{32})(W_{21} + A_{21})} \qquad ...(1.129)$$

Dividing Eq. (1.126) with Eq. (1.128), we get

$$\frac{N_2 - N_1}{N_2 + N_1} = \frac{W_p(S_{32} - A_{21}) - S_{32}A_{21}}{2W_{21}(W_p + S_{32}) + (W_p + A_{21})S_{32} + W_p A_{21}}$$

Below threshold for laser oscillation, W_{21} is very small and the above equation may be approximated to

$$\frac{N_2 - N_1}{N_2 + N_1} \approx \frac{W_p\, S_{32} - S_{32}A_{21}}{W_p\, S_{32} + S_{32}\, A_{21} + W_p\, A_{21}}$$

$$= \frac{(W_p - A_{21})\, S_{32}}{(W_p + A_{21} + W_p A_{21} / S_{32})\, S_{32}}$$

As $S_{32} >> A_{21}$, the term $W_p A_{21}/S_{32}$ may be neglected.

$$\therefore \qquad \frac{N_2 - N_1}{N_2 + N_1} = \frac{W_p - A_{21}}{W_p + A_{21}} \qquad ...(1.130)$$

or

$$W_p N_1 = A_{21} N_2 \qquad ...(1.131)$$

The condition necessary for laser oscillation to occur is that $(N_2 - N_1)$ must be positive. It requires that $W_p > A_{21}$.

Let us now estimate the threshold pumping power required to start laser oscillations. The number of atoms pumped per unit volume per unit time from E_1 to E_3 is W_pN_1. If the pump frequency is denoted by ν_p, then the power required per unit volume will be

$$P = W_pN_1h\nu_p \qquad ...(1.132)$$

Threshold pump power can be written as

$$P_{th} = A_{21}N_1h\nu_p \qquad ...(1.133)$$

As there will be very few atoms in E_3, $N_3 \approx 0$ and

$$N_0 \approx N_1 + N_2$$

$$N_0 >> N_2 - N_1$$

We can therefore assume $N_1 \approx N_2 \approx N_0/2$ and write Eq. (1. 133) as

$$\boxed{P_{th} = \frac{N_0 h\nu_p}{2\tau_{sp}}} \qquad ...(1.134)$$

1.48.3 Four Level Laser

Let us consider the ideal four level laser shown in Fig. 1.23. Level E_1 is the ground level, E_4 the pumping level and E_3 and E_2 are the upper and lower lasing levels. Atoms are pumped from the ground level E_1 to the pumping level E_4. They make quick non-radiative transition to the metastable level E_3. We assume that $E_2 >> kT$ so that the level is not populated at ordinary temperatures by thermal process. It is further necessary that level E_2 has very small lifetime such that atoms are not accumulated in E_2 and spoil the population inversion condition between levels E_3 and E_2. Thus, the lifetime t_2 of atoms in level E_2 is short compared to the lifetime τ_3 of atoms in energy level E_3. We are interested only in the rate of change of the atomic density N_3 in level E_3 and that of N_2 in E_2.

Let R_2 denote the rate at which atoms are pumped to level E_4 and from there the atoms make a quick nonradiative transition to the level E_3. In effect, R_2 represents the rate at which atoms are arriving in E_3. Let R_1 denote the rate at which atoms are pumped into level E_2. Process R_1 is detrimental to laser action as it tends to reduce the population inversion condition between the levels E_3 and E_2. In gas lasers, pumping into the lower laser level E_2 is unavoidable and consequently, we take R_1 into consideration. The decrease in the number of atoms in level E_3 has the following components:

(i) The stimulated transition to E_2 which produces the lasing light given by $W_{32}(N_3 - N_2)$,

(ii) The spontaneous emission to the level E_2 given by N_3A_{32}.

(iii) Process R_2 populates the level.

The rate equation for N_3 may be written as

$$\frac{dN_3}{dt} = R_2 - W_{32}(N_3 - N_2) - N_3A_{32} \qquad ...(1.135)$$

The change in the number of atoms in level E_2 has the following components :

(i) The stimulated emissions from level E_3 given by $W_{32}(N_3 - N_2)$;

(ii) The spontaneous emission from level E_3 given by N_3A_{32}

(iii) The process R_1 which pumps atoms from level E_1 ;

(iv) The spontaneous emission to level E_1 given by N_2A_{21}.

The rate equation for N_2 can be written as

$$\frac{dN_2}{dt} = R_1 + W_{32}(N_3 - N_2) + N_3A_{32} - N_3A_{21} \quad ...(1.136)$$

In the steady state condition, $\frac{dN_3}{dt} = 0$ and $\frac{dN_2}{dt} = 0$

Thus from equations (1.135) and (1.136) we obtain

$$R_2 - W_{32}(N_3 - N_2) - N_3A_{32} = 0 \quad ...(1.137)$$

$$R_1 + W_{32}(N_3 - N_2) + N_3A_{32} - N_2A_{21} = 0 \quad ...(1.138)$$

Adding Eq. (1.137) to equation (1.138) we get

$$R_2 + R_1 = N_2 A_{21}$$

Since $R_2 >> R_1$, we write

$$R_2 = N_2 A_{21}$$

$$\therefore \quad N_2 = \frac{R_2}{A_{21}} \quad ...(1.139)$$

Using Eq. (1.139) into Eq. (1.137) we obtain

$$R_2 = (W_{32} + A_{32}) N_3 - W_{32}R_2/A_{21}$$

or

$$N_3 = R_2\left[1 + \frac{W_{32}}{A_{21}}\right]\left[\frac{1}{W_{32} + A_{32}}\right] \quad ...(1.140)$$

$$\therefore \quad N_3 - N_2 = R_2\left[\frac{(A_{21} + W_{32})}{A_{21}(W_{32} + A_{32})} - \frac{1}{A_{21}}\right]$$

$$\therefore \quad N_3 - N_2 = R_2\left[\frac{1 - A_{32}/A_{21}}{(W_{32} + A_{32})}\right] \quad ...(1.141)$$

From the above equation it is evident that

$$N_3 - N_2 > 0 \text{ if } A_{21} >> A_{32}$$

As $A_{21} = \frac{1}{\tau_{21}}$ and $A_{32} = \frac{1}{\tau_{32}}$,

the necessary condition for population inversion is thus

$$\tau_{21} < \tau_{32}$$

The above condition implies that atoms dropping to E_2 level by spontaneous emission must be removed at a faster rate than the arrival rate. If this condition is not satisfied atoms accumulate at E_2 and however hard we pump, population inversion cannot be attained between E_3 and E_2 levels. Hence lasing action does not occur.

Below the threshold, the stimulated transition rate

$$W_{32} = 0$$

$$\therefore \quad N_3 - N_2 = R_2\left[\frac{1 - A_{32}/A_{21}}{A_{32}}\right]$$

This condition continues upto the threshold level. Therefore, the threshold value is given by

$$N_{th} = (N_3 - N_2)_{th} = R_{th}\left[\frac{1 - A_{32}/A_{21}}{A_{32}}\right]. \quad ...(1.142)$$

Since $A_{32} << A_{21}$, $1 - A_{32}/A_{21} \approx 1$,

$$\therefore \quad R_{th} = N_{th} A_{32}$$

or
$$R_{th} = \frac{N_{th}}{\tau_{32}} \qquad ...(1.143)$$

Using Eq. (1.143) into Eq. (1.90), we get

$$N_{th} = \frac{8\pi \nu_0^2 \, \tau_{32} \gamma_{th} \, \Delta\nu}{\upsilon^2} \qquad ...(1.144)$$

This is the stage at which the gain at ν_0 due to the population inversion is large enough to balance the cavity losses. Under steady state condition, $(N_3 - N_2)$ remains then on equal to N_{th} even if the rate of pumping is made greater than the pumping threshold. If an increase in $(N_3 - N_2)$ would occur, it would lead to an increase in stimulated emissions thereby an increase of stored energy with time in the cavity. This obviously violates the steady state assumption. Hence $(N_3 - N_2)$ remains equal to N_{th}.

Each atom raised into level E_3 absorbs an amount of energy E_4 so that the total pumping power per unit volume required at threshold is

$$P_{th} = \frac{N_{th}}{\tau_{32}} E_4 \qquad ...(1.145)$$

Using Eq. (1.144) into the above, we get

$$\boxed{P_{th} = \frac{8\pi \nu_0^2 \, \gamma_{th} \, \Delta\nu}{\upsilon^2} E_4} \qquad ...(1.146)$$

Comparison of three level and Four Level Lasers

In the case of a three-level laser, we can write

$$N_{th} = (N_2 - N_1)$$

and
$$N_0 = N_2 + N_1$$

For the laser to begin lasing $N_2 > \frac{N_0}{2} + \frac{N_{th}}{2}$

As $N_0 >> N_{th}$, $N_2 > N_0/2$ ....(1.147)

On the other hand in a four-level laser,

$$N_2 = N_1 \, e^{-(E_2 - E_1)/kT}$$

Assuming that $(E_2 - E_1)/\mathrm{kT} >> 1$, $N_2 \approx 0$

For the laser to begin lasing $(N_3 - N_2) > N_{th}$ *i.e.* $N_3 > N_{th}$

$$\therefore \quad \frac{(N_{th})_{3-level}}{(N_{th})_{4-level}} = \frac{N_0}{2N_{th}} = a \text{ very large quantity}$$

It is obvious that it is much easier to pump a four-level laser than a three-level laser. This is the reason why most of the lasers are of four-level.

1.49 OPTIMUM OUTPUT POWER

The population inversion $(N_3 - N_2)$ is given by Eq. (1.141) as

$$N_3 - N_2 = \frac{R_2 (1 - A_{32} / A_{21})}{W_{32} + A_{32}} \qquad ...(1.148)$$

or
$$N_3 - N_2 = \frac{R_1}{W_{32} + A_{32}} \quad ...(1.149)$$

where
$$R = R_2(1 - A_{32}/A_{21}) \quad ...(1.150)$$

We rewrite Eq. (1.149) as

$$N_3 - N_2 = \frac{R / A_{32}}{1 + W_{32} / A_{32}} \quad ...(1.151)$$

Since the gain coefficient γ (ν) is proportional to $(N_3 - N_2)$, we use Eq. (1.151) to write

$$\gamma = \frac{\gamma_0}{1 + W_{32} / A_{32}} \quad ...(1.152)$$

where $\gamma_0 = R/A_{32}$ is the gain coefficient in the absence of feedback.

The power emitted by the laser is given by

$$P_e = N_{th} V W_{32}\, h\nu \quad ...(1.153)$$

where V is the active volume of the lasing material.

The amount of spontaneous light generated by the lasing material when it is just at the threshold, but not lasing, is given by

$$P_s = N_{th} V A_{32} h\nu = \frac{8\,\pi\mu_0^3\, h\Delta\nu V}{\lambda^3 t_c} \cdot \frac{\tau_{sp}}{\tau_{32}} \quad ...(1.154)$$

P_s is called *critical fluorescence.*

Using eqs.(1.153) and (1.154) into eq.(1.152), we get

$$\gamma = \frac{\gamma_0}{1 + P_e / P_s} \quad ...(1.155)$$

The expression (1.154) gives us the total power generated within the cavity by the atoms due to stimulated emission. However, only a fraction of the total emitted power P_0 is coupled out of the cavity as useful output laser beam through the output mirror. As far as the oscillation condition is concerned, the output power is a loss to the cavity. We would like to extract more power from the cavity, and it could be done by increasing the transmission coefficient of the output mirror.

If the transmission coefficient of the output mirror is increased, the light output increases but it means an increase in the cavity losses. Further, increasing transmission reduces mirror reflectivity. If the mirro reflectivity is smaller, the cavity losses exceed the gain and the laser ceases oscillating. On other hand, if the output mirror reflectivity is increased to say 100%, the laser oscillates but the output power will become zero. It means that for a given pumping rate, there exists an optimum output coupling which yields the maximum output power.

We recall that the oscillation condition is given by

$$r_1 r_2\, e^{[\gamma_t - \alpha]L} = 1 \quad ...(1.156)$$

We rewrite the above equation as

$$e^{\gamma_t L}\,(1 - l_c) = 1 \quad ...(1.157)$$

where $l_c = 1 - r_1 r_2 e^{-\alpha L}$ is the fractional loss per pass.

The term l_c consists of two types of losses : one due to the useful power output and the other due to the inherent losses. Thus,

$$l_c = T_0 + l_i \quad ...(1.158)$$

$$P_0 = P_e \frac{T_0}{T_0 + \ell_i} \quad ...(1.159)$$

From Eq. (1.155), we can write

$$P_e = P_s\left[\frac{\gamma_0}{\gamma}-1\right]$$

Using Eq. (1.157), we write

$$P_e = P_s\left[\frac{2\gamma_0 L}{l_c}-1\right] \quad ...(1.160)$$

Using Eq. (1.160) into Eq. (1.159) we get

$$P = P_s\left[\frac{T_0}{T_0+\ell_i}\right]\left[\frac{2\gamma_0 L}{T_0+\ell_i}-1\right] \quad ...(1.161)$$

It is seen from the above expression that as $T_0 \to 0$, $P_0 \to 0$. On the other hand, as $T_0 \to \infty$, P_0 decreases.

We write for P_s the expression (1.154).

$$P_s = \frac{8\pi\mu_0^3\, h\nu_0 \Delta\nu V}{\lambda^3 t_c}\frac{\tau_{sp}}{\tau_{32}} \quad ...(1.162)$$

But $t_c = nL/cl_c$ and $V/L = A$. Using these relations in the above expression (1.159) we get

$$P_0 = \frac{8\pi\mu_0^3\, h\nu_0 \Delta\nu A}{\lambda^2(\tau_{32}/\tau_{sp})}\, T_0\left[\frac{2\gamma_0 L}{T_0+\ell_i}-1\right]$$

or

$$P_0 = I_s A T_0\left[\frac{2\gamma_0 L}{T_0+\ell_i}-1\right] \quad ...(1.163)$$

where $I_s = \dfrac{8\pi\mu_0^2\, h\nu_0 \Delta\nu}{\lambda^2(\tau_{32}/\tau_{sp})}$ is called the *saturation intensity.*

To find out the optimum value of P_0, we set

$$\frac{\partial P_0}{\partial T_0} = 0$$

$$P_0 = I_s A\left[\frac{2\gamma_0 L T_0}{T_0+\ell_i}-T_0\right]$$

$$\frac{\partial P_0}{\partial T_0} = I_s A\left\{\frac{(T_0+\ell_i)2\gamma_0 L - 2\gamma_0 L T_0}{(T_0+\ell_i)^2}-1\right\} = 0$$

$$\therefore \quad (T_0 + l_i)\, 2\gamma_0 L - 2\gamma_0 L T_0 = (T_0 + l_i)^2$$

$$2\gamma_0 L l_i = (T_0 + l_i)^2$$

$$\therefore \quad (T_0)_{opt} = \sqrt{2\gamma_0 L l_i} - l_i \quad ...(1.164)$$

This is the condition for the mirror transmission that yields the maximum power output. The power output at optimum coupling is obtained by using Eq. (1.164) into Eq. (1.163).

$$(P_0)_{opt} = I_s A\left[-\ell_i + \sqrt{2\gamma_0 L\ell_i}\right]\left[\frac{2\gamma_0 L}{\sqrt{2\gamma_0 L\ell_i}}-1\right]$$

$$= I_s A\left[-2\sqrt{2\gamma_0 L\ell_i} + \ell_i + 2\gamma_0 L\right]$$

$$\therefore \quad (P_0)_{opt} = I_s A\left(\sqrt{2\gamma_0 L} - \sqrt{\ell_i}\right)^2 \qquad ...(1.165)$$

Useful power output P_0 versus mirror transmission T_0 for different values of internal loss l_i in a Helium-Neon laser is shown in Fig. 1.41. It is evident from the plot that there exists a maximum power output for each l_i value.

It is important to note here that the power levels are higher within the cavity than outside. In a typical case where the output mirror transmits about 1.5% of light, the balance 98.5% light is reflected back into the cavity. Consequently the power output is about 1.5% of the power existing inside the cavity. If the output power is assumed to be 1 mW, the power within the cavity would be as large as 67 mW.

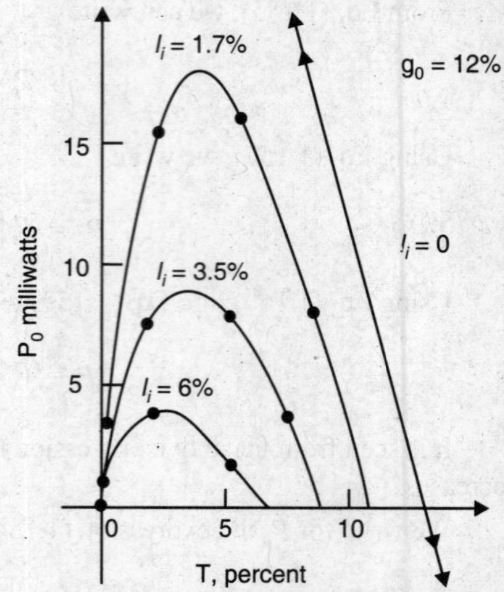

Fig. 1.41. Power output versus mirror transmission for various values of l_i for a He-Ne laser.

Example 1.35 *Calculate threshold pumping power per unit volume in case of Ruby laser. Given : $N = 1.6 \times 10^{25}$ atoms/m³, frequency $\nu_p = 6.25 \times 10^{14}$ and $t_{sp} = 3 \times 10^{-3}$ sec.*

Solution. In three level laser scheme, as in the case of Ruby laser, let N_2 be the number of atoms in the upper laser level E_2 (metastable state) and t_{sp} the spontaneous time at this level, then

Number of atoms decaying per unit time from upper laser level $= \frac{N_2}{t_{sp}}$. For each atom lifted from the ground state E_1 to E_2.

Energy required = $h\nu_p$ where ν_p is average pump frequency. Therefore, to maintain N_2 atoms in the level E_2,

the minimum power, $$P = \frac{N_2 h\nu_p}{t_{sp}}$$

Now, If N is the total number of atoms per unit volum, then

$$N_2 = \frac{N}{2}$$

$\therefore$ Minimum power, $$P = \frac{N}{2}.\frac{h\nu_p}{t_{sp}}$$

$$= \frac{1.6\times10^{25}}{2} \times \frac{6.6\times10^{-34}\times6.25\times10^{14}}{3\times10^{-3}}$$

$$= \mathbf{1100 \times 10^6\ W/m^3}$$

1.50 PROPERTIES OF LASER MODES

In general, several longitudinal and transverse modes exist simultaneously within the laser cavity. The characteristics of laser modes may be summarized as follows.

Spatial Dependence :

Each longitudinal mode is a distinct standing wave as shown in Fig. 1.29 and has a value of the electric field of zero at the mirrors. The longitudinal modes are defined by

$$\nu = \frac{c}{2\mu L}$$

where $m = 1, 2, 3, \ldots$ gives the mode number. Each longitudinal mode is associated with a number of transverse modes characterized by the integers (m, n). The standing wave pattern for each mode corresponds to a distinct spatial distribution of laser intensity within the cavity. The spatial distribution of each mode is slightly different from that of the other modes. Consequently, different modes cover different regions of the medium.

Frequency Dependence :

Two transverse modes may belong to the same longitudinal mode but they may differ slightly in frequency, since they may have different optical path lengths within the cavity. In contrast, adjacent longitudinal modes will have a greater frequency difference.

Mode Competition :

Let us consider a medium located inside a cavity. The pumping is increased slowly from less than threshold to above the threshold. For much-below-threshold case spontaneous emission is present. With an increase in pumping, cavity modes will build up. With further pumping, the amplitude of the cavity modes increases until the threshold is reached. The waves associated with different modes within the same region compete for the same upper laser level species (excited atoms) and the various modes develop. Each mode tries to grow towards reaching its saturation intensity by stimulating more emissions at its frequency. At threshold, oscillations at ν_o only start as it corresponds to the mode with the highest round trip gain. Let us now examine separately the cases of homogeneously broadened and inhomogeneously broadened lasers.

In lasers with homogeneously broadened transitions, an increase in pumping cannot increase the population inversion beyond the threshold value. As all atoms in an ideal homogeneously broadened laser behave identically and are characterized by identical spectral line shape function g(ν), the oscillations build up only at the central frequency ν_o where the gain is highest. The gain at other frequencies will remain below threshold. As it will be difficult for other modes to stimulate emissions into them, only one mode will lase. However more than one transverse mode can excite along with the strongest longitudinal mode because they access different spatial regions.

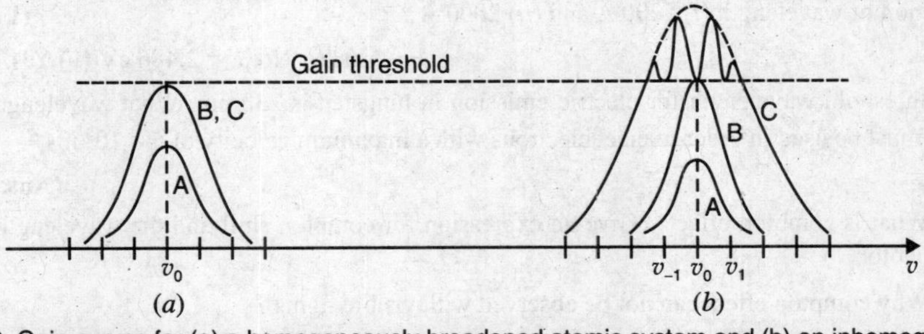

Fig. 1.42. Gain curves for (a) a homogeneously broadened atomic system and (b) an inhomogeneously broadened atomic system (A - below threshold; B - at threshold' C - above threshold). In (b), different groups of atoms respond to the stimulating radiation at different cavity mode frequencies. The gain is saturated at each of these frequencies independently, creating 'holes' in the gain curve.

In case of inhomogeneous broadening, individual atoms behave differently from one another and the population inversion and gain profile can increase above threshold values at frequencies other than ν_o. The gain at ν_o remains clamped at the threshold value due to gain saturation. Further pumping may increase the gain at other frequencies until oscillations commence at those frequencies also. It means that different longitudinal modes can operate independently as they are not competing for the same upper laser level species. However, transverse modes associated with the same longitudinal mode, can be close enough in frequency that they may compete for the same upper laser level species.

Spectral hole Burning :

Different populations within the upper laser level contribute to different frequency components of the gain profile. The population in the upper laser level will be reduced only at those frequencies where the modes are developing. Therefore, the gain decreases to the threshold value at the corresponding frequencies only and the gain curve exhibits depressions at these oscillating frequencies. This effect is known as *spectral hole burning.*

Spatial Hole Burning :

When the standing wave pattern of a single longitudinal mode develops within the homogeneously broadened laser cavity, the laser intensity pattern along the length of the medium is regularly spaced. The spacing occurs at half-wavelength intervals, with the intensity being zero at the cavity mirrors and at every half-wavelength interval between the mirrors. At these points of zero intensity, there is no stimulated emission and thus no reduction in gain. The gain profile therefore exhibits *a periodic spatial variation* within the gain medium. This effect is known as *spatial hole burning. Such a spatial distribution leads to wasted gain. If higher order transverse modes are present there will use part of this gain, but otherwise it is wasted energy.*

EXERCISE-1

1. What is LASER ? How does it differ from an ordinary source of light ? Mention a few engineering applications of LASER ? **(Pune 2010, N.U. 2002)**
2. Discuss the interaction of light with matter.
3. Discuss the importance of Planck's quantum theory.
4. Explain briefly quantum theory of radiation. What is a photon ? State its properties. **(Pune U. 2010)**
5. Express linear momentum of a photon in terms of a wave vector $|\vec{K}|$ and energy of photon in terms of angular velocity *w*. **(G.N.D.U. 2001, M.D.U. 2000)**
6. What is photo-electric effect ? Derive Einstein's photoelectric equations ? **(H.P.U. 2002, Meerut U. 2001, Luck. U. 2001, Nagpur U. 2011, 2002)**
7. Is it possible to liberate an electron from a metal surface having work function 4.8 eV with an incident radiation of wavelength (*i*) 5000 Å and (*ii*) 2000 Å ? **(Luck U. 2003)**

   (Ans. (*i*) No, $E = 2.486$ eV **(*ii*)** Yes, $E = 6.21$ eV**)**
8. The threshold wavelength for electric emission in tungsten is 230 nm. What wavelength of incident light must be used in order to eject electrons with a maximum velocity of 5×10^5 ms^{-1}. **(P.U. 2004)**

   (Ans. $\lambda = 203$ nm**)**
9. **(a)** What is compton effect ? Give an expression for compton shift and the wavelength of scattered photon.

   (b) Why compton effect can not be observed with visible lights ?

   (Pbi.U. 2003, 2000; G.N.D.U. 2004; H.P.U. 2001)
10. An X-ray photon is found to have doubled its wavelength on being scattered by 90°. Find the energy and wavelength of incident photon. **(Ans.** E = 0.51 MeV**)**
11. In an experiment on compton scattering; X-rays of wavelength 10^{-10} m are incident. Calculate the wavelength of X-rays scattered in the backward direction. **(Gharwal U. 2000) (Ans.** 1.0484 Å**)**
12. Explain de-Broglie hypothesis ? Derive expression for de-Broglie wavelength.
13. What is the de-Broglie hypothesis ? Show that the wavelength λ associated with an electron of mass *m* and kinetic energy *E* is given by

$$\lambda = \frac{h}{\sqrt{2mE}}$$

(Nagpur U. 2011)

14. The work function of photoelectric material is 4 eV and the wavelength of incident light is 3000 Å. Calculate the velocity of a photoelectron. **(Nagpur U. 2011)**

(Ans. 2.1×10^5 m/s)

15. What should be the wavelength of incident photon so that increase in wavelength of scattered photon at a scattering angle $\phi = \pi/2$ is 1 % ? **(Ans.** 2.43 Å)

16. State and explain Heisenberg's uncertainty principle.

17. Show that an electron can not exist in the Nucleus of radius in the range 10^{-14} m, on the basis of uncertainty principle. **(Nagpur U. 2011, Pbi. U. 2002, 2000; M.D.U. 2003, Meerut 2003; Gharwal U. 2000; Purwanchal 2002)**

18. What is the relation between work function and threshold frequency. **(Nagpur U. 2011)**

19. Discuss quantum physics briefly and deduce equation. **(Third Sem. B.Tech. G.B.T.U. Lucknow 2010)**

20. Ultra violet light of wavelength 1500Å is incident on Molybdenum whose work function is 4.15 eV. Find the de-Broglie wavelength associated with the photo-electron. **(Bang.U. 2006)**

(Ans. $\lambda = 6$Å)

21. Lifetime of nucleus in the excited state is 10^{-12} s. Calculate the probable uncertainty in energy and frequency of a γ-ray photon emitted by it. **(Kanpur U. 2003)**

(Ans. 1.054×10^{-22} J; 1.59×10^{11} Hz)

22. Calculate the voltage through which an electron may be accelerated so that it has, de-Broglie wavelength of 0.5Å associated with it. **(Pune U. 2010)**

(Ans. 614.4 V)

23. What do you understand by a wave function ψ of a moving particle ? **(Meerut U. 2002)**

24. Give the physical significance of wave function. What does the square of wave function signify ? **(Nagpur U. 2002; M.D.U. 2003; 2000; Meerut U. 2003)**

25. Derive Schrodinger's time dependent and time independent wave equations. **(Nagpur U. 2011. B.Tech Third sem G.B.T.U. Luck. 2010)**

26. A proton and electron have equal to de-Broglie wavelength. Compare their velocities. **(Nagpur U. 2007)**

27. State and explain Heisenberg's uncertainty principle and its two applications. **(Mumbai U. 2006)**

28. With the help of uncertainty principle, prove the existence of protons and neutron's in the nucleus. **(Mysore Uni. 2006)**

29. **(a)** Explain the concept of wave packet.

(b) Define wave velocity and group velocity. Find the relation between them. **(Pune U. 2010)**

30. Write short note on basic principle and efficiency of laser. **(Nagpur U. 2008)**

31. What do you mean by lineshape function ? How it is normalized ?

32. Explain spontaneous emission and stimulated emission. Describe coherent absorption. **(B.Tech, Third sem. G.B.T.U. Luck 2010)**

33. Describe the following terms:

(*i*) Population inversion, **(Pune U. 2010)**

(*ii*) Pumping

(*iii*) Gain of laser. **(B.Tech, Third sem. G.B.T.U. Luck 2010)**

34. 'As long as the material is in thermal equilibrium, the population inversion of the higher state cannot exceed the population of the lower state', Explain.

35. What are the Einstein's coefficients ? Discuss in detail.

(B.Tech, Fourth sem. G.B.T.U. Luck 2010)

36. What are Einstein's postulates and Einstein's coefficients ? Derive the Einstein's relations.

(Pune U. 2010)

37. Explain absorption. Differentiate between spontaneous emission and stimulated emission.

(Kerala U. 2006)

38. Explain (*i*) Absorption
(*ii*) Spontaneous emission and
(*iii*) Stimulated emission. **(Nagpur U. 2007)**

39. Explain light amplification. Why do we say light is Amplified in a LASER ? Establish the condition for light amplification. **(Pune U. 2010)**

40. What are optical cavities and pumping techniques ?
Explain different types of pumping techniques in different types of lasers.

(B.Tech, Third sem. G.B.T.U. Luck 2010)
(P.U. 2007, 2006, G.N.D.U. 2007, 2006)

41. How many methods do you know for optical pumping ? Discuss various methods citing examples.

(B.Tech, Fourth sem. G.B.T.U. Luck 2010)

42. What do you mean by 'population inversion'. Why does it necessary ?

(Pune U. 2010, Nagpur U. 2006)

43. Illustrate basic differences between spontaneous emission and stimulated emission of radiation.

(Pune U. 2010)

44. What do you mean by 'Pumping'? Give methods of pumping briefly.

(Pune U. 2010, Nagpur U. 2005)

45. Explain Laser-Action. Illustrate all levels of Laser-Actions.

46. Explain optical gain and pumping techniques. **(Nagpur U. 2004)**

47. **(a)** What do you understand by spectral line width of a Laser ? **(Pune U. 2010)**
(b) Explain the term pulse-width of a Laser.

48. What are metastable states ? Discuss their role in laser action ?

(B.Tech, Fourth sem. G.B.T.U. Luck 2010; P.U. 2008, 2007)

49. What are laser resonators ? Discuss them briefly. **(Agra U. 2007)**

50. Show that in optical region, stimulated emission is negligible as compared to spontaneous emission.

(Guhati U. 2006)

51. Explain quality factor and losses inside the cavity.

52. What is amplification factor ? How is it related to population inversion ? What is critical population inversion ? **(Pune U. 2010)**

53. Explain the terms: (a) line broadening , (b) natural broadening and (c) collision broadening of a spectral line. **(Pune 2010)**

54. What is Doppler broadening ? Obtain the relation

$$g_G(\nu_0) = \frac{2}{\Delta\nu}\left(\frac{\log_e{}^2}{\pi}\right)^{\frac{1}{2}}$$

Give its interpretation.

55. Explain the meaning of gain bandwidth. How it depicts the energy loss level in a cavity ?

(Guhati U. 2005)

56. What are modes ? Explain different types of modes ? Describes a technique for obtaining single longitudinal mode. **(Kanpur U. 2004)**

57. Compare the three level and four level lasers ? Obtain the relation for optimum output power.

58. Enunciate the properties of laser modes exists within the laser cavity.

59. **(a)** What is a resonator in a Laser ?

(b) Discuss the working of resonators of laser systems. How many types of such resonators you know? Explain the working of one resonator in detail. **(B.Tech, Fourth sem. G.B.T.U. Luck 2010)**

60. Find the intensity of a laser beam of 20mW power and having a diameter of 1.3 mm. Assume uniform intensity across the beam. **(B.Tech, Fourth sem. G.B.T.U. Luck 2010)**

(Ans. 15.09 mW/mm**)**

61. Calculate the ratio of spontaneous emission to stimulated emission if the wavelength of radiation emission is 5500Å at 2000 K (Given $h = 6.602 \times 10^{-34}$ JS and $K = 1.38 \times 10^{-23}$ SI units).

(Ans. 4.62×10^5**)**

62. Find the coherence length of He-Ne line (1150 Å) if coherence time is 26.7 ns. **(Ans.** 8m**)**

63. Find the intensity of laser beam of 20 mW power having a diameter of 0.13 cm. Consider the intensity to be uniform across the beam. **(Ans.** 1.5 kW/m^2**)**

64. The ratio of population of two energy levels out of which upper one corresponds to metastable state is 1.059×10^{-30}. Calculate the wavelength of light emitted at 330 K.

65. In a He-Ne laser beam, the two plane mirrors form the resonant cavity are at a distance of 0.5 m. Find the mode seperation of longitudinal cavity in terms of frequency. **(Ans.** 3×10^8 Hz**)**

66. Calculate the energy difference in eV between two energy levels of Ne atoms of He-Ne gas laser, the transitions between which result in the emission of a light of wavelength 6328 Å. Also calculate the number of photons emitted/sec if optical power output is 1 mW. **(Ans.** 3.182×10^{15}**)**

67. **(a)** Explain the concept of Doppler broadening of spectral lines.

(b) Explain the terms (*i*) Coherence length and (*ii*) Coherence time **(Nagpur U. 2010)**

68. Define: **(a)** Population inversion and

(b) Temporal coherence. **(Nagpur U. 2010)**

2 CHAPTER Types of Lasers

2.1 INTRODUCTION

There are several ways in which we can classify the different types of lasers. It can be done according to what material or element is used as the active medium. It can be done according to the operation of laser in a pulsed mode or in a continuous wave (CW) mode. The classification may be done basing on other parameters such as gain of the laser medium, power delivered by the laser, efficiency or applications. We prefer here to classify the lasers on the basis of the material used as active medium. Accordingly, lasers are broadly divided into four categories-solid lasers, gas lasers, liquid lasers and semiconductor lasers. The subject of laser systems is a vast topic and it is not possible to cover each and every system. We describe only some of the important and practical lasers.

2.2 CLASSIFICATION OF LASERS (on the basis of medium)

As mentioned above, lasers can be classified in several ways as per convenience. Now a days, a large number of lasers are available and many new lasers are coming out regularly. Basically all these lasers can be classified into two categories : (*i*) high density gain medium laser and, (*ii*) low density gain medium laser. High density gain medium lasers are further sub-divided into solid and liquid lasers. Om the basis of medium, the lasers are classified into three catergories : Solid, liquid and gas lasers. In brief they can be put as under:

Solid lasers : Solid state layer was the first known laser. They belong to the laser system involving high density gain media. Different examples of solid state laser are :

- Ruby
- Nd :- YAG laser
- Nd :- Glass laser
- Tunable solid state laser
- Alexaandrite laser
- Titanium-Sapphire laser
- Colour-centre laser
- Fibre laser

Amongst all, Rubby laser is the most important and widely used solid state laser. The active material used is C_r^{3+} doped in Al_2O_3 crystal. Neodymium is a typical example of an optically pumped rare-earth laser system. Neodymium ion (Nd^{3+}) is doped in Yttrium alluminum garnet (YAG) with nearly 0.725% concentration by weight.

Liquid lasers : There are certain unavoidable draw backs in same solid state lasers. For examples, crystals have *optical strain*. Hence, they crack down or shatter. Moreover, they are prone to internal damages and susceptical to (crystal) defect and imperfections. Homogeneous liquid overcomes these problems. The active media of the liquid laser are formed by solutions of certain dyes dissolved in liquids such as water or alcohols. Dye lasers are lunable and have high gain as compared with solid state lasers. Dyes used are of the type

- Polymethene dye
- Xanthene dye
- Coumarin dye
- Scintillator dye.

Gas lasers : Gas lasers belong to low density gain media. Depending upon the characteristics of the active medium, the gas lasers are further categorized into (*i*) atomic, (*ii*) ionic and (*iii*) molecular lasers. The optical quality of gas lasers is much better as it does not suffer from optical inhomogenities. The gas lasers commonly used are Helium-Neon (He – Ne) laser, Argon ion laser, Krypton ion laser, Helium-Cadmium laser, Copper vapour laser, Gold vapour laser, Carbon-diOxide (CO_2) laser, Excimer laser, HF laser, Nitrogen laser.

Semiconductor lasers : Semiconductor laser is a special designed p-n junction dioe which emits coherent radiation when it is forward biased. A small size devoice capable of giving output of wavelength in gigahertz and has an ability to modulate the optical output. Two types of semiconductor lasers namely homojunction laser and heterojuction lasers are observed. They may also be classified in terms of bond gap properties of the gain media as Direct band-gap medium laser and Indirect band-gap medium laser.

We describe some important and usually used lasers as under.

2.3 SOLID STATE LASERS

The term solid state has different meanings in the field of electronics and lasers. A solid state laser is one in which the active centres are fixed in a crystal or glassy material. Solid state lasers are electrically non-conducting. They are also called *doped insulator lasers* to avoid connotation of semiconductor.

The basic principles that underlie the operation of solid state lasers are the same. A generic solid state laser is illustrated in Fig. 2.1.

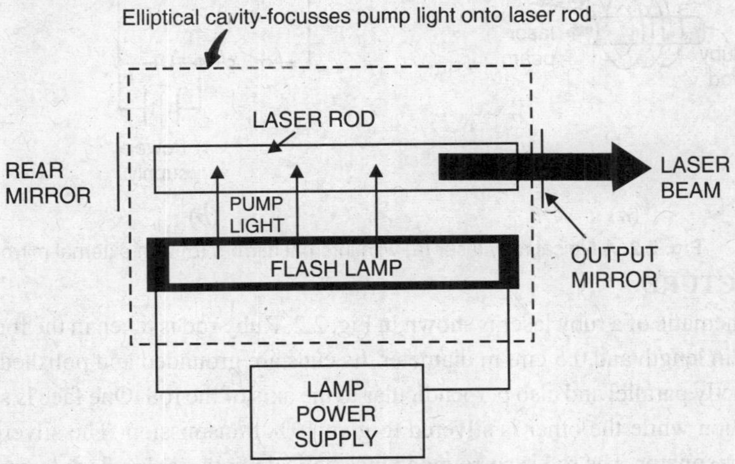

Fig. 2.1. A typical solid state laser

The active centres are dispersed in a dielectric crystal or a piece of glass. The crystal atoms do not participate directly in the lasing action but act as a host lattice to the active centres which are present in concentrations of around 1%. The crystal is usually shaped into a rod with reflecting mirrors placed at each end. Light from an external source such as a flashlamp excites the active centres in the rod. The flashlamp is usually linear in shape. The linear lamp and laser rod are placed close to each other in a reflective cylinder, which focusses pump light onto the rod. The mirrors on the sides of laser rod form a resonant cavity and provide the necessary feedback to generate laser beam.

It is required that the host material should be transparent to the pump light and should not absorb light at the laser wavelength. Most of the excitation energy ends up as heat rather than light. Excess heat damages the laser crystal. Therefore, the crystal should have good thermal conductivity. Normally, cooling with forced air or circulating water needs to be provided.

The active centres are ions of metallic elements : chromium (Cr), neodymium (Nd), erbium (Er), holonium (Ho), cerium (Ce), cobalt (Co) and titanium (Ti). Chromium is the active centre in ruby and alexandrite lasers while neodymium is the active centre in common type of solid state lasers.

These lasers are rugged, simple to maintain, and capable of generating high peak powers.

2.3.1 Ruby Laser

GENERAL DESCRIPTION

Historically, the ruby laser was the first laser. The ruby laser rod consists of a synthetic ruby crystal, Al_2O_3, doped with chromium ions at a concentration of about 0.05% by weight. At this concentration there are about 1.6×10^{25} Cr^{3+} ions per cubic meter. The chromium ions constitute the active centers as they have a set of three energy levels suitable for realizing lasing action whereas the aluminium and oxygen atoms are inert.

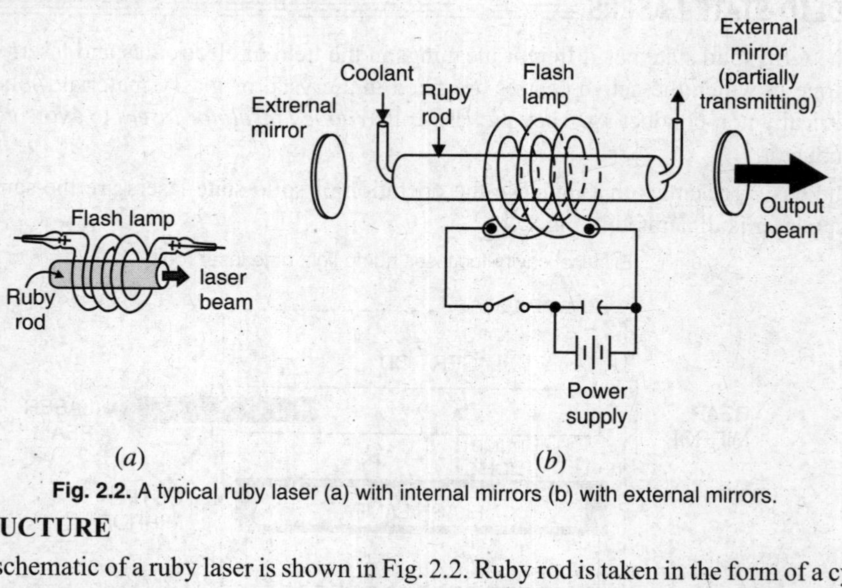

Fig. 2.2. A typical ruby laser (a) with internal mirrors (b) with external mirrors.

STRUCTURE

The schematic of a ruby laser is shown in Fig. 2.2. Ruby rod is taken in the form of a cylindrical rod of 4 cm. in length and 0.5 cm. in diameter. Its ends are grounded and polished such that the end faces are exactly parallel and also perpendicular to the axis of the rod. One face is silvered to achieve 100% reflection while the other is silvered to give 10% transmission. The silvered faces form the Fabry-Perot resonator. The rod is surrounded by a helical photographic flash lamp filled with xenon. The lamp produces flashes of white light whenever activated by the power supply. The system is cooled with the help of a coolant circulating around the ruby rod.

In practical lasers, flash lamps of helical design are no longer used. The most common types are linear lamps. The laser rod is generally placed in an elliptical cavity with a linear flash lamp mounted at the focus of internally reflecting cavity, as shown in Fig. 2.3, to ensure an efficient transfer of pumping energy to the ruby crystal. The mirrors are arranged outside.

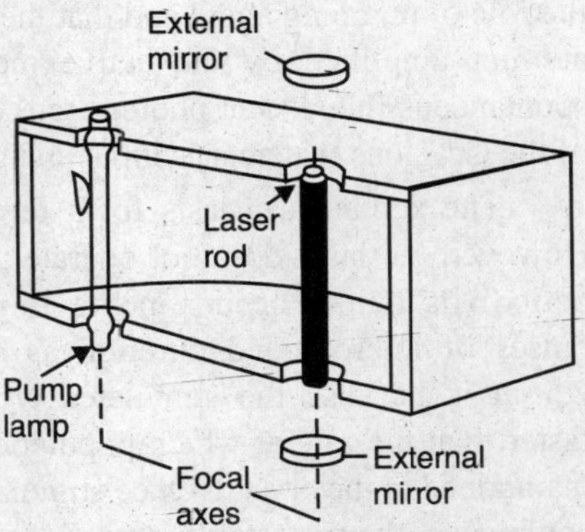

Fig. 2.3. Elliptic cylinder housing for concentrating light onto the laser rod.

WORKING

Ruby is a three-level laser system and the energy levels of Cr^{3+} ions in the crystal lattice are shown in Fig. 2.4. There are two wide energy bands E_3 and E_3', and a pair of closely spaced levels at E_2. When the ruby rod is irradiated with an intense burst of white light from the xenon lamp, the ground state Cr^{3+} ions absorb light in two pump bands one centered near 5500 Å and the other at about 4000 Å, and are excited to the broad upper bands. The energy levels in these bands have a very small lifetime ($\leq 10^{-9}$ sec). Hence, the excited Cr^{3+} ions rapidly lose some of their energy to the crystal lattice and undergo non-radiative transitions to the pair of adjacent levels denoted as E_2. These levels are metastable states having a lifetime of 3×10^{-3} sec. The transition from E_2 to E_1 is radiative and under normal population condition produces spontaneous, incoherent, red fluorescence typical of ruby with a peak near 6943 Å.

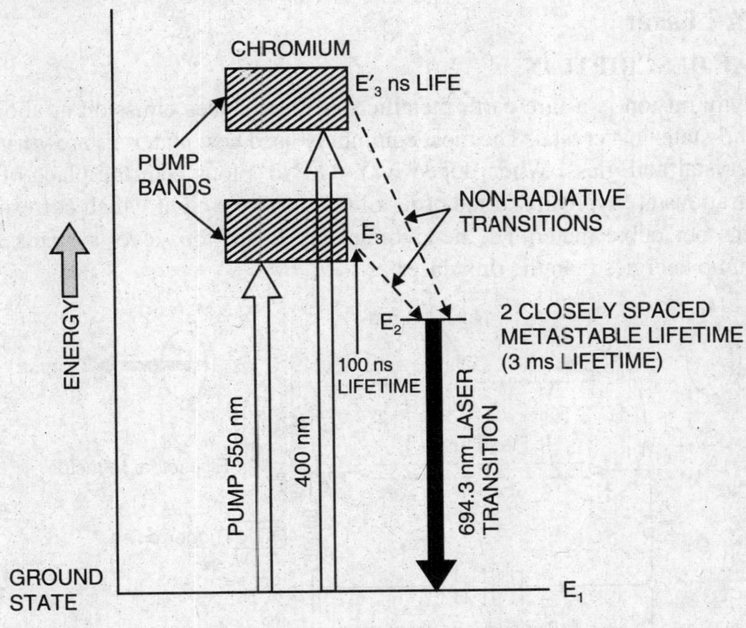

Fig. 2.4. Energy levels of chromium atoms in ruby laser.

Under very intense excitation where pumping energy increases above a critical threshold, population inversion occurs in E_2 with respect to the ground level E_1. Then one of the spontaneously emitted fluorescent photons travelling parallel to the axis of the ruby rod would initiate stimulated emissions. The photons make many passes through the medium building up the stimulated emissions in a large way and the system lases. The photons travelling in any other direction would be lost after a few reflections. The laser beam ensuing from the ruby rod is red in colour and corresponds to a wavelength of 6943 Å. It is to be remembered that the green and blue components of light play the role of pumping agent and that these components are not amplified by the active medium. It is a spontaneous fluorescent photon (red) emitted by one of the Cr^{3+} ions that acts as input and gets amplified.

The xenon flash lasts for a few milliseconds. However, the laser does not operate throughout this period. Its output occurs in the form of irregular pulses of microsecond duration as shown in Fig. 2.5. It is because the stimulated transitions occur faster than the rate at which population inversion is maintained in the crystal. Once stimulated transitions commence, the metastable state E_2 gets depopulated very rapidly and at the end of each small pulse, the population at E_2 has fallen below the threshold value required for sustained emission of light. As a result the lasing ceases and laser becomes inactive. The next pulse appears after the population inversion is once again restored. The process repeats.

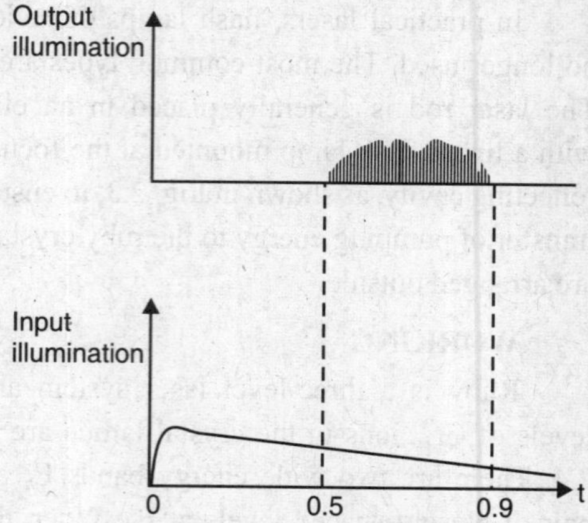

Fig. 2.5. Ruby laser light pulses

The ruby laser has high energy-storage capability because of long upper laser level lifetime of 3 ms. Therefore, pulse energies of upto 100 J are possible. The ruby laser is relatively inefficient having a typical efficiency of 0.1 to 1%.

The ruby laser is used in a variety of applications, the most important one being in holography.

2.3.2 Nd : YAG Laser

GENERAL DESCRIPTION

The neodymium ion is a rare earth metallic ion. It produces emission at about 1 μm, when doped into a solid-state host crystal. The most commonly used host materials are yttrium aluminium garnet (YAG) crystal and glass. When doped in YAG, Nd^{3+} ions take the place of yttrium ions. Doping concentrations are typically of the order of 0.725% by weight which corresponds to about 1.4×10^{26} atoms per cubic meter. The neodymium lasers are four-level systems and therefore, require lower pump energies than the ruby laser.

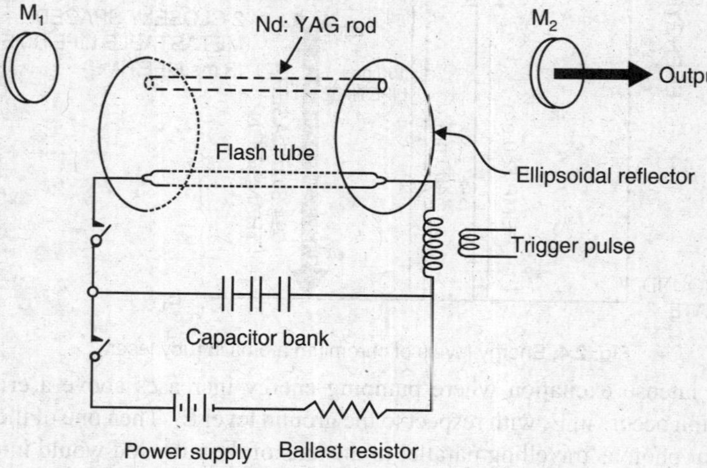

Fig. 2.6. A typical arrangement for the Nd : YAG laser

STRUCTURE

Fig. 2.6 illustrates the typical design of Nd:YAG laser. The laser rods are typically of 10 cm

in length and a diameter of 12 mm. The YAG rod and a linear flash lamp are housed in a reflector cavity of elliptical cross-section. The light issuing from the lamp is closely coupled to the laser rod as they are located at the foci of the ellipse. The ends of the YAG rod are ground flat and parallel. The optical cavity may be formed by silvering the ends of the rod. In practice, external mirrors are used as shown in Fig. 2.6. One mirror is made 100% reflecting while the output mirror is about 90% reflecting. The system is cooled by circulating air. High power lasers are cooled by water circulation.

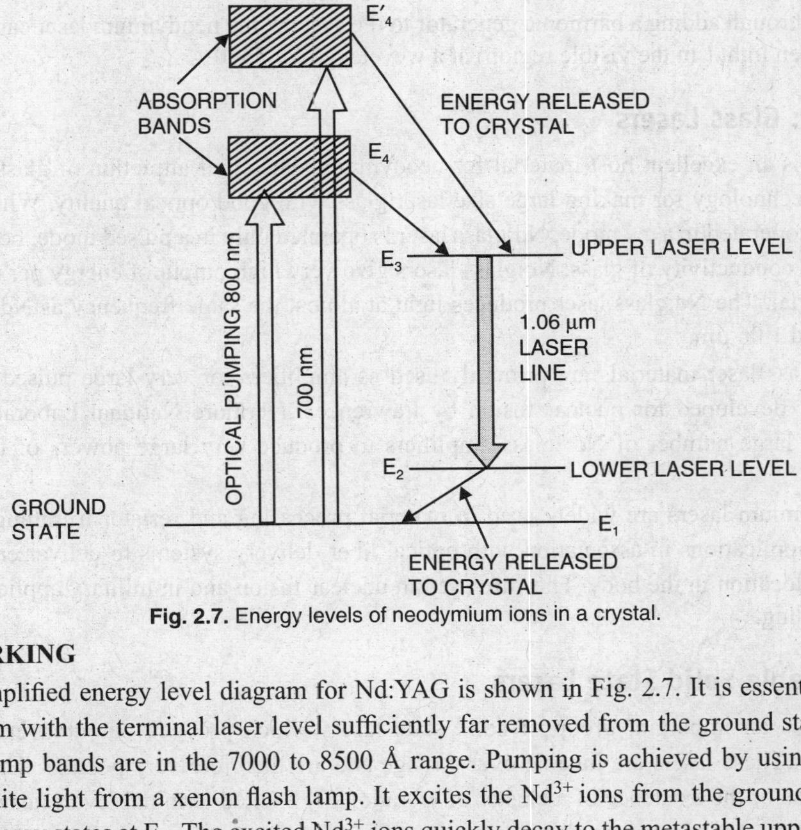

Fig. 2.7. Energy levels of neodymium ions in a crystal.

WORKING

A simplified energy level diagram for Nd:YAG is shown in Fig. 2.7. It is essentially a four-level system with the terminal laser level sufficiently far removed from the ground state. The two primary pump bands are in the 7000 to 8500 Å range. Pumping is achieved by using an intense flash of white light from a xenon flash lamp. It excites the Nd^{3+} ions from the ground state to the multiple energy states at E_4. The excited Nd^{3+} ions quickly decay to the metastable upper laser level E_3, releasing their excess energy to the crystal lattice. As the lower laser level E_2 is located at 0.25 eV above the ground state E_1, it cannot be populated by Nd^{3+} ions through thermal transitions from the ground level. Thus, it is sparsely populated at normal operating temperature. The population inversion can be readily achieved between E_3 and E_2 levels. In the E_3 state, Nd^{3+} ions are stimulated to emit on the main 1.064 μm laser transition, and drop to the lower laser level E_2.

From the level E_2, Nd^{3+} ions quickly drop to the ground state again by transferring energy to the crystal.

The energy level scheme of Nd^{3+} ions in YAG crystal as shown in Fig 2.6 is an oversimplified diagram. In reality, the upper and lower laser levels are also split. Because of the splitting several other laser transitions occur in the near infrared, but all of them are weaker than the main 1.064 μm transition. The only other strong transition occurs at 1.318 μm and produces about 20% as much power as the 1.064 μm transition. It is useful in fibre-optic systems.

The laser output is in the form of pulses occurring at a very high rate. The overall efficiency is within 0.1 to 1% range.

A xenon flash tube is used for pulsed output and tungsten-halide incandescent lamps are used for cw operation. Thereshold pumping levels of about 3J at 3% conversion efficiency can be achieved. Continuous output powers of over 1 kw are obtainable. The lasers can be pumped by a diode laser also. The GaAs laser diode has the ideal pumping wavelength for the Nd^{3+} ion arround 0.8 μm.

The primary wavelength of neodymium laser lies in the infrared. It is possible to double the frequency by second harmonic generation. Twice the frequency corresponds to half the wavelength. Therefore, through adding a harmonic generator to the system, the neodymium laser can be made to produce green light (in the visible region) at a wevelength of 5320 Å .

2.3.3 Nd : Glass Lasers

Glass is an excellent host material for neodymium. The chief attraction of glass is the well developed technology for making large size laser glass with good optical quality. While Nd:YAG laser can be operated in a cw mode, Nd:glass laser is operated only in a pulsed mode, because of the low thermal conductivity of glass. Nd:glass lasers give very high output of energy per unit volume of the material. The Nd:glass laser produces light at almost the same frequency as Nd:YAG laser, *i.e.* at around 1.06 μm.

Nd: glass laser materials are primarily used as amplifiers for very large pulsed lasers. The NOVA laser developed for nuclear fusion by Lawrence Livermore National Laboratories, USA employed a large number of Nd: glass amplifiers to produce very large powers of the order of 10^{13} W.

Neodymium lasers are widely used in material processing and resistor trimming. It is used in medical applications in association with optical fiber delivery systems to deliver energy to the appropriate location in the body. They are used in nuclear fusion and in military applications such as range finding.

2.3.4 Tunable Solid State Lasers

There is an important group of solid state lasers which produce output over a range of wavelengths. These lasers are tunable over a range and are therefore very popular. The tunability arises because of the existence of a cluster of vibrationally excited terminal levels near the ground state. Therefore, these lasers are also known as *vibronic lasers.*

In liquid lasers, we find that dye lasers are tunable. However, the dye lasers suffer from dye degradation and other limitations. In contrast solid state tunable lasers have unlimited shelf life and operational life. Therefore, they are useful in applications like remote sensing and in space craft.

2.3.4.1 Alexandrite Laser

GENERAL DESCRIPTION

The alexandrite laser is a three-level solid state laser similar to a ruby laser. While a ruby laser is a fixed wavelength laser, the alexandrite laser is a tunable laser and lases in the wavelength range of 7000 Å to 8200 Å. The mineral alexandrite ($BeAl_2O_4$) is doped with chromium in concentrations of about 0.1% corresponding to a chromium ion density of 3.5×10^{25} per cubic meter. Alexandrite has pump bands at 3800 Å and 6300 Å. Therefore, pumping can be done using a flash lamp.

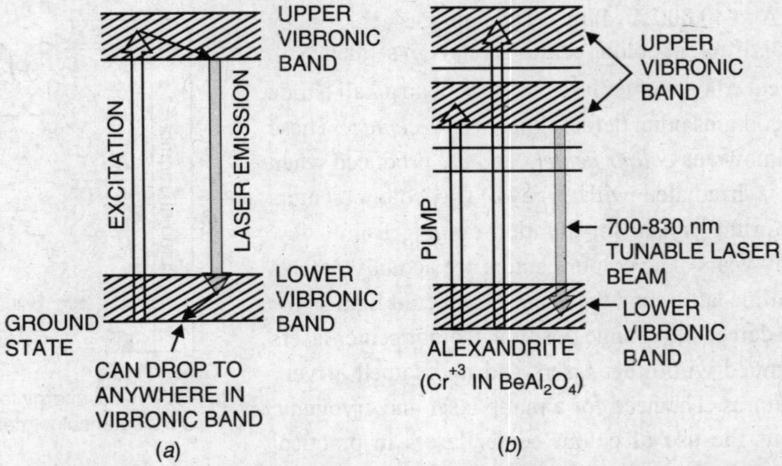

Fig. 2.8. Energy level scheme (a) for a typical vibronic laser (b) Cr^{3+} ions in Alexandrite crystal.

STRUCTURE

The alexandrite laser rod is typically of 0.10 m long and 6 mm dia and is pumped using a flash lamp. Linear flash lamps and a double elliptical pumping cavity are used.

WORKING

The energy levels of Cr^{3+} in alexandrite are shown in Fig. 2.8(*b*). In this case both the upper and lower laser levels are a band of energy levels. The bands consist of vibrational sublevels of a single electronic energy level which arise due to vibrations of the crystal lattice. When Cr^{3+}ions drop from the upper laser level to the lower band of energy levels, they undergo a compound transition where both electronic and vibrational energy change. This compound transition is called a vibrational-electronic transition or in short a *vibronic* transition. The interesting feature of the vibronic transition is that it can occur over a range of energies, because the excited ion can drop from the upper level to any level within the lower vibronic band. It implies that a vibronic laser can be made to operate over a range of wavelengths within its gain bandwidth. Therefore vibronic laser is tunable to any desired wavelength within its emission spectrum. This characteristic of vibronic laser makes it possible to produce wavelengths that are not obtained from other solid-state lasers. An alexandrite laser emits light in the range of 7000 Å to 8300 Å. Alexandrite rod can store more energy than an equal sized Nd:YAG rod. Alexandrite laser can operate in a pulsed or cw mode.

The alexandrite laser is widely used in cancer therapy. Other applications are in pollution detection and kidney stones removal.

2.3.4.2 Titanium-Sapphire Laser

The titanium-sapphire laser uses a titanium doped sapphire crystal designated as $Ti:Al_2O_3$ where Ti^{3+} ions replace some of the Al^{3+} ions. Doping concentrations are about 0.1% by weight. The titanium-sapphire laser is the most widely used tunable solid-state laser which is tunable between 6600 Å in the red and 11,800 Å in the near infrared. These lasers can be operated in pulsed or cw mode. The lifetime of the upper laser level is too short, only 3.8 μs. Therefore, they cannot be pumped with flash lamp. They are pumped with argon laser for cw operation or with frequency-doubled Nd:YAG lasers for pulsed operation. These lasers are used in laser radar (lidar), range finders, and remote sensing.

2.3.4.3 Colour-Center Lasers

Colour center lasers are broadly tunable solid state lasers. They operate in the wavelength range 8000 Å to 40,000 Å (4 μm). The tuning is achieved by using several different colour center crystals in sequence.

A typical colour center laser consists of an alkali halide crystal that contains point defects known as *F centers*. These centers are known as *colour centers* and are produced when the crystal is irradiated with X-rays. The colour centers remain within the crystal for duration ranging from a few days to many years. The colour centers are actually defects in the crystalline lattice and they absorb and emit light as the atoms at the defect site change position. Colour center lasers must be pumped with other lasers and maintained at very low temperatures. The need for a pump laser and cryogenic cooling limits the use of colour center lasers in practical applications.

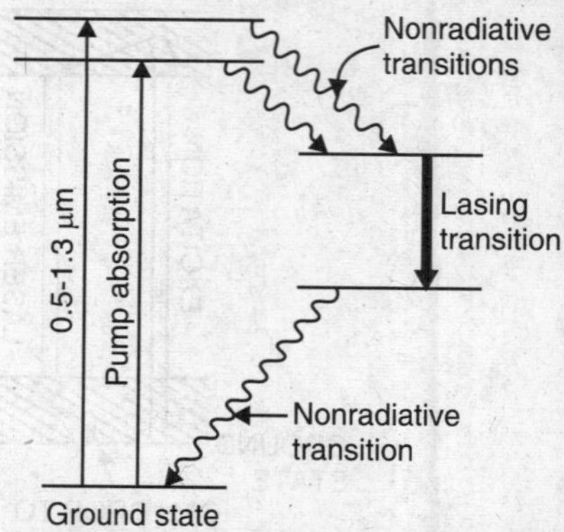

Fig. 2.9. Typical energy level diagram of a F-center colour center laser

2.3.5 Fiber Lasers

Erbium in a glass host forms a three-level laser with a wavelength range centered around 1.55 μm. Erbium lasers and amplifiers assume importance in fibre-optic communications because optical signals of 1.55 μm travel longer distances through optical fibres with least loss. In long distance fibre-optic communications, amplifiers are required after each specific distance to increase the strength of the signal. They are known as repeaters. Earlier, optical signals are converted into electrical signals and are amplified in each repeater station. It can be alternately done by passing the weak signal through a length of the erbium-doped optical fibre. If the erbium atoms are pumped by a laser diode pump laser at a different wavelength, the weak signal can stimulate the atoms to emit light at the signal wavelength, 1.55 μm, causing an increase in the signal strength. These erbium-amplifiers are highly useful in under-sea communication and long haul communication links.

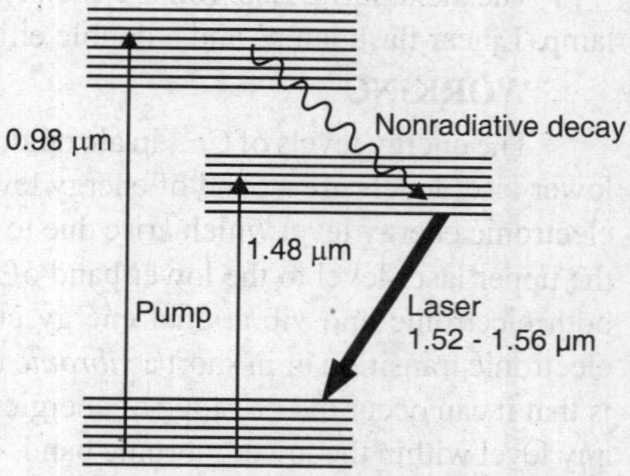

Fig. 2.10. Energy level diagram of an erbium-doped fibre laser

2.4 GAS LASERS

Gas lasers are the most widely used lasers and the most varied. They range from the low power helium-neon (He-Ne) laser used in college laboratories to very high power carbon dioxide laser used in industrial applications. These lasers operate with rarefied gases as their active media and excited by an electric discharge. There are three different types of gas laser: ion lasers, neutral atom lasers and molecular lasers. In gases, unlike in crystals, the energy levels of atoms involved in the lasing process are well defined and narrow. Broad pump bands do not exist and the pump levels are also narrow. In order to excite atoms, sources with sharp wavelength are required. Finding an appropriate optical source for pumping poses a problem. Therefore, optical pumping is not used in gases. The most common method of exciting gas laser medium is by passing an electric discharge through the gas. Electrons in the discharge transfer energy to atoms in the laser gas by collisions.

Gas lasers, in general can be classified as (*i*) Atomic, (*ii*) Ionic and (*iii*) Molecular lasers.

(*i*) Atomic lasers: He-Ne (Helium-Neon) lasers are of this type. In atomic lasers the active medium is noble gas *viz.* Helium, neon, argon, krypton and considered in the neutral state. The atomic lasers are characterised by mixing active gas with other gases which increases the excitation efficiency of the system.

***(ii)* Ionic lasers:** In iônic lasers, ionized gas is used as lasing medium. For this purpose, ionized noble gases like A_r^+ (Argon ion) and K_r^+ (Krypton ion) are prefered. The most commanly used ionic lasers are A_r^+ (Argon) and K_r^+ (Krypton) lasers which mostly operate in CW mode. They are operated in Ultra-Violet (UV) Spectral region. Large amount of heat is generated and hence cooling arrangement is essentially needed. These lasers produce a wide range of colours and hence are popularly used in art and entertainment to give visual effects.

***(iii)* Molecular lasers :** All the gas lasers are based on *electronioc levels* however in molecular lasers other energy levels are also taken into account. The atoms comprising the molecular vibrate about their mean position giving rise to *vibrational energy levels* and also can rotate as a whole. The gives rise to *rotational energy levels.*

The first gas laser was He-Ne laser which was demonstrated in 1961 at Bell Telephone Laboratories, U.S.A. by Ali Javan, William R. Bennett, Jr., and Donald R. Herriott. The generic gas laser is shown in Fig. 2.11.

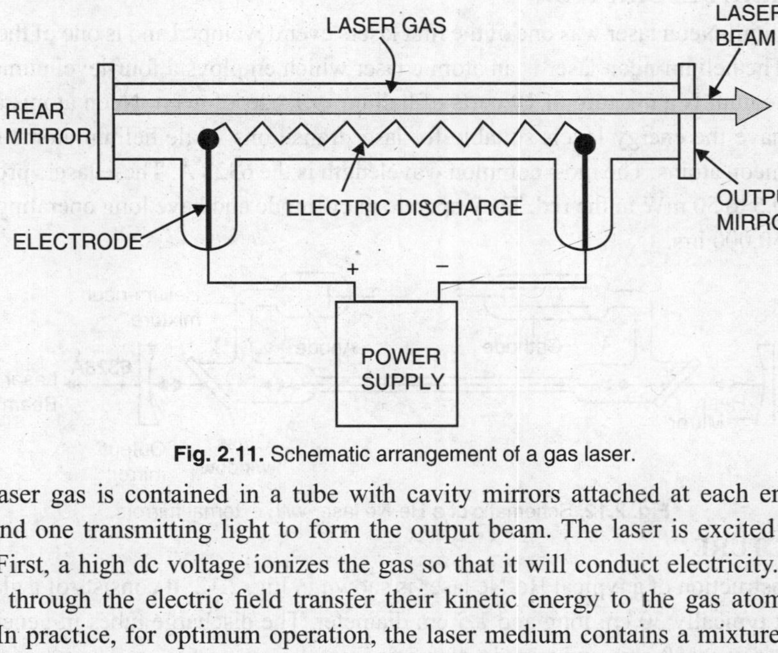

Fig. 2.11. Schematic arrangement of a gas laser.

The laser gas is contained in a tube with cavity mirrors attached at each end, one totally reflecting and one transmitting light to form the output beam. The laser is excited by an electric discharge. First, a high dc voltage ionizes the gas so that it will conduct electricity. The electrons accelerated through the electric field transfer their kinetic energy to the gas atoms by inelastic collisions. In practice, for optimum operation, the laser medium contains a mixture of two gases, say *A* and *B*, at low pressure. Atoms of kind *A* are initially excited by electron impact and they in turn transfer their energy to atoms of kind *B* which are the actual active centers.

The cavity mirrors can be either inside the gas container or outside. If they are inside, than the output light is generally unpolarized. For the outside case, to minimize reflection loss, discharge tube edges are cut at the Brewster angle. The Brewster angle is given by

$$\theta_B = \tan^{-1}\sqrt{\frac{\mu_2}{\mu_1}}$$

where μ_2 and μ_1 are the refractive indices of glass and the gas mixture respectively. It is known that light incident at Brewster angle is polarized parallel to the plane of incidence and has a transmission coefficient of 1. The passage of light through a glass container at Brewster angle does not involve any transmission losses and thus parallel polarization component has a higher effective gain. Therefore, for the outside mirror arrangement, light is plane polarized in general.

Gas lasers vary widely in their characteristics. Their output wavelengths range from near UV to far IR . Some of them operate in pulsed mode and others in CW mode. The output power varies from less than a milliwatt (mW) to over 10 kW.

Important feature of gas lasers. The gas lasers having following important features :

1. The line width of the spectral transition is relatively smaller as compared to solid state lasers.
2. The laser emission is narrower.
3. No optical inhomogeneties. Hence, yields the good quality laser.
4. The conerence properties are also superior.
5. The monochromaticity is very high.
6. The quality factor (Q) of the modes are high we shall describe them in detail.

2.4.1 Helium-Neon Laser

GENERAL DESCRIPTION

The Helium-Neon laser was one of the first lasers ever developed and is one of the most widely used lasers. The helium-neon laser is an atomic laser which employs a four-level pumping scheme. The active medium is a mixture of 10 parts of helium to 1 part of neon. Neon atoms are the active centers and have the energy levels suitable for laser transitions while helium atoms help efficient excitation of neon atoms. The most common wavelength is the 6328 Å. These lasers produce powers in the range 0.5 to 50 mW in the red. They operate in cw mode and have long operating life times of the order of 50,000 hrs.

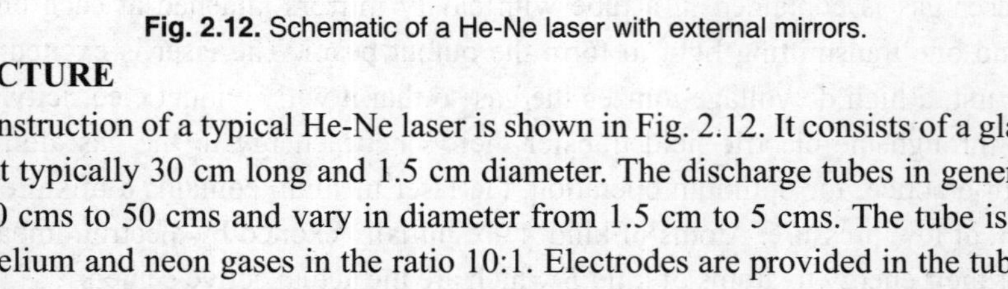

Fig. 2.12. Schematic of a He-Ne laser with external mirrors.

STRUCTURE

The construction of a typical He-Ne laser is shown in Fig. 2.12. It consists of a glass discharge tube of about typically 30 cm long and 1.5 cm diameter. The discharge tubes in general may have lengths of 10 cms to 50 cms and vary in diameter from 1.5 cm to 5 cms. The tube is filled with a mixture of helium and neon gases in the ratio 10:1. Electrodes are provided in the tube to produce a discharge in the gas. They are connected to a high voltage power supply. The tube is hermetically sealed with glass windows orientated at the Brewster angle to the axis of the tube. The cavity mirrors are arranged externally.

WORKING

The energy level schemes of helium and neon are shown in Fig. 2.13.

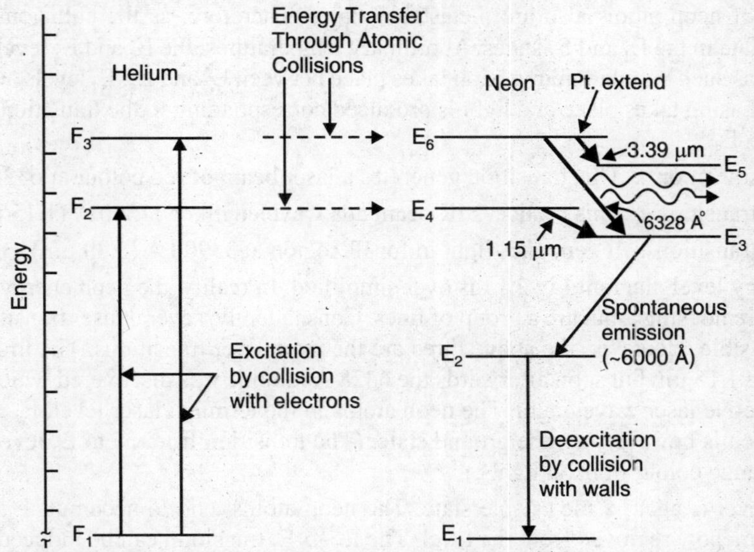

Fig. 2.13. Energy levels of helium and neon atoms and transitions between the levels.

When the power is switched on, a high voltage of about 10 kV is applied across the gas. It is sufficient to ionize the gas. The electrons and ions produced in the process of discharge are accelerated towards the anode and cathode respectively, as shown in Fig. 2.14.

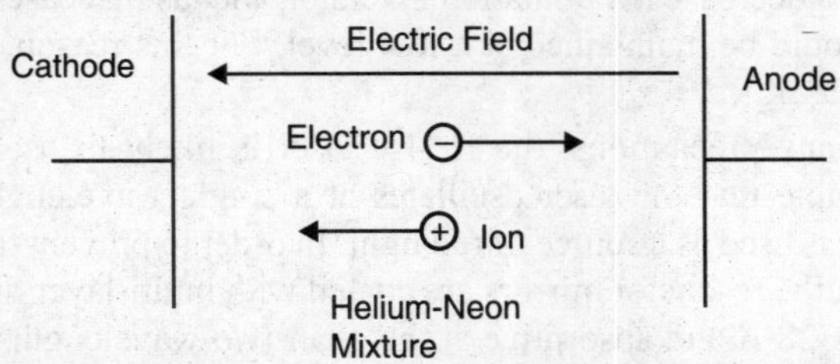

Fig. 2.14. Migration of electrons and positive ions in a discharge tube.

Since the electrons have a smaller mass, they acquire a higher velocity. They transfer their kinetic energy to helium atoms through inelastic collisions. Helium atoms are much more readily excited by electron impact because they are fairly light. Thus, the initial excitation affects only the helium atoms, with the result that these atoms are excited to the levels F_2 and F_3 which lie at 19.81 eV and 20.16 eV respectively above the ground sate. These two levels are metastable levels and the excited helium atoms cannot return to the ground state through spontaneous emission. Though a radiative transition is not possible, the excited helium atoms can return to the normal state by transferring their energy to neon atoms through collision. Such an energy transfer can take place when the two colliding atoms have identical energy states. It is called *resonant transfer* of energy. The neon energy levels, namely E_6 level at 20.66 eV and E_4 level at 18.7 eV nearly coincide with the F_3 and F_2 levels respectively of helium atom. Therefore, the resonant transfer of energy can occur readily. The additional 0.05 eV energy required is provided by the kinetic energy of helium atoms. When a helium atom in the metastable state collides with a neon atom in the ground state, the neon atom is excited to E_4 or E_6 level and the helium atom drops back to the ground state. This is the pumping mechanism in He-Ne laser. The neon atoms which are much heavier could not be pumped up efficiently without the helium atoms. The role of helium atoms is thus to excite neon atoms and cause population inversion. The probability of energy transfer from helium atoms to neon atoms is more as there are 10 atoms of helium per 1 neon atom in the gas mixture. For this reason, the probability of reverse transfer of energy from neon to helium atom is extremely small. The E_4

and E_6 levels of neon atoms also are metastable states. Therefore, as the collisions go on, neon atoms accumulate in the E_4 and E_6 states. At ordinary temperatures, the E_5 and E_3 levels are sparsely populated and as such population inversion takes place between E_6 and E_5, E_3 levels and between E_4 and E_3 levels. Lasing takes place and light is produced corresponding to the transitions, $E_6 \rightarrow E_5$, $E_6 \rightarrow E_3$ and $E_4 \rightarrow E_3$.

(*i*) $E_6 \rightarrow E_3$ transition : This transition generates a laser beam of red colour at 6328 Å.

(*ii*) $E_4 \rightarrow E_3$ transition : This produces IR beam at a wavelength of 11500 Å (1.15 μm).

(*iii*) $E_6 \rightarrow E_5$ transition : It generates light in for IR region at 33900 Å (3.39 μm)

The energy level diagram Fig 2.13 is over simplified. In reality, the neon energy levels E_6, E_5, E_4, E_3 and E_2 are not single but are a group of lines. Consequently, several laser transitions, as many as 150, are possible. However, the above three are the main laser transitions. The first He-Ne laser was operated at 1.15 μm but soon afterwards the 6328 Å red line was discovered which has become the standard He-Ne laser wavelength. The neon atoms in the terminal laser level, E_3, decay rapidly to E_2 level which is however, not the ground state. (The transition from E_3 to E_2 level gives rise to the familiar orange colour of neon lights.)

The E_2 level is again a metastable state. The neon atoms tend to accumulate at this level if they are not somehow removed from the level. The $E_2 \rightarrow E_1$ transition can be induced by collisions with the walls of the discharge tube. To enhance the probability of atomic collisions with the walls, the discharge tube is made as narrow as possible. Once the atoms reach the ground state, they will be again available for pumping. E_2 level is more likely to be populated by the electric discharge itself. It may be recalled here that electric discharge is used to produce neon lighting. An increase in population at E_2 leads to a decrease in population inversion and lasing ceases. Therefore, the current in the discharge tube should be maintained at a low level. For this reason, high powers cannot be obtained from a He-Ne laser.

In the absence of any precautions, the He-Ne laser is likely to oscillate at the three main wavelengths. It is desirable that the laser oscillates at a single wavelength only. In most of the applications He-Ne laser is used as a source of red light. In order to prevent the laser from oscillating at the other wavelengths, the resonator mirrors are coated with multi-layer dielectric coatings which are highly reflective at 6328 Å and absorptive at the other two wavelengths. The laser oscillates at the wavelength which is reflected more by cavity mirrors.

Without the Brewster windows, the light output is unpolarized. The Brewster windows cause the laser output to be linearly polarized.

Although, 6328 Å is the standard wavelength of He-Ne laser, other visible wavelengths 5430 Å (green), 5940 Å (yellow-orange), 6120 Å (red-orange) can also be produced.

The He-Ne laser operates in CW mode and the over all gain is very low and is typically about 0.01% to 0.1%. However, the laser is simple, practical and less expensive. The laser beam is highly collimated, coherent and monochromatic.

He-Ne lasers are widely used in laboratories as a monochromatic source, in interferometry, laser printing, bar code reading etc. These lasers provide a very stable single transverse mode reference beam to identify optical properties of materials such as surface smoothness. Most supermarket and other types of stores now use these lasers as scanners to read the digitally encoded bar codes located on products. They are also used as a reference beam in surveying, for alignment in pipe laying etc works.

2.4.2 Ion Gas Lasers

Helium, neon, argon, xenon and krypton are noble gases and they have electronic states capable of laser transitions. However, except for neon, noble gases are difficult to pump and thus are not of practical interest. However, if these noble gases are first ionized by electron collisions then they are easy to pump.

2.4.2.1 Argon Laser

GENERAL DESCRIPTION

The argon laser belongs to the group of ion lasers. It is a four-level laser which operates in visible region over a wavelength range from 3510 Å to 5200 Å. It is the most powerful CW laser operating in visible region. The importance of argon and krypton lasers is that they are extensively used in laser light shows. The argon ion laser can provide approximately 25 visible wavelengths ranging from 4089 Å to 6861 Å and more than 10 ultraviolet wavelengths ranging from 2750 Å to 3638 Å. In the visible region, CW powers of about 100 W are available. The discharge tubes have an operating life of 2000 to 5000 hrs.

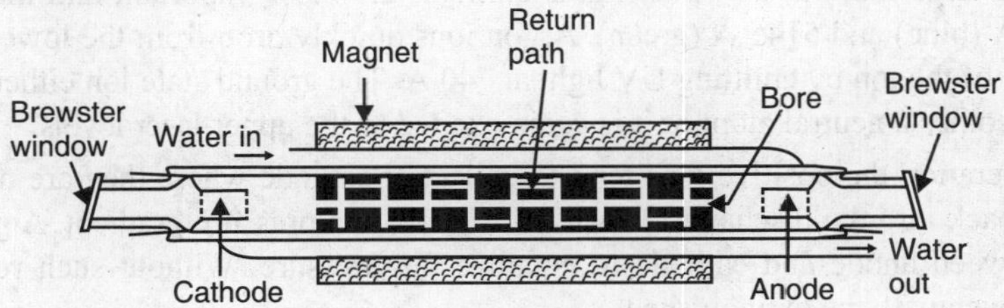

Fig. 2.15. Schematic of a typical ion laser tube

STRUCTURE

In argon lasers, the active medium is argon gas and the active centres are ionized argon atoms. A typical design of the discharge tube is shown in Fig 2.15. It consists of a narrow water-cooled ceramic tube in which an arc discharge takes place. The electrodes are arranged at the ends of the capillary tube. The anode and cathode spaces communicate through a return gas path which ensures free circulation of the gas. A magnet surrounds the discharge tube. Its function is to constrict the discharge area and increase the concentration of ions along the axis of the tube. This increases the output power and efficiency.

WORKING

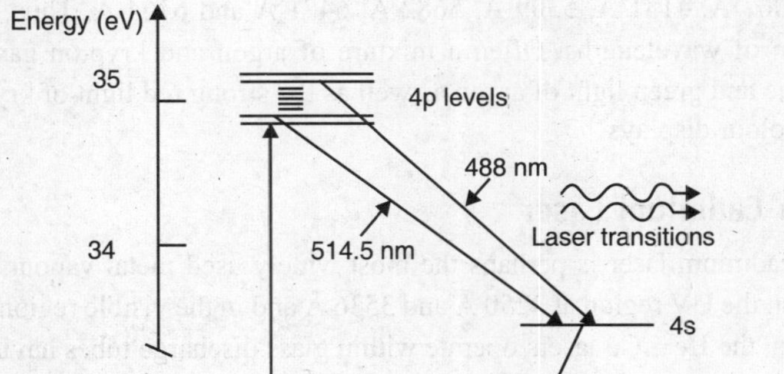

Fig. 2.16. Energy level scheme for an argon atom.

As in He-Ne laser, an initial high voltage pulse ionizes the gas so that it conducts current. Electrons in the current transfer energy directly to argon atoms, ionizes them and raises the ions to a group of high energy levels, shown in Fig 2.16, which are about 35 eV above the ground state of neutral argon atom. Different processes populate the metastable upper laser level. Three possible processes are : (*a*) electron collisions with Ar+ ions in the ground state, (*b*) collision with ions in metastable states and (*c*) radiative transitions from higher states. The life time of upper laser level is 10^{-8}s while that of the lower laser level is of the order of 10^{-9}s. Therefore, the condition for population inversion is satisfied. Transitions can occur between many pairs of upper and lower lasing levels and therefore many laser wavelengths are emitted.

Out of the large number of wavelengths emitted, the most important and the most intense lines are 4881 Å (blue) and 5145 Å (green). Argon ions quickly drop from the lower laser level to the ground state of the ion by emitting UV light at 740 Å. The ground state ion either recaptures an electron and becomes a neutral atom or it is again excited to the upper laser levels.

During operation the positive ions tend to collect at cathode where they are neutralized and diffuse slowly back into the discharge. However, it leads to a pressure gradient. A gas return path is provided between anode and cathode to equalize the pressure. Without such return path, the discharge may eventually be extinguished.

The laser needs active cooling. Any desired wavelength can be selected through proper cavity optics. For example, a small prism may be inserted into the optical cavity and the position of the end mirror is rotated such that it comes into a position normal to the path of the light of desired wavelength. Then, only the desired wavelength is sustained in to and fro reflections while other wavelengths are lost from the cavity after a few reflections.

2.4.2.2 Krypton Ion Laser

The krypton ion laser resembles the argon ion laser in energy level arrangement and in operation. Krypton laser provides different laser wavelengths than the argon laser. The dominant output occurs at 4067 Å, 4131 Å, 5309 Å, 5682 Å, 6471 Å and 6764 Å. Thus, these lasers offer a broader spectrum of wavelengths. Often a mixture of argon and krypton gases is used which provides strong blue and green light of argon as well as the strong red light of krypton. Such lasers are used for multicolour displays.

2.4.2.3 Helium-Cadmium Laser

The helium-cadmium laser is perhaps the most widely used metal vapour laser. It produces continuous output in the UV region at 3250 Å and 3536 Å and in the visible region at 4416 Å (blue). Like He – Ne lasers, the He – Cd lasers operate within glass discharge tubes having a bore of 1 – 2 mm. The cadmium metal is to be heated to about 250 °C and vapourized at the anode end in order to produce cadmium vapour needed for laser operation. Helium gradually leaks out and therefore a helium reservoir is to be included in the system to replace helium that is lost. He – Cd lasers need discharge voltages around 1500 V and currents in the range of 60- 100 mA. The lasers typically have lifetimes of 4,000 to 5,000 hrs.

The electrons in the gas discharge excite helium atoms to upper levels. The excited helium atoms transfer energy to cadmium atoms through ionization process and raise them to upper laser levels. The transition from one of the upper metastable states produces UV lines and the transition from the other state produces blue line.

The He–Cd laser is used in many applications which include photolithography, stereolithography, inspections of electronic circuit boards, CD–ROM mastering, fluorescence analysis etc.

2.4.2.4 Copper-Vapour Laser

Copper – vapour laser is one of the important metal vapour lasers. It operates only in pulsed mode and produces pulsed energies of the order of 1mJ leading to average powers of 10 – 100W for pulse repetition rates of 10 – 100 kHz. Copper vapour laser produces light at 5105 Å (green) and 5782 Å (yellow). The metallic vapour is contained within a cylindrical discharge tube filled with helium or neon as a buffer gas. The optimum pressure for the laser is of the order of 1 torr of metal vapour which requires the metallic copper to be heated upto 1500 ^{0}C. However, the copper metal will last typically for 500 – 100 hrs of operation, after which it is to be reloaded.

When an electric discharge passes through copper vapour, copper atoms are excited to one of the two upper laser levels. If the vapour pressure is optimum, the atoms stay long enough at the upper laser level to produce stimulated emission. With the onset of stimulated emissions, the atoms quickly drop to lower laser levels. The lower laser levels are metastable levels and as atoms accumulate at these levels, the laser operation ceases. All this takes less than 100 ns. Depopulation of lower laser level takes place by collision of atoms with the walls of discharge tube. Again the laser gets ready for producing next pulse of light.

Copper vapour lasers have high gain and can operate without resonant mirrors.

2.4.2.5 Gold-Vapour Laser

The gold vapour laser operates in a similar way to that of the upper vapour laser. It produces orange beam at a wavelength of 6278 Å.

Copper vapour lasers are used to pump tunable dye lasers, in high – speed flash photography and material processing. The gold vapour laser is used in photodynamic therapy for destroying the cancerous tissue.

2.4.3 Carbon Dioxide (CO_2) Laser

GENERAL DESCRIPTION

The carbon dioxide laser is one of the most powerful and efficient lasers. It is a four-level molecular laser and operates on a set of vibrational-rotational transitions. The laser output is in the middle infrared at 10.6 μm and 9.4 μm wavelength region. Both cw and pulsed output occurs. The laser produces CW powers of greater than 100 kW and pulsed energies of as much as 10 kJ.

Energy levels of CO_2 molecule :

In case of isolated atoms, the electron energy levels are discrete and narrow. On the other hand, the energy spectrum of molecules is complex and includes many additional features. Each electron level is associated with a number of equispaced vibrational levels and each vibrational level in turn has a number of rotational levels.

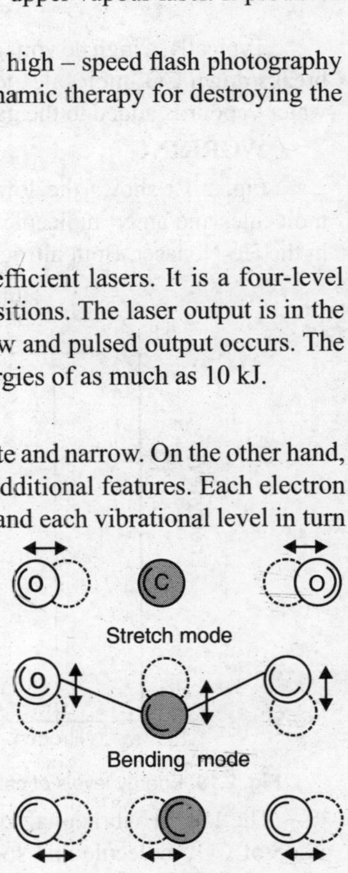

Fig. 2.17. Vibrational modes of CO_2 molecule.

The CO_2 molecule is basically a linear molecule consisting of a central carbon atom with two oxygen atoms linked one on either side. It can undergo three independent vibrational oscillations (Fig. 2.17) known as the *vibrational modes.* The modes are termed as stretching mode, the bending mode and the asymmetric stretching mode. Each mode is quantized, so the molecule can have 0, 1, 2, or more units of vibrational energy in each mode. At any one time, a CO_2 molecule can vibrate in any linear combination of these three fundamental modes. The energy states of the molecule is then represented by three quantum numbers (m n q). These numbers represent the amount of energy associated with each mode. For example, the number (020) indicates that the molecule in this energy state is in the pure bending mode and has two units of energy. Each vibrational state is associated with rotational states corresponding to the rotation of the molecule about its centre of mass. The separations between vibrational – rotational states are much smaller on the energy scale compared to the separations between electron energy levels.

STRUCTURE

The schematic of a typical CO_2 laser is shown in Fig 2.18. It is basically a discharge tube having a bore of cross-section of 1.5 cm^2 and a length of about 26 cm. The discharge tube is filled with a mixture of carbon dioxide, nitrogen and helium gases in 1:4:5 proportions respectively.

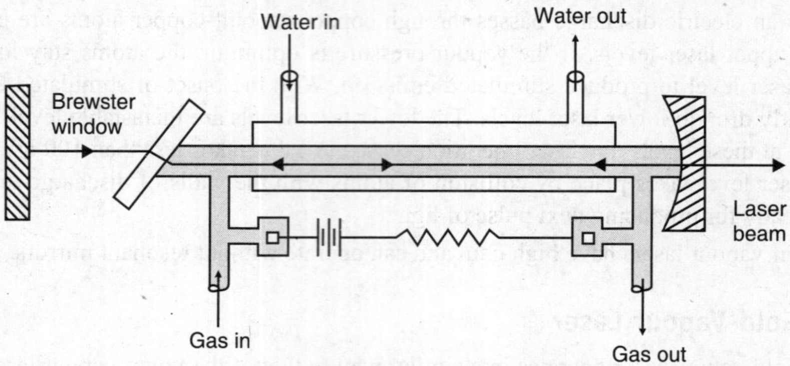

Fig. 2.18. A schematic of a typical CO_2 laser.

Typically, a high dc voltage causes an electric discharge to pass through the tube. The discharge breaks down CO_2 molecules to form oxygen and carbon monoxide. Therefore, a small amount of water vapour is added to the gaseous mixture which regenerates CO_2.

WORKING

Fig. 2.19 shows the lowest vibrational levels of the ground electron energy state of CO_2 molecules and an N_2 molecule also. In the CO_2 laser, nitrogen plays a similar role to that of helium in the He-Ne laser. Both nitrogen and CO_2 absorb energy from electrons in the discharge.

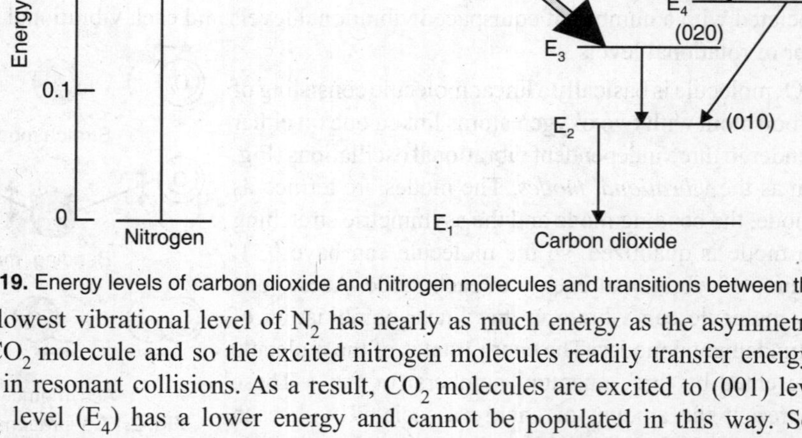

Fig. 2.19. Energy levels of carbon dioxide and nitrogen molecules and transitions between the levels.

The lowest vibrational level of N_2 has nearly as much energy as the asymmetric stretching mode of CO_2 molecule and so the excited nitrogen molecules readily transfer energy to the CO_2 molecules in resonant collisions. As a result, CO_2 molecules are excited to (001) level (E_5). The (100) CO_2 level (E_4) has a lower energy and cannot be populated in this way. So population inversion is established between the (001) level, (E_5) and the (100) level, (E_4). Simultaneously, in the same manner population inversion is created between (001) level, E_5, and (020) level E_3. The lasing transitions $E_5 \rightarrow E_4$ and $E_5 \rightarrow E_3$ produce IR radiation at wavelength 10.6 μm and 9.6 μm respectively. The E_3 and E_4 levels are metastable states. The CO_2 molecules at these levels fall to the lower bending level E_2, through inelastic collisions with helium atoms. The decay from the E_2 level to the ground state should be very fast. Otherwise there would be accumulation of molecules in this level which leads to decrease in population inversion. The helium atoms deplete the level E_2 through collisions. There is yet another function of helium. The E_2 level is very close to the ground state E_1 and tends to be populated through thermal excitations. It becomes therefore necessary to keep the temperature of CO_2 low. Helium has a high thermal conductivity and conducts heat away to the walls and keeps CO_2 cold. Thus, while nitrogen helps to increase the population of the upper laser level, helium depopulates the lower laser level.

Some of the other structures used for CO_2 lasers are briefly described here.

Wave-guide laser

This structure is the most efficient way to produce a compact CW CO_2 laser. It consists of two transverse RF electrodes separated by insulating sections that form the bore region. The lateral dimensions of the bore are up to a few mm. An rf power supply is connected to the electrodes to provide a h.f. field across the electrodes within the bore region. This type of laser produces cw powers of about 100 W.

Gas-dynamic laser

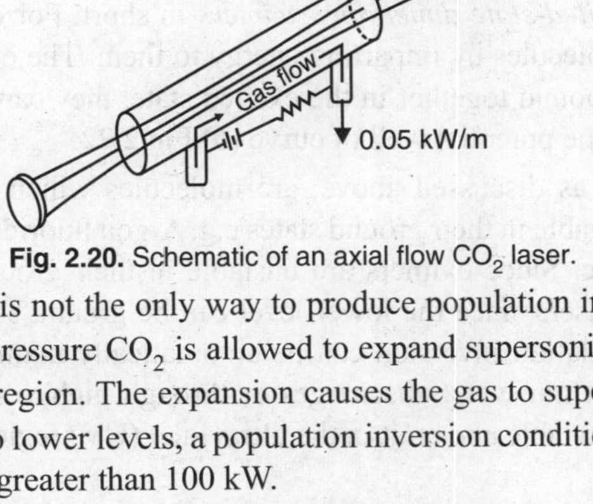

Fig. 2.20. Schematic of an axial flow CO_2 laser.

An electric discharge is not the only way to produce population inversion in carbon dioxide. Rapidly flowing hot, high pressure CO_2 is allowed to expand supersonically through an expansion nozzle into a low pressure region. The expansion causes the gas to supercool. In the process as all the molecules do not drop to lower levels, a population inversion condition is attained. Lasers of this design produce cw outputs greater than 100 kW.

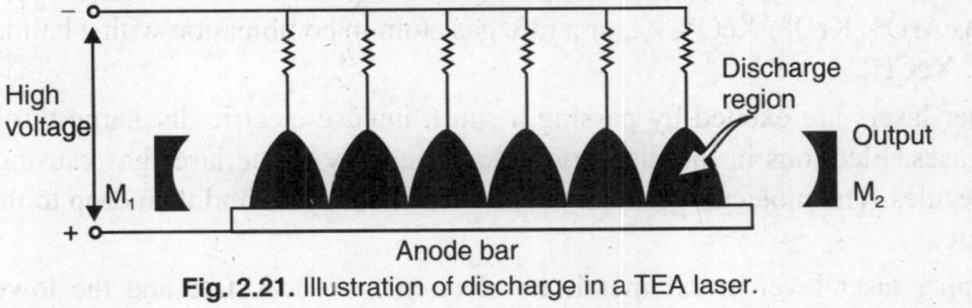

Fig. 2.21. Illustration of discharge in a TEA laser.

TEA CO_2 Lasers

This laser operates at gas pressure of around one atmosphere with a pulsed electric discharge passing through the gas. It works better if the electric discharge is transverse to the laser axis. Such lasers are called TEA lasers (Fig. 2.21).

The CO_2 lasers are largely used in the field of material processing, which includes cutting, drilling, material removal, welding, etching, melting, annealing, hardening etc. In medical applications, it is employed for cutting and cauterizing.

2.4.4. Excimer Lasers

Excimer lasers constitute an interesting and important class of molecular lasers. Excimer laser was first demonstrated in mid - 1970s and their importance lies in that they are the most powerful UV lasers. The active medium of these lasers consists of diatomic molecules that can be bound into a single system when they are in the excited state only. These diatomic molecules exist only as monomers in the ground state, as they repel one another at inter atomic distance.

Fig. 2.22 shows that potential energy curve (*a*) as a function of distance between the monomers. An excitation modifies the state of the atom so that there appears an attractive force with other atoms. The potential energy curve (*b*) in Fig. 2.22 shows the potential energy variation in the excited state of atoms. The curve exhibits a dip indicating that at that interatomic distance, the atoms are bound together because they have a minimum energy at that separation. Such molecules which exist only in excited state are called *excited-state dimers or excimers* in short. For example, the atoms of inert gases can be bound into molecules by imparting energy to them. The excimer state is a metastable state. When the atoms are bound together in the excited state, they can occupy several vibrational levels (horizontal lines) in the potential well of curve (*b*) Fig 2.22.

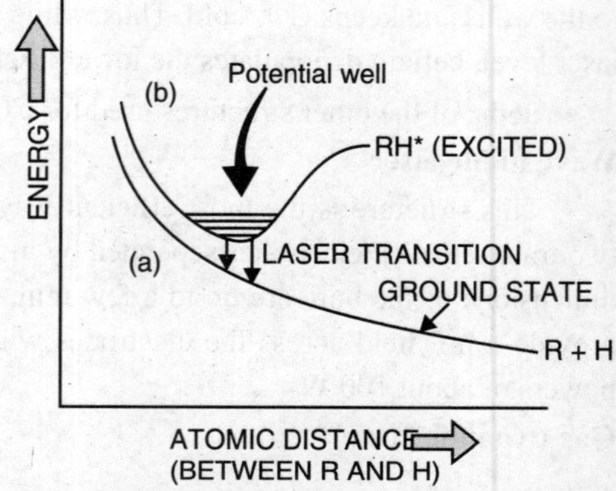

Fig. 2.22. Internal energy of a rare gas halide molecule in excited and ground states.

Principle. Excimers, as discussed above, are molecules which are bound in their excited electronic states but are unstable in their ground states e.g. Argon fluoride, krypton chloride, krypton fluoride, xenon bromide etc. Since eximers are unstable in their ground state, they are ideal for forming active media for lasers since the lower level can be ground state in which the molecules is unstable and hence population inversion can be automatically maintained. Pumping is usually accomplished either by fast transverse discharges or through high voltage, high current electron sources. Eximer lasers are tunable and emit in the ultraviolet (UV) region.

Method,

The active medium can be an excited rare gas dimer such as Ar_2*, Kr_2*, Xe_2*, ..., a rare gas oxide such as ArO*, KrO*, XeO*, ..., or a rare gas atom in combination with a halide atom such as ArF*, KrF*, XeCl*, ...

Excimer lasers are excited by passing a short, intense electric discharge through a mixture of desired gases. Electrons in the discharge transfer energy to the laser gas causing formation of excited molecules. The molecules remain excited for about 10 ns and then drop to the ground state and dissociate.

The upper laser level in excimer is its electronic excited state and the lower level is the electronic ground state. Therefore, population inversion will occur as soon as some atoms are bound to form molecules in the excited state. Once a molecule drops to the lower laser level, it separates out into atoms. Therefore, the lower level in excimer lasers is always vacant.

As excimer lasers have high gain, the cavity mirrors are not required. However, one fully reflective mirror is used in the rear and unsilvered transparent window is used as output mirror.

Excimer lasers are mainly used in refractive surgery on the eye and in the manufacture of semiconductor devices. They are also used in material processing, photolithography and pumping of dye lasers. Table - 2.1 lists some excimer lasers and the wavelengths at which they operate.

Table 2.1. Excimer Lasers

Type	λ (Å)
Ar_2*	1260
ArCl*	1750
ArF*	1930
KrCl*	2220
KrF*	2490
XeCl*	3080
XeF*	3500

2.4.5. Chemical Lasers

Chemical lasers are pumped by the energy liberated in a chemical reaction. The most well known chemical lasers are the HF (hydrogen fluoride) laser and DF (deuterium fluoride) laser. They operate on vibrational transitions and therefore the output wavelength is always in the infrared. Chemical lasers have been developed primarily for military and space applications where pumping power, in the form of electrical energy, may not be available. Chemical lasers can produce powers of several megawatts. The HF laser emits light in the wavelength range of 2.6 to 3.3 μm while DF laser operates in the wavelength range of 3.5 to 4.2 μm.

HF Laser

The ingredients of HF laser are molecular hydrogen, and fluorine gas. Helium is added as a buffer gas. These molecular species can react at normal temperatures under external excitation such as UV radiation. The reaction is highly exothermic and a large quantity of chemical energy is

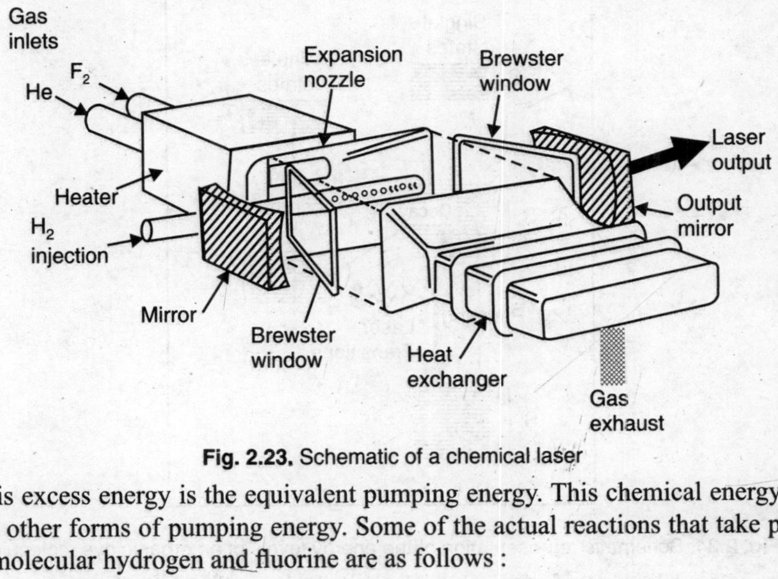

Fig. 2.23. Schematic of a chemical laser

released. This excess energy is the equivalent pumping energy. This chemical energy is enormous compared to other forms of pumping energy. Some of the actual reactions that take place between atomic and molecular hydrogen and fluorine are as follows :

$$F + H_2 \longrightarrow HF^* + H \qquad \Delta H = -31.7 \text{ k cal/mol.}$$

$$H + F_2 \longrightarrow HF^* + F \qquad \Delta H = -97.9 \text{ k cal/mol.}$$

Each step produces a vibrationally excited HF* molecule and such excited molecules participate in the lasing action. The end product of the reactions shown above contains atomic hydrogen and fluoride. Hence, once the reaction starts it continues until all the molecular H_2 and F_2 are consumed. The schematic of a chemical laser is shown in Fig. 2.23.

The main application of chemical lasers is for high-power weapons on battle field or in space.

2.5 TUNABLE DYE LASERS

Dye lasers belong to the family of liquid lasers. The active material is a dye dissolved in a host medium of a liquid solvent, such as methyl alcohol, ethyl alcohol or ethylene glycol. Their sturctures are

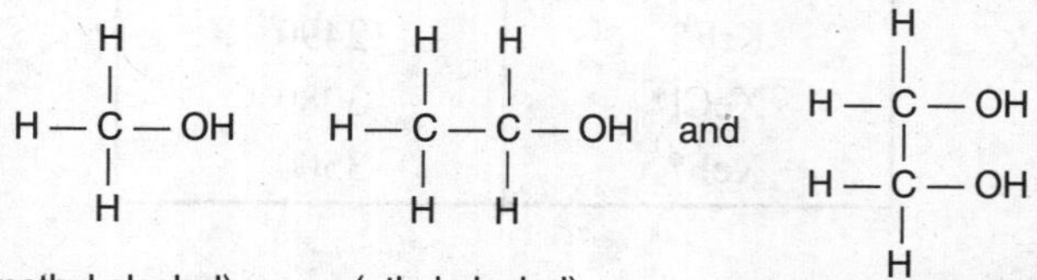

(methyl alcohol) (ethyl alcohol) (ethylene glycol)

The situation is very similar to solid state lasers where Cr^{3+}, Nd^{3+} or Ti^{3+} ions are used in a solid host. The advantage of a liquid host is that the concentration of the active ions can be easily varied. Typical dye concentration of 10^{-4} to 10^{-3} molar solution is used which corresponds to 10^{24} and 10^{25} dye molecules per cubic meter. There are over 200 laser dyes, the most important one being rhodamine 6G. When used sequentially, they can produce tunable laser output over wavelengths ranging from 0.32 to 1.2 μm. Dye lasers operate both in CW and pulsed modes. Pulsed dye lasers are pumped by a flash lamp or other lasers and can produce upto 400 J output in a 10 μs-pulse. CW dye lasers are pumped by other CW lasers such as argon ion laser and produce powers of about 2 W. The gain of the dye lasing medium is very high. Consequently, a small volume of dye solution is sufficient to sustain the lasing action.

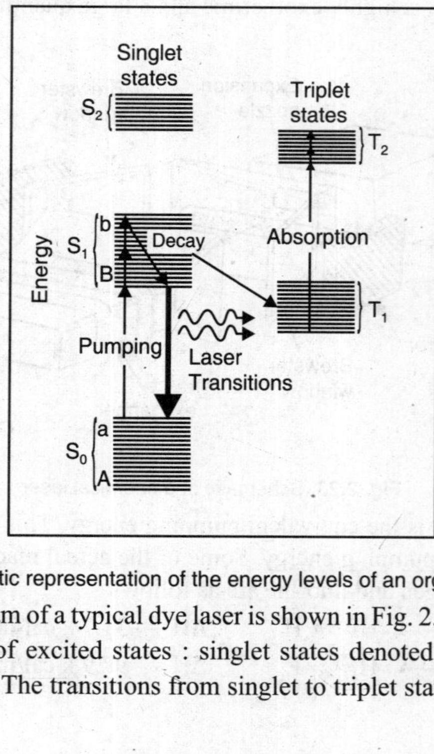

Fig. 2.24. Schematic representation of the energy levels of an organic dye molecule.

The energy band diagram of a typical dye laser is shown in Fig. 2.24. The organic dye molecule is known to have two sets of excited states : singlet states denoted by S_0, S_1 and S_2 and triplet states denoted by T_1 and T_2. The transitions from singlet to triplet states are forbidden. The optical pumping excites the dye molecules from the lowest vibrionic level of the ground state S_0 to one of the upper vibronic level of the excited state S_1. From there, they undergo a non-radiative transition to the lower vibronic level of S_1 state which acts as the upper laser level. The role of lower laser level is played by one of the upper vibronic levels of S_0. The levels at S_0 are so closely spaced that they

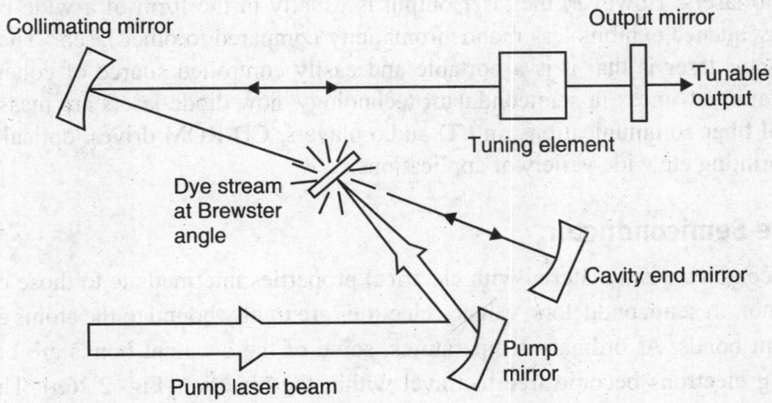

Fig. 2.25. Schematic diagram of a laminar-flow dye laser.

form a continuum. It may be seen from Fig. 2.24 that the laser transition can be to various levels within a range defined by the vibrationally excited sublevels of the ground state. Consequently, the laser operates over a broad range of wavelengths. Thus dye lasers differ from other lasers that emit light at fixed wavelengths. The schematic of a laminar-flow dye laser is shown in Fig. 2.25.

2.6 COMPARISON BETWEEN SOLID, LIQUID AND GAS LASERS

No.	Solid Lasers	Liquid Lasers	Gas Lasers
1.	The active medium is a soild.	The active medium is a liquid or a dye.	The active medium is a gas at low pressure .
2.	Either crystalline or amorphous.	In the form of dye dissolved in organic or inorganic solvent.	Either atomic, ionic or molecular.
3.	Pumping is done usually by optical method.	Liquid laser is excited optically usually by another laser source.	It is excited by electrical discharge method.
4.	Emits a narrow range of wavelengths.	Emits a broad-range of wavelengths.	Emits a narrow range of wavelengths.
5.	Due to imperfections and defects in crystals, inhomogeneties in the output.	Output is optically homogeneous.	Output is highly homogeneous.
6.	Example : Ruby, Nd-YAG, Nd-Glass Fibre laser etc.	Pulse dye laser, Tunable pulsed and CW dye laser etc.	He-Ne, Argon ion, Krypton ion, CO_2 laser etc.

2.7 SEMICONDUCTOR LASERS

A semiconductor laser is a specially fabricated pn-junction device that emits coherent light when it is forward biased. The first semiconductor laser was made in 1962 by R. N. Hall and his coworkers of General Electric R & D Laboratories, U.S.A. It is made from gallium arsenide (GaAs) which operated at low temperature and emitted light in the near IR. Semiconductor lasers working at room temperature and in continuous wave mode are produced by 1970. Now pn-junction (diode) lasers are made to emit light almost anywhere in the spectrum from the UV to the IR. Diode lasers are remarkably small in size (0.1 mm long). They have high efficiency of the order of 40%. Modulating the biasing current easily modulates the diode laser output. They operate at low powers. Inspite of their small size and low power requirement, they produce power outputs equivalent to that of He-Ne lasers. However, the laser output is usually in the form of a wide beam, 5 to 15° and as a consequence exhibits less monochromaticity compared to other lasers. The advantage of a semiconductor laser is that it is a portable and easily controlled source of coherent radiation. Because of rapid advances in semiconductor technology, now diode lasers are mass produced for use in optical fiber communications, in CD audio players, CD-ROM drives, optical reading, high speed laser printing etc wide variety of applications.

2.7.1. Pure Semiconductor

A semiconductor is a material with electrical properties intermediate to those of a conductor and an insulator. In semiconductors, valence electrons are tightly bound to the atoms and are locked up in covalent bonds. At ordinary temperatures, some of the covalent bonds are broken and the corresponding electrons become free to travel within the material (Fig. 2.26*a*). The situation is described in the following alternate way. The allowed energy values of the valence electrons in semiconductors occur within two well defined energy bands, separated by a forbidden energy gap, as shown in Fig. 2.26(*b*). A pure semiconductor crystal has exactly enough electrons to fill all the states in the lower band, namely *valence band*. However, a few electrons gain enough energy E_g to jump into the upper band, namely *conduction band*. In case of silicon and germanium, the number of electrons that can jump into the conduction band at room temperature is small. The electron jumping to the conduction band leaves behind a quantum vacancy in the valence band. The vacancy is called a *hole* and is assigned a positive charge along with the other attributes of a particle. In a pure semiconductor the number of electrons in the conduction band and holes in the valence band are equal. When a conduction electron falls into the valence band, it recombines with a hole there. The excess energy E_g is given out in the form of heat or light. If the recombination energy is released in the form of heat as in silicon and germanium crystals, the materials are of no use for laser action.

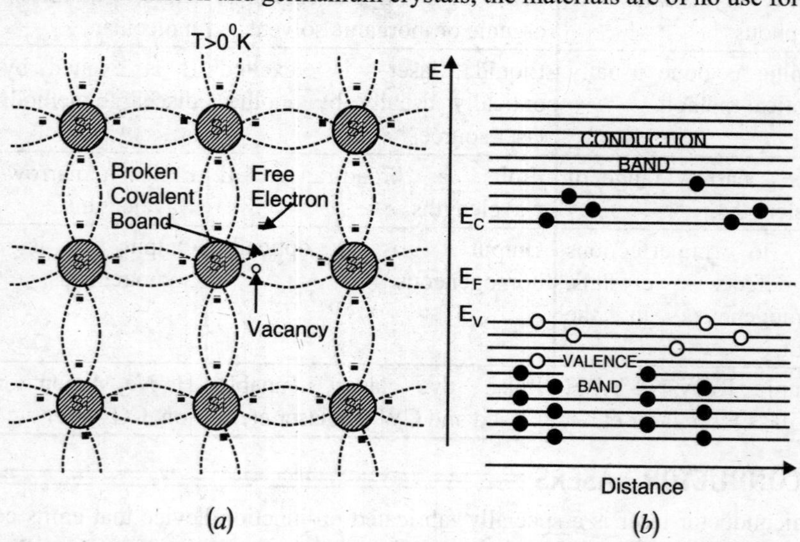

Fig. 2.26. Silicon crystal at temperatures above OK. (a) Thermal vibrations of atoms lead to breaking of covalent bonds. Consequntly, a free electron and a vacancy are simultaneously produced. (b) energy band representation. An energy $E_g = (E_c - E_v)$ causes the transition of electrons from valence band to conduction band leaving behind vacancies

Now, if a pure semiconductor material is illuminated by photons of energy $hv = E_g$, a photon can be absorbed by an electron in the valence band. The excited electron will jump into the conduction band. If, on the other hand, the photon hits an electron in the conduction band, it can induce electron to make a downward transition to the valence hand where the electron recombines with a hole and a quantum of radiation with energy equal to E_g is released. At ordinary temperatures, the number of electrons at the bottom of the conduction band is very small compared to the number of electrons at the top of the valence band. Therefore, the process of light absorption dominates over the process of stimulated emission. In order to make stimulated emissions to dominate over absorption of light in a semiconductor, the concentration of electrons at the bottom of the conduction band should be made higher than that at the top of the valence band. To create such a situation, pure semiconductors are to be doped with certain impurity atoms.

2.7.2. Impurity Semiconductors

The characteristics of pure semiconductors can be drastically altered in very important ways by the introduction of small amounts of impurities. Pentavalent or trivalent atoms are used as *dopants*, which easily substitute themselves in place of some of the host atoms in the semiconductor crystal. Consider the silicon crystal. Each silicon atom has four valence electrons and therefore each atom has four atoms bonded to it. If the crystal is doped with phosphorous, the phosphorous atoms will fit into the crystal structure, but they have five valence electrons out of which four participate in bonding to neighbour atoms. The fifth electron is superfluous to the structure and therefore is loosely bound to the phosphorous atom. It is readily excited into the conduction band by the thermal energy. Thus, pratically every phsophorous atom introduced into the silicon lattice contributes a conduction electron *without* creating a positive hole. Thus, the addition of the phosphorous greatly increases the number of condution electrons, which become the *majority carriers* in the material. In this case phosphorous is called a *donor* impurity and it makes the silicon crystal an *n-type semiconductor* (Fig. 2.27).

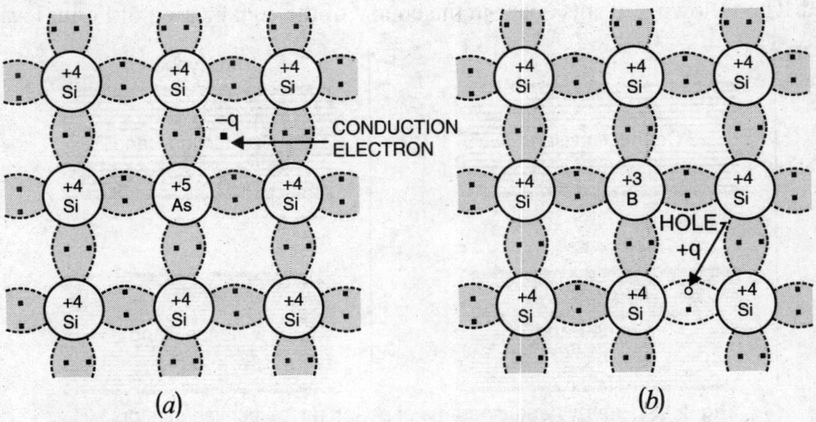

Fig. 2.27. Schematic bond pictures for (a) n-type Si with donor (arsenic) and (b) p-type Si with acceptor (boron).

Boron, on the other hand, has three valence electrons. If it is used as a dopant in silicon crystal, it can supply only three of the four electrons necessary to fit into the silicon lattice. However, thermal excitation is sufficient to cause silicon valence electrons to complete the bonding by attaching themselves to boron atoms. It leaves holes in the valence band. Thus boron is an *acceptor impurity*, which causes as many positive holes as there are boron atoms. The acceptor impurity makes silicon a *p-type semiconductor*.

If the concentration of impurities in the semiconductor material is very high, the semiconductor is said to be heavily doped.

2.7.3. Population Inversion

Population inversion is required for producing stimulated emission and consequent amplification of light. The way in which population inversion is achieved in semiconductors is very different from the way it is established in other types of lasers. A semiconductor is not a two-level atomic system, but consists of electrons and holes distributed in the respective energy bands. Therefore, laser action in semiconductors involves energy bands rather than discrete levels. Secondly, in many types of lasers, population inversion is obtained by exciting electrons in spatially isolated atoms. In semiconductors, the excited electrons are not associated with specific atoms but are injected into the conduction band from the external circuit. Therefore, the conduction band represents an excited state while the valence band plays the role of the ground state.

Population inversion requires that there is a large concentration of electrons in the conduction band and a large concentration of holes in the valence band. A simple way of achieving population inversion is to use a semiconductor in the form of a *pn*-junction diode with heavily doped *p* and *n* type semiconductors.

2.7.4. *pn*-Junction

Let us first understand the formation of an ordinary *pn* junction. Fig. 2.28 shows the energy band diagrams of *p*-type and *n*-type semiconductors. The Fermi level E_{Fp} of *p*-type material is nearer to the valence band and the Fermi level E_{Fn} of *n*-type material is nearer to the bottom of the conduction band. Fermi level on each side is the reference energy level, analogous to the liquid level in a container. When both the *p* and *n* materials are joined, equilibrium is attained only when equalisation of Fermi levels E_{Fp} and E_{Fn} takes place. When two containers filled with liquid to two different levels are connected, the liquid level in one container falls down while the level in the other rises till both levels are equalized. In a similar manner, when *p* and *n*-type materials are joined, the energy levels in *p*-region move up and those in *n*-region move down till the Fermi levels E_{Fp} and E_{Fn} are equalized. The following events occur in the course of the equalization of Fermi levels.

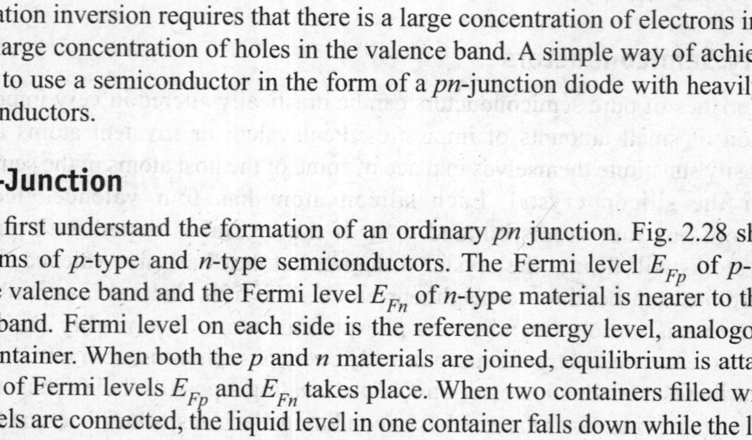

Fig. 2.28. Energy band diagrams of p- and n-type semiconductors.

(i) The electrons are larger in number (majority carriers) on *n*-side and smaller (minority carriers) on *p*-side. Similarly holes are larger in number on *p*-side and smaller on *n*-side. Thus, there is a concentration gradient for majority carriers across the junction. Hence, they tend to diffuse across the junction.

(ii) The electrons and holes diffuse in opposite directions and recombine in the vicinity of junction. Therefore, a narrow region around the junction is left with only immobile impurity atoms and neutral host atoms.

(iii) The immobile ions are negative on the *p*-side and positive on *n*-side. These layers of opposite charges create a potential barrier for the diffusion of majority charge carriers.

The potential barrier halts movement of electrons from *n*-side and that of holes from *p*-side. As a result, a narrow depletion region devoid of mobile charges forms.

The mutual displacement of the energy levels on both sides of the junction causes a bending of the energy bands around the junction region, as shown in Fig. 2.29.

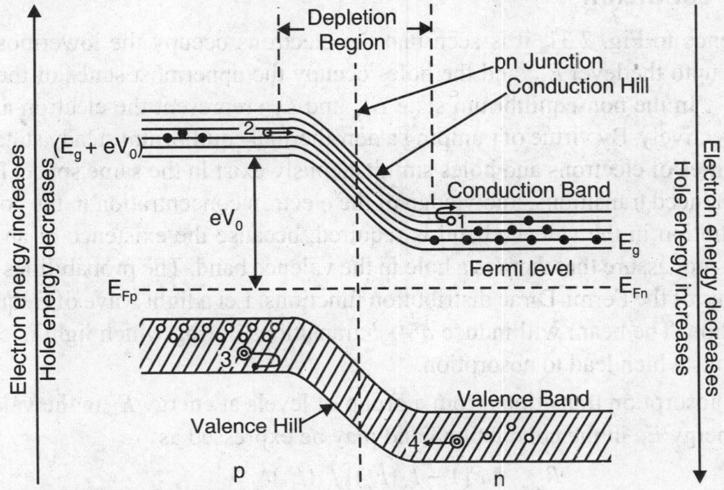

Fig. 2.29. Energy band diagram of pn-junction at equilibrium.

When the doping levels are very high, the Fermi level in *p*-type region will be below the valence band edge and will be above the conduction band edge in *n*-type region.

The energy band diagram of a *pn* junction formed between two heavily doped semiconductors is shown in Fig. 2.30.

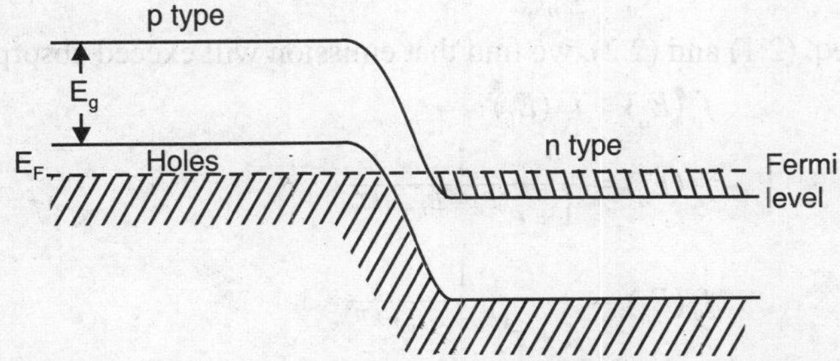

Fig. 2.30. Energy band diagram of a heavily doped *pn*-junction in equilibrium.

The probability of stimulated emission event occurring is low even now because there must be a region of the device where a large number of conduction electrons and a large number of holes would be *simultaneously* present. This condition is never realized in thermal equilibrium. However, when the junction is forward biased with a large enough voltage, electrons and holes (minority carriers) are injected into the junction region in considerable concentrations. In other words carriers are pumped by the dc source. As a result, the region 'd' (Fig. 2.31) contains a large concentration of electrons within the conduction

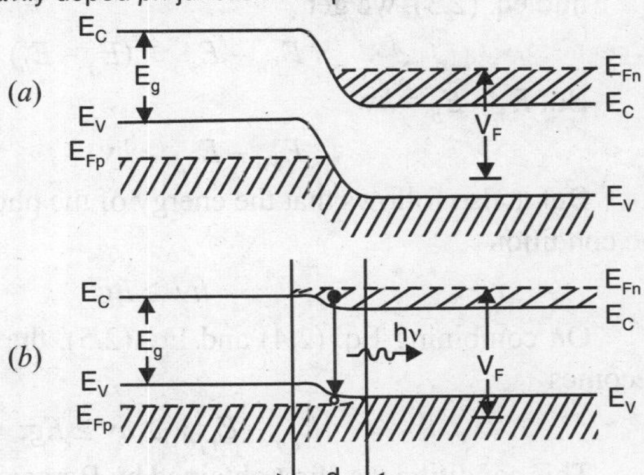

Fig. 2.31. Energy band diagram of a heavily doped *pn*-junction under
(a) forward bias of smaller magnitude
(b) larger forward bias leading to lasing.

band and large concentration of holes within the valence band. This is the condition of population inversion. The narrow zone over which population inversion occurs is called an *inversion region* or *active region*. The electrons and holes recombine in the active region emitting photons.

2.7.5. Lasing Condition

With reference to Fig. 2.31, it is seen that the electrons occupy the lowermost states of the conduction band upto the level E_{Fn} and the holes occupy the uppermost states of the valence band upto the level E_{Fp}. In the non-equilibrium state E_{Fn} and E_{Fp} represent the electron and hole quasi-Fermi levels respectively. By virtue of pumping a non-thermal equilibrium steady state is obtained in which large densities of electrons and holes simultaneously exist in the same space. For writing the probabilities of induced transitions, knowledge of the electron concentration in the conduction band and hole concentration in the valence band is required, because the existence of an electron at an excited level does not assure that there is a hole in the valence band. The probabilities for transitions are written in terms of the Fermi-Dirac distribution functions. Let a light wave of frequency ν_0 travel through the junction. The beam will induce $a \to b$ transitions during which light is emitted as well as $b \to a$ transitions which lead to absorption.

The rate of absorption transitions from a group of levels at energy E_v in the valence band to a group levels at energy E_C in the conduction band may be expressed as

$$R_{abs} = P[1 - f_c(E_a)] f_v(E_b) N \qquad \text{...(2.1)}$$

where f_c and f_v are Fermi functions in conduction and valence bands respectively, N is the density of photons of energy E, and P is the transition probability. The stimulated emission rate is given by

$$R_{st} = P f_c E_a [1 - f_v(E_b)] N \qquad \text{...(2.2)}$$

For emission of light from the *pn*-junction, we require that

$$R_{st} > R_{abs}$$

On comparing eq. (2.1) and (2.2), we find that emission will exceed absorption if

$$f_c(E_a) > f_v(E_b) \qquad \text{...(2.3)}$$

Using,

$$f_c(E_a) = \frac{1}{1 + e^{(E_a - E_{Fn})/kT}}$$

and

$$f_v(E_b) = \frac{1}{1 + e^{(E_b - E_{Fp})/kT}}$$

into eq. (2.3), we get

$$E_{Fn} - E_{Fp} > (E_a - E_b)$$

But $E_a - E_b = h\nu$

$\therefore$

$$E_{Fn} - E_{Fp} > h\nu \qquad \text{...(2.4)}$$

But it also follows that the energy of the photon emitted due to electron transition must satisfy the condition

$$h\nu \geq Eg. \qquad \text{...(2.5)}$$

On combining Eq. (2.4) and Eq. (2.5), the necessary condition for lasing in the *pn*-junction becomes

$$\{E_{Fn} - E_{Fp}\} \geq h\nu \geq Eg. \qquad \text{...(2.6)}$$

This condition was first obtained by Bernard and Duraffourg. E_g and $(E_{Fn} - E_{Fp})$ constitute the outside limits of the laser spectra. In order to fulfill the condition (2.6), the *p* and *n*-regions of the diode has to be doped heavily such that the Fermi levels penetrate the main bands as shown in the Fig. 2.27.

2.7.6. Gain in a Semiconductor

The fractional increase in intensity of radiation of frequency ν per unit distance travelled within the depletion layer is defined as the gain coefficient per unit length of the semiconductor laser. The eq. (1.51) for the gain coefficient γ is valid here also. Thus,

$$\gamma = (N_2 - N_1)\frac{\upsilon^2 g(\nu)}{8\pi\nu^2}A_{21} \quad ...(2.7)$$

If τ_{sp} denotes the time for spontaneous emission of electrons,

$$A_{21} = \frac{1}{\tau_{sp}} \quad ...(2.8)$$

$$\therefore \quad \gamma = (N_2 - N_1)\frac{\upsilon^2 g(\nu)}{8\pi\nu^2\tau_{sp}} = \frac{n\upsilon^2 g(\nu)}{8\pi\nu^2\tau_{sp}} \quad ...(2.9)$$

In semiconductor laser, population inversion is not strictly the quantity $(N_2 - N_1)$. It is actually caused by an injection current. *n* represents the density of carriers. Now, let the current density through the diode be J. The number of electrons injected per second per unit volume of the active region is equal to *J*/*ed*, where *d* is the width of the depletion layer. This number of carriers is removed in the steady state by recombination. The rate equation governing the steady state is

$$\frac{dn}{dt} = \frac{J}{ed} - \frac{n}{\tau_{sp}} = 0 \quad ...(2.10)$$

$$\therefore \quad \frac{n}{\tau_{sp}} = \frac{J}{ed} \quad ...(2.11)$$

It means that under steady state condition the rate at which carriers are injected into the active region must equal the electron-hole combination rate.

On combining eq. (2.11) and eq. (2.9), we get

$$\therefore \quad \gamma = \frac{J\upsilon^2 g(\nu)}{8\pi e\nu^2 d} \quad ...(2.12)$$

2.7.7. Optical Cavity

For lasing to take place, optical cavity is required. In a semiconductor laser, the Fabry-Perot optical cavity is formed by *cleaving* both sides of the diode. The cleaved edges being optically flat to near perfection serve as the mirrors of the cavity because of the large refractive index difference between the semiconductor material and air. Thus, external mirrors are not needed in a diode laser. The reflectance at the boundary of semiconductor-air is given by

$$R = \left(\frac{\mu_2 - \mu_1}{\mu_2 + \mu_1}\right)^2$$

where μ_2 and μ_1 are the refractive indices of the semiconductor and air respectively. Taking $\mu_2 = 3.6$, $\mu_1 = 1$, $R = 0.32$. It means that above 32% of incident light gets reflected into the medium, providing enough feed back for laser action.

The initial light output occurs due to spontaneous emission transition of carriers between the two energy bands and it is selectively amplified as the light travels in the cavity between the end mirrors and is "fed back". During the round trip there is amplification as well as absorption in the light intensity. Lasing occurs when the gain provided by the medium overcomes the loss in the cavity for a round trip.

2.7.8. Threshold Condition for Lasing

Let us consider a *pn* junction diode, where the laser cavity is of length L and has sides of mirror-polish. Let the reflectivities of the two ends be R_1 and R_2 respectively. Now, if I_0 denotes the intensity of light at the beginning of the round trip and α is the absorption coefficient within the medium, then the intensity of radiation after a round trip is given by

$$I = I_0 R_1 R_2 e^{2(\gamma - \alpha)L} \quad \text{...(2.13)}$$

The diode will start functioning as a laser if $I \geq I_0$. It requires that $\gamma > \alpha$. At the threshold, when the round trip gain exactly equals the losses

$$I = I_0 \quad \text{...(2.14)}$$

$$\therefore \quad R_1 R_2 e^{2(\gamma_{th} - \alpha)L} = 1 \quad \text{...(2.15)}$$

where γ_{th} is the threshold gain.

The eq. (2.15) yields

$$\gamma_{th} = \alpha + \frac{1}{2L} \ell n \left(\frac{1}{R_1 R_2} \right) \quad \text{...(2.16)}$$

After a few oscillations laser action reaches a steady state and the gain maintains at a steady state value equal to γ_{th}.

2.7.9. Threshold Population Inversion

The threshold population inversion can be expressed as

$$(N_2 - N_1)_{th} = \frac{8\pi \nu_0^2 \gamma_{th} \, \tau_{sp} \mu^2}{c^2 g(\nu)} \quad \text{...(2.17)}$$

or

$$n_{th} = \frac{8\pi \nu_0^2 \gamma_{th} \, \tau_{sp} (\Delta\nu) \mu^2}{c^2} \quad \text{...(2.18)}$$

The above equation contains terms that are either physical constants or parameters that can be determined for a real laser.

2.7.10. Threshold Current Density

Population inversion occurs in semiconductors over an active thickness d. However, the thickness over which the optical mode extends is d_m which is usually larger than d.

The threshold density of injected carriers can be written from eq. (2.18) as

$$n_{th} = \frac{d_m}{d} \left[\frac{8\pi \nu_0^2 \gamma_{th} \, \tau_{sp} (\Delta\nu) \mu^2}{c^2} \right] \quad \text{...(2.19)}$$

From eq. (2.11), we get

$$J_{th} = \frac{n_{th} e d}{\tau} \quad \text{...(2.20)}$$

Using eq. (2.19) into eq. (2.20), we obtain

$$J_{th} = \frac{e d_m}{\tau} \left[\frac{8\pi \nu_0^2 \mu^2 \gamma_{th} \, t \Delta\nu}{c^2} \right] \quad \text{...(2.21)}$$

Substituting the expression (2.16) for γ_{th} into the above equation, we get

$$J_{th} = \frac{8\pi \nu_0^2 \mu^2 e d_m \tau_{sp} \Delta\nu}{\tau c^2} \left[\alpha + \frac{1}{2L} \ell n \left(\frac{1}{R_1 R_2} \right) \right] \quad \text{...(2.22)}$$

From the above equation, it is seen that the threshold current density J_{th} does not depend on the thickness of the active layer, d. On the other hand, it is directly proportional to d_m.

2.7.11. Power Output

The optical power output into modal volume due to a current $I\ (> I_{th})$ can be written as

$$P_e = \left(\frac{I - I_{th}}{e}\right)\eta_i h\nu \qquad ...(2.23)$$

where η_i denotes the internal quantum efficiency of the semiconductor. Part of this power is dissipated inside the cavity while overcoming the cavity losses and a fraction of P_e is coupled out through the end facets as useful light output. The output power of the laser can be written as

$$P_o = \frac{(I - I_{th})\eta_i h\nu}{e}\left[\frac{(1/2L)\ell n\,(1/R_1R_2)}{\alpha + (1+2L)\ \ell n\,(1/R_1R_2)}\right] \qquad ...(2.24)$$

2.7.12. Efficiency

The overall efficiency of the laser in converting electrical power into optical output power may be defined as

$$\eta = \frac{P_0}{VI}$$

$$\eta = \left(\frac{I - I_{th}}{I}\right)\frac{\eta_i h\nu}{eV}\ \frac{\ell n\,(1/R_1R_2)}{2\alpha L + \ell n\,(1/R_1R_2)} \qquad ...(2.25)$$

As the applied voltage is approximately equal to $h\nu/e$, and for current $I >> I_{th}$

$$\eta \sim \eta_i$$

η_i is very high in GaAs leading to an efficiency of 70% for the laser.

2.7.13. Basic Laser Structure

The simplest *pn* junciton structure is called a *homojunction* laser, which is a basic *pn* junction. The same semiconductor material, say GaAs, is on both sides of the junction. Starting with a heavily doped large area *n*-type sample, a *p* region is formed on its top by diffusing zinc atoms into *n*-type GaAs. A heavily zinc doped layer constitutes the heavily doped *p*-layer. Grooves are cut along the

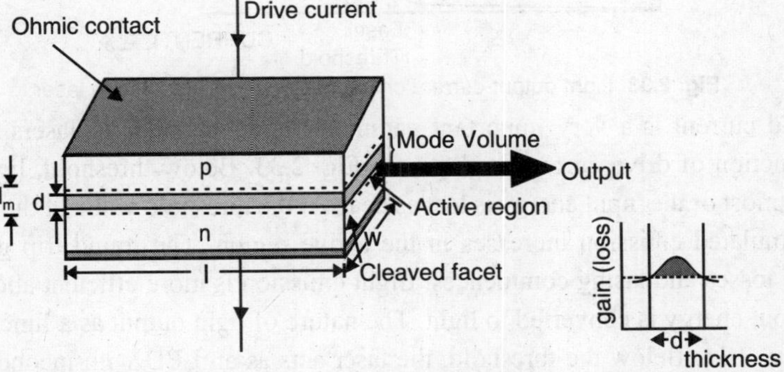

Fig. 2.32. Structure of a typical homo-junction semiconductor laser.

length of the sample, and small *pn*-junction chips are broken apart. For forming optical resonator cavity, external mirrors are not needed in a diode laser. A pair of parallel planes cleaved at the two ends of the *pn*-junction provides the required reflection to form the cavity. The two remaining sides of the diode are roughened to eliminate lasing in the directions other than the main ones. This structure is the Fabry-Perot cavity. The top and bottom faces are metallized and ohmic contacts are provided to pass current through the diode. The structure of a homojunction diode laser is shown in Fig. 2.32. Typical diode chips are 300 to 500 μm long and about 100 μm wide and thick. The entire structure is packaged in small cases which look like the metal cases used with discrete transistors.

2.7.14. Diode Laser Operation

A *pn* junction formed from very heavily doped p and n materials provides the active medium. When forward bias is applied to the junction diode, the bias current performs the role of pumping agent. As a result of forward bias, electrons and holes are injected into the junction region. At low forward current level, electron-hole recombination causes spontaneous emission of photons and the junction acts as a light emitting diode (LED). When the current reaches a threshold value, the carrier concentrations reach to very high values within the depletion layer and the state of population inversion is established. There is simultaneously a large concentration of electrons within the conduction band and a large concentration of holes within the valence band of the active region. The spontaneous photons propagating in the junction plane induce the conduction electrons to jump into the holes of valence band. These stimulated electron-hole recombinations cause emission of coherent radiation. The stimulated emission is confined to a particular direction by the reflective surfaces formed perpendicular to the junction. The reflective surfaces are the smooth "facets" formed by cleaving the chip. Semiconductors have high refractive indices and the facets reflect about 30% of the light back into the material providing enough feedback for laser action.

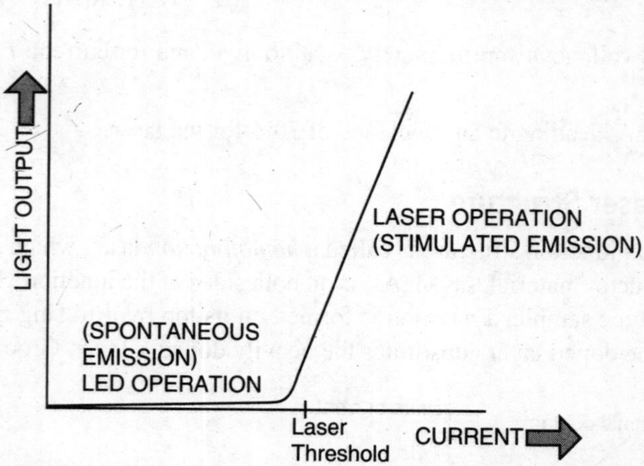

Fig. 2.33. Light output-current characteristic of an ideal diode laser.

Threshold current is a very important parameter in semiconductor lasers. A plot of output power as a function of drive current is shown in Fig. 2.33. Below threshold, light output is very inefficient and most of the input energy is lost as heat. At the threshold value of the injection current, the rate of stimulated emission increases in the active region. The round trip gain in the cavity overcomes the losses and lasing commences. Light emission is more efficient above threshold and more of the input energy is converted to light. The nature of light output as a function of frequency is shown in Fig. 2.34. Below the threshold, the laser acts as an LED with incoherent spontaneous emission dominating the output light (Fig. 2.34*a*). At threshold, the laser modes emerge and coherent light is emitted, as shown in Fig. 2.34(*b*). Beyond threshold, a single dominant mode is observed, as in Fig. 2.34(*c*). Homojunction lasers are operated in pulsed mode at room temperature because the threshold pumping current density requirement is very high.

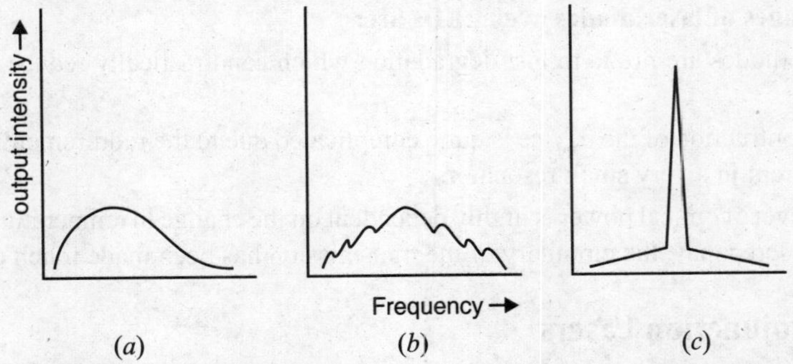

Fig. 2.34. Light output as a function of frequency for a diode laser (a) below threshold (b) with the laser modes at threshold, and (c) with the dominant laser mode above threshold.

While other lasers emit round cylindrical beams, the diode laser beams are long thin ovals, because the cross section of the active region is approximately rectangular. The beam from a diode laser diverges at different rates in the two perpendicular directions as shown in Fig. 2.35. Typical values are 10^0 in the direction parallel to the layer and 40^0 in the direction normal to the active layer. Therefore, collimating lenses are to be used along with diode lasers.

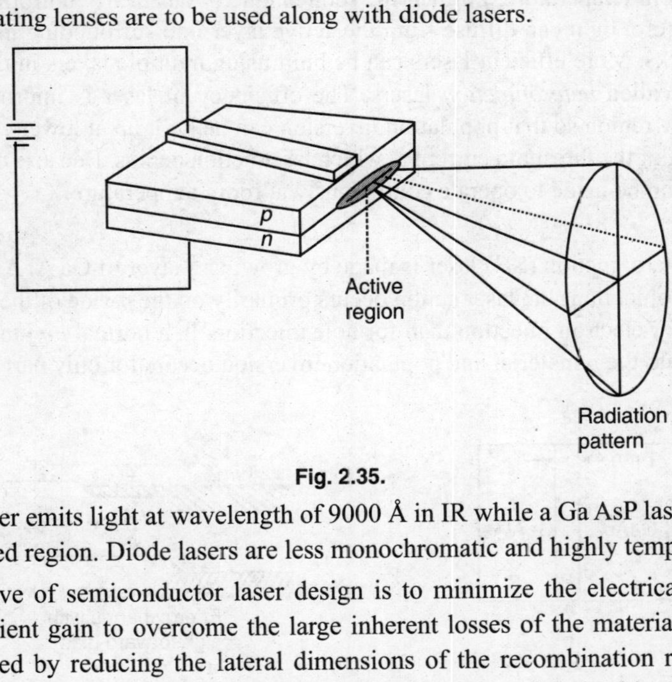

Fig. 2.35.

The Gas laser emits light at wavelength of 9000 Å in IR while a Ga AsP laser rediates at 6500 Å in the visible red region. Diode lasers are less monochromatic and highly temperature sensitive.

The objective of semiconductor laser design is to minimize the electrical current required to produce sufficient gain to overcome the large inherent losses of the material. Minimizing this current is achieved by reducing the lateral dimensions of the recombination region, both in the direction of layers and orthogonal to the layers. Current can be minimized in the direction of the layers by reducing the active layer region to dimensions of 0.1-0.2 μm as in heterojunction devices. Quantum well devices have been made to operate at threshold currents as low as 0.5 mA.

Advantages of laser diodes over LEDs are:

(*i*) The laser can be coupled to a optic fibre, which can give a large distance transmission of signals.

(*ii*) Laser diodes have faster response-time which give much greater modulation rates. It means data transmission rates is higher in laser diodes.

(*iii*) Laser diodes give a much narrow spectral width of the output which means that the distribution caused by induced dispersion can be minimised.

Disadvantages of laser diodes over LEDs are:

(*i*) Laser diodes are prone to fast degradation which can drastically reduce the life-time of device.

(*ii*) The contruction of the device is quite complicated due to the requirement of confinement of current in a very small resonator.

(*iii*) The level of optical power is highly dependent on the change in temperature. To overcome this discrepancy, the circuitary of the transmission has been made much complex.

2.7.15. Heterojunction Lasers

The device described above is a homojunction laser, since the *p*- and *n*-regions are made of the same semiconductor. In homojunction lasers, the active region '*d*' and the thickness d_m denoting the mode volume are not well defined. The active region *d* is ill defined owing to the diffusion length of the carriers. As the semiconductor has nearly uniform refractive index throughout, light can diffuse from active layer into the surrounding medium. Therefore, d_m can far exceed *d*. As a result the cavity losses will increase. Therefore, high threshold currents are required and the laser cannot be operated continuously at room temperature. Further, the semiconductor has nearly uniform refractive index throughout. Therefore, light can diffuse from the active layer into surrounding medium. These are the major drawbacks. More efficient lasers can be built using multiple layers in the laser structure. Such devices are called *heterojunction* lasers. The efficiency of laser is improved by confining carriers to a narrow region so that population inversion can be built up at lower current levels. The result is a lowering of the threshold current at which laser action begins. Because of lower threshold current, the laser can be made to operate continuously at room temperature.

SH Laser

A single - heterojunction (SH) laser is made by growing a layer of Ga Al As on the GaAs. In a GaAs laser (homojunction) the laser action occurs primarily on the *p*-side of the junction due to a higher efficiency for electron injection than for hole injection. In a normal *pn* junction the injected electrons diffuse into the *p* material and population inversion occurs for only part of the electrons

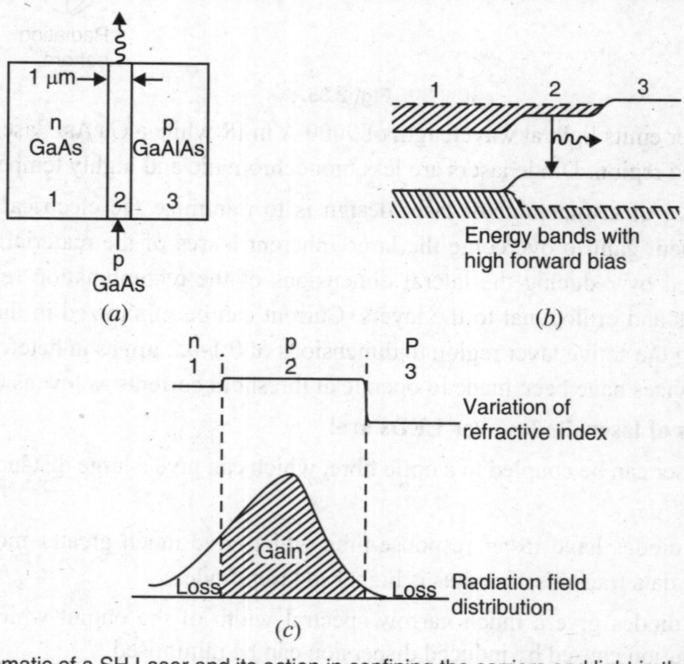

Fig. 2.36. Schematic of a SH Laser and its action in confining the carriers and light in the active region.

near the junction. If the p material is made narrow and terminated in a barrier, the injected electrons can be confined near the junction. Referring to Fig. 2.36 the electrons injected from the n-GaAs layer to the p-GaAs layer is blocked from diffusing over a large distance by the potential barrier provided by the higher band gap p-GaAlAs layer. Now, the active region thickness is the thickness of p-GaAs layer. If its thickness is made small, a smaller drive current can lead to population inversion and lasing. Further, the change of refractive index at the heterojunction provides a waveguide effect which confines light. However, SH lasers do not confine light well enough to emit light continuously at room temperature.

DH Laser

A further improvement is obtained by sandwiching the active GaAs layer between two layers of GaAlAs, which has a wider energy gap and a lower refractive index than GaAs. This double-heterojunction (DH) structure further confines injected electrons to the active region and reduces the threshold current required for lasing. Similarly, the step change in refractive index provides a much more efficient waveguide structure for light than the SH laser.

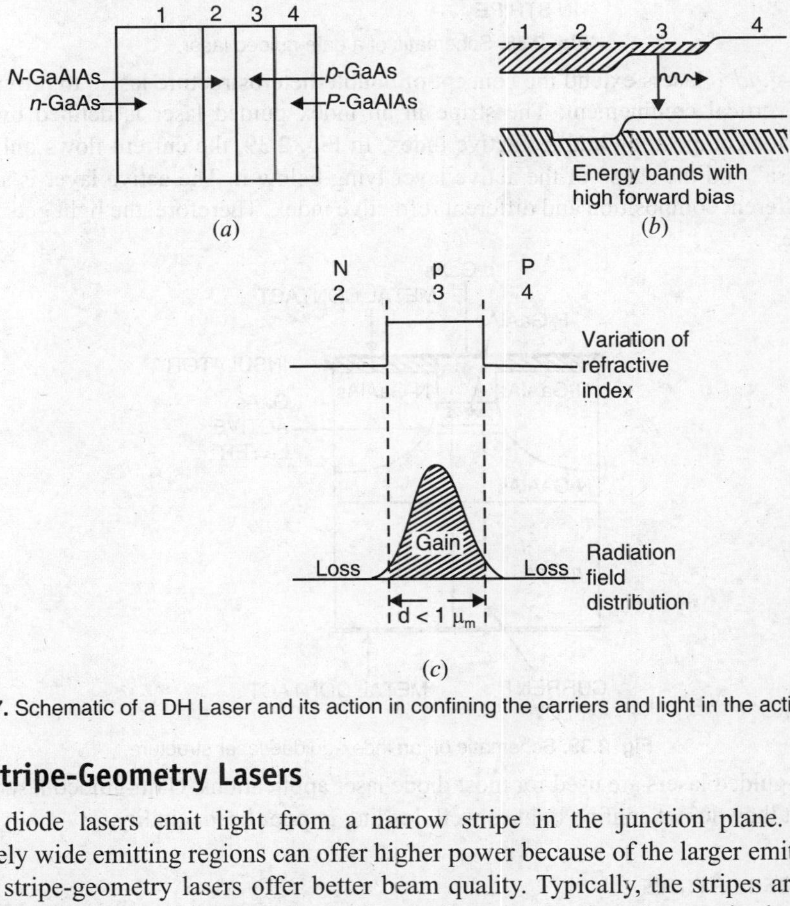

Fig. 2.37. Schematic of a DH Laser and its action in confining the carriers and light in the active region.

2.7.16. Stripe-Geometry Lasers

Most diode lasers emit light from a narrow stripe in the junction plane. Lasers with comparatively wide emitting regions can offer higher power because of the larger emitting volume, but narrow stripe-geometry lasers offer better beam quality. Typically, the stripes are only a few micrometers wide. The stripes are limited either by boundaries where laser gain drops off, or by changes in the refractive index of the material itself.

The basic concept of a *gain-guided* laser is shown in Fig. 2.38. An insulating layer at the top of the laser chip blocks current flow at the sides, and confines it to a narrow stripe along the length of the chip. Current flows to produce a population inversion and the right conditions for laser gain within the narrow stripe. The sides do not emit light, even though no physical boundary separates the stripe from the rest of the active layer.

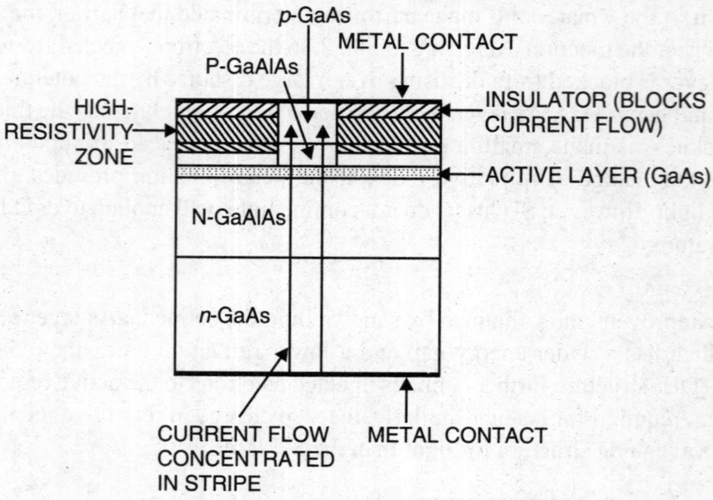

Fig. 2.38. Schematic of a gain-guided laser.

Index-guided lasers extend the concept of double-heterostructure lasers to provide horizontal as well as vertical confinement. The stripe in an index guided laser is defined by a change in semiconductor composition and refractive index. In Fig. 2.39, the current flows only through the central "mesa" and the stripe of the active layer lying below it. The active layer is surrounded by layers of different composition and different refractive index. Therefore, the light gets trapped in the active stripe.

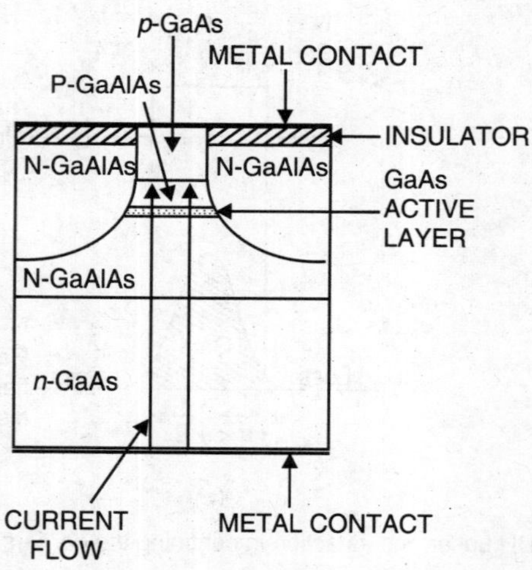

Fig. 2.39. Schematic of an index-guided laser structure.

Index-guided lasers are used for most diode laser applications. Gain-guided lasers are simpler to make, but they do not confine light as well, leading to poor beam quality.

2.7.17. Laser Arrays

Many parallel laser stripes can be fabricated on the same semiconductor substrate, which can produce much higher power than single laser stripes, although their beam quality tends to be poorer.

Arrays with tens or even hundreds of parallel laser stripes can generate continuous-wave powers above 10 W.

So far most high-power arrays have been of gallium aluminum arsenide, emitting near 800 nm used for such applications as pumping solid-state neodymium lasers.

2.7.18. Quantum Well Laser

Semiconductor layers of only a few nanometers thick can serve as "quantum wells". The basic idea of a quantum well is to sandwich a thin semiconductor layer with a small band gap between two thick semiconductor layers with larger band gaps, as shown in Fig. 2.40. Electrons passing through the semiconductor can be captured in the quantum well. They have enough energy to remain free in the small-band-gap confinement layer. This makes quantum wells good traps for electrons, which are confined very tightly in the thin layer.

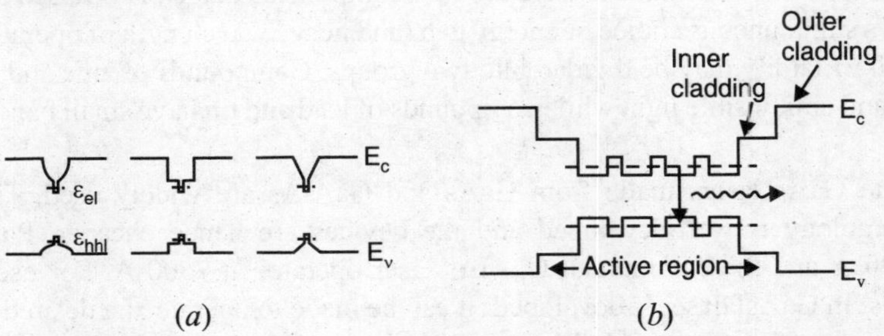

Fig. 2.40. Band diagrams of the active region of (a) single quantum well (b) multiquantum well lasers.

The tight confinement is a big advantage if the quantum well is put at the junction of a semiconductor laser. By concentrating the electrons as they recombine with holes, it improves the laser efficiency and decreases its threshold current. The confinement layers also have a different refractive index than the quantum well, so they can confine light as well.

The volume of a single quantum well is small, which can limit the amount of light it can generate. Therefore, quantum wells are stacked in alternating layers with confinement layers, producing a multiple quantum well structure that contains a larger volume of light-emitting material. This allows the production of higher powers.

2.7.19. DFB Lasers

Conventional semiconductor lasers have broad bandwidth and hence contain several longitudinal modes. As the length of the optical cavity is shorter, the modes are widely spaced, and often one mode dominates. But because of changes in temperature and operating conditions, a diode laser tends to from one dominant mode to another adjacent mode, leading to instability in laser power and output wavelength.

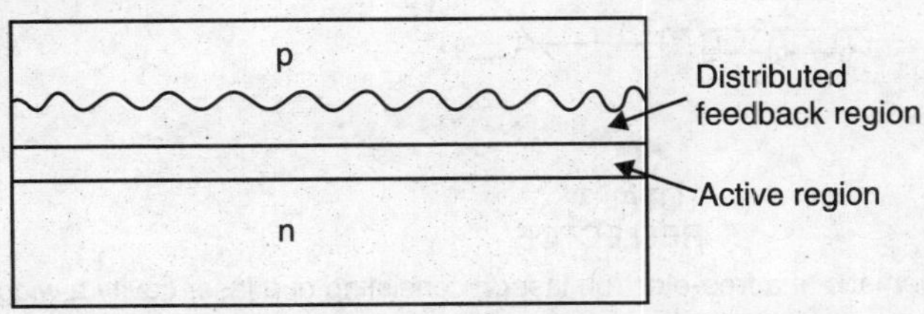

Fig. 2.41. Schematic of a distributed feedback laser structure

Distributed feed back (DFB) laser is designed to create wavelength stability and strong mode selectivity. It incorporates a periodic grating into the laser structure. Feed back from the grating causes interference effects as a result of which laser oscillation is limited to a single longitudinal mode of narrow bandwidth and high stability.

2.7.20. Materials

In choosing a material for junction laser fabrication, it is necessary that electron-hole recombination occurs directly rather than indirectly as in silicon and germanium. Thus, the elemental semiconductor silicon is not a suitable material, as it is a very poor light emitter. Compound semiconductors are used for all practical semiconductor lasers. The most important are the III-V semiconductors followed by II-VI compounds. The simplest of III-V materials is the binary compound, gallium arsenide, GaAs. By replacing some of the gallium atoms in GaAs with aluminium atoms, the energy band gap is changed. The "ternary" compound GaAlAs and the "binary" compound GaAs have identical lattice constants and are said to be lattice matched. Lattice matching is necessary for growing heterostructures. The "quaternary" alloy In GaAsP allows considerable choice of wavelength. As there are four components that can be varied in composition, In GaAsP allows simultaneous choice of energy gap (and hence wavelength of operation) and lattice constant. The II-VI family may be divided into two groups. Compounds of zinc and cadmium have large band gap and emit visible light while compounds of lead and tin have small band gaps and emit in the IR.

As on date GaAs lasers made from GaAs and GaAlAs are widely used. The GaAs laser fabrication technology is well-developed and the devices are hence cheaper. Pure GaAs laser operates at 9040 Å and the commercial GaAlAs laser operates at 7500 Å. For use in fibre optic communications, In GaAsP laser is developed. It can be made to operate at a desired wavelength in the range 1.1 μm to 1.65 μm, by carefully controlling the concentrations of the constituent elements.

2.8 FREE-ELECTRON LASER

There are a few lasers which do not fall into the categories we have discussed till now. Free electron laser is one such laser.

The active medium in free electron laser is a beam of free electrons. Therefore, the basic requirement is a high velocity high current beam and the electrons are to be periodically accelerated along their path. One way of doing this is to set up a periodic array of permanent magnets with alternating polarity, as shown in Fig. 2.42, across the path of the electron beam.

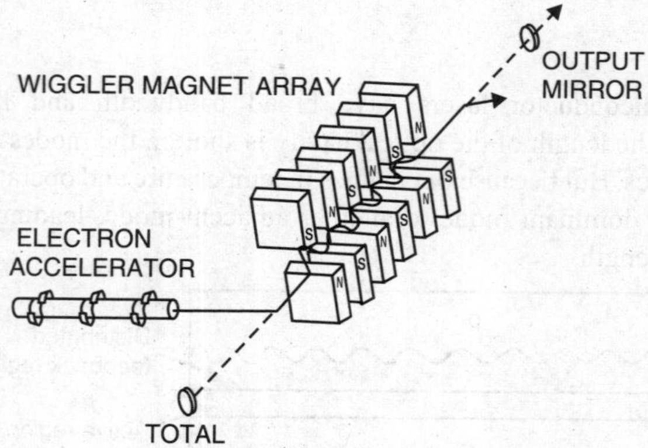

Fig. 2.42. Schematic of a free-electron laser n consisting of a laser cavity a wiggler magnet.

Such an arrangement is called a *wiggler*. Electrons wiggling through the wiggler magnetic field generate some energy as light. Much of the energy is reabsorbed when the electrons wiggle back in the opposite direction. Light is amplified if the following resonance criterion is satisfied.

$$\lambda = \frac{p}{2\,(1 - \upsilon^2 / c^2)} \qquad \text{...(2.26)}$$

where p is the period of the wiggler magnet and v is the velocity of electron along the axis. The importance of this laser lies in the possibility of generating coherent radiation from millimetre to visible wavelengths. Their main disadvantage is that they are enormously bulky and expensive.

EXERCISE–2

1. Enumerate the types of lasers on the basis of medium. Discuss their relative merits and demerits. **(B.Tech. Fourth sem., G.B.T.U. Lucknow 2010)**
2. Name the active material in Ruby laser. Draw a neat labelled diagram of a Ruby laser. **(Pune U. 2010)**
3. Draw schematic diagram to explain construction of CO_2 laser. **(Pune U. 2010)**
4. What is the role of Ne atoms in He-Ne laser ? Explain the working of He-Ne laser with a suitable diagram. **(Pune U. 2010)**
5. What is an excimer laser ? **(Pune U. 2010)**
6. What are solid-state lasers ? **(Pune U. 2010)**
7. Which is the prominent wavelength emitted by Ruby laser ? **(Pune U. 2010)**
8. Explain the principle and construction and working of a ruby laser. **(Garhwal U. 2004; P.U. 2007, 2006, 2003; G.N.D.U.2007, 2006; Kerala U. 2001; Indore U. 2001, Meerut U. 2001 M.D.U. 2007, 2000)**
9. (*a*) Discuss the construction and working of helium-neon laser.
   (*b*) Why does excited helium prefer energy transfer to neon atoms instead of returning to its ground state ? Explain process of lasing in He-Ne laser. What role is played by resonant cavity ?
   (P. U. 2008, 2006; G.N.D.U. 2007, 2004; Meerut U. 2002; K.U. 2001, 2000; Garhawal U. 2004)
10. Explain with the help of an appropriate energy level diagram, how stimulated emission results from electron impact pumping in He-Ne gas laser. **(N.U., 2001)**
11. What is the reason for monochromatically of laser beam ? Explain the working of He-Ne laser and give its engineering applications. **(N.U., 2002)**
12. Explain how lasing action is achieved in a semiconductor laser. With the help of energy band diagram discuss the working of a semiconductor laser. **(N.U., 2000)**
13. Explain in brief the characteristics of laser beam. **(N.U., 2000)**
14. Explain the reason for the monochromaticity of laser beam. **(N.U., 2000)**
15. Can we obtain light amplification in the absence of stimulated emission ?
    (G.N.D.U. 2007, Pbi, U., 2003, P.U. 2003) (Ans. No)
16. Discuss the essential requirements and various steps involved for producing laser action. Laser is basically a three-component device. Explain.
    (P.U. 2007, 2006; Meerut U. 2003, 2001; Pbi. U. 2000, G.N.D.U. 2001)
17. Calculate threshold pumping power per unit volume in case of Ruby laser. Given : N = 1.6×10^{25} atoms/m^3; frequency $\nu_p = 6.25 \times 10^{14}$ Hz, and $t_{sp} = 3 \times 10^{-3}$ sec. **(Pbi. U. 2008) (Ans.** 1100×10^6 W/m^3)
18. Explain the main features and conditions for laser action in a semi conductor.
    (G.N.D.U. 2008, 2006, 2005; P.U. 2008, 2007, 2005; Pbi. U. 2006; K.U. 2002; M.D.U. 2002)
19. Discuss the construction and working of a CO_2 laser.
    (P.U. 2008, 2006; G.N.D.U. 2007, 2006; Pbi. U. 2006)
20. Write a short note on Nd : YAG laser. **(G.N.D.U. 2007, 2006, 2005)**
21. Enunciate the advantages and disadvantages of laser diode over LEDs.
22. Discuss dye laser and its relative advantages. Why is it known as tunable laser ?
    (G.N.D.U. 2007, 2006; P.U. 2006)
23. What do you think that energy conservation is violated in a laser or not ? Expalin.
    (Pbi. U. 2006, P.U. 2005) (Ans. No)
24. Why gas lasers emit light which is more unidirectional and monochromatic as compared to solid state lasers ?
    (Hints: because of absene of effect such as crystalline imperfectios, thermal distortion and scattering which is present in solid state lasers).
25. Discuss the working of a 'Nitrogen laser'. What are the techniques used for population inversion ? Show the energy level diagram to explain your answer. **(B.Tech. Fourth Sem. G.B.T.U. Luck. 2010)**

3 CHAPTER

Laser Beam Characteristics

3.1 INTRODUCTION

Light from a laser differs from light from conventional sources in a number of ways. The most striking features of a laser beam are (*i*) directionality, (*ii*) high intensity (*iii*) high degree of coherence and (*iv*) extraordinary monochromaticity. Laser light can also be produced as polarized light. It can be generated in the form of very short pulses, at high powers.

3.2 DIRECTIONALITY

The conventional light sources emit light in all directions. When we need a narrow beam, we obtain it by placing an aperture in front of the source. In case of a laser, the active material is in a cylindrical resonant cavity. Any light that is travelling in a direction other than parallel to the cavity axis is eliminated and only the light that is travelling parallel to the axis emerges from the cavity and becomes the laser beam. Hence, the light is emitted by a laser only in one direction. The directionality of a laser beam is expressed in terms of *beam divergence*.

3.2.1. Divergence

Light from a laser diverges very little. Upto a certain distance, the beam shows little spreading and remains essentially a bundle of parallel light rays. The distance from the laser over which the light rays remain parallel is known as *Rayleigh range.* The laser beam diverges beyond the Rayleigh range as shown in Fig. 3.1. There are two parameters which cause beam divergence. They are (*i*) the size of the beam waist and (*ii*) diffraction. The *divergence angle* is measured from the centre of the beam to the edge of the beam, where the edge is defined as the location in the beam where the intensity decreases to $1/e$ of that at the centre.

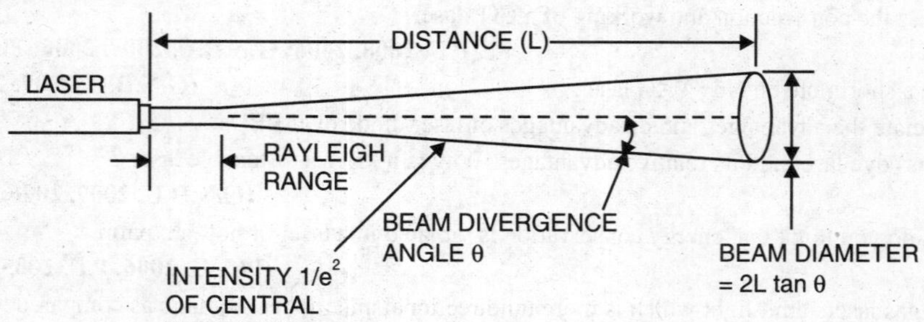

Fig. 3.1. Divergence of a laser beam

Twice the angle of divergence is known as the *full angle beam divergence.* This angle tells us how much the beam will spread as it travels through space. The full angle divergence is given by

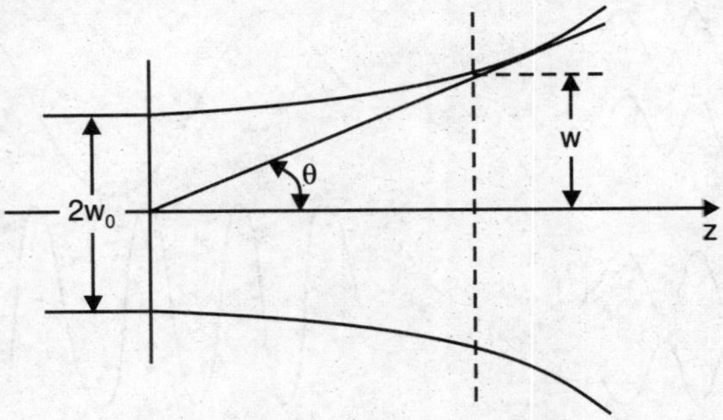

Fig. 3.2. Relationship between beam waist and divergence of a laser beam.

$$2\theta = \frac{4\lambda}{\pi d_0} \qquad \text{...(3.1)}$$

where $d_0 = 2W_0$ is the diameter of the beam waist (Fig. 3.2). It is seen from Eq. (3.1) that the divergence is inversely proportional to d_0. Thus, divergence is large for a beam of small waist.

The beam divergence due to diffraction is determined from Rayleigh's criterion

$$\theta = 1.22\frac{\lambda}{D} \qquad \text{...(3.2)}$$

where D is the diameter of the laser's aperture.

In case of gas lasers, the diffraction divergence is about twice as large as the beam-waist divergence. A typical value of divergence for a He-Ne laser is 10^{-3} rad. It means that the diameter of the laser beam increases by about 1 mm for every metre it travels. Beam divergence of large lasers may be as small as a micro-degree. A laser beam of 5 cm. diameter (divergence = 10^{-6} degree) when focussed from Earth will have spread to a diameter of only about 10 m on reaching the surface of the moon. This extreme collimation of the beam makes lasers a very useful tool for surveying.

3.3 INTENSITY

The power output of a laser may vary from a few milliwatts to a few kilowatts. But this energy is concentrated in a beam of very small cross-section. The intensity of a laser beam is approximately given by

$$I = \left(\frac{10}{\lambda}\right)^2 P \qquad W/m^2 \quad \text{...(3.3)}$$

where P is the power radiated by the laser. In case of 1 mW He-Ne laser,

$\lambda = 6328 \times 10^{-10}$ m and

$$I = \frac{100 \times 10^{-3} W}{\left(6328 \times 10^{-10}\right)^2 m^2} = 2.5 \times 10^{11} W/m^2$$

To obtain light of same intensity from a tungsten bulb, it would have to be raised to a temperature of 4.6×10^6 K. The normal operating temperature of a bulb is about 2000 K.

3.4 COHERENCE

Light waves are said to be coherent if they are in phase with each other – for example, if they maintain crest-to-crest and trough-to-trough correspondence, as in Fig.3.3(a). Two things are necessary for light waves to be coherent. First, they must start with the same phase at the same position. Second, their wavelengths must be the same or they will drift out of phase because the crests of the higher frequency will arrive ahead of the crests of the lower frequency wave.

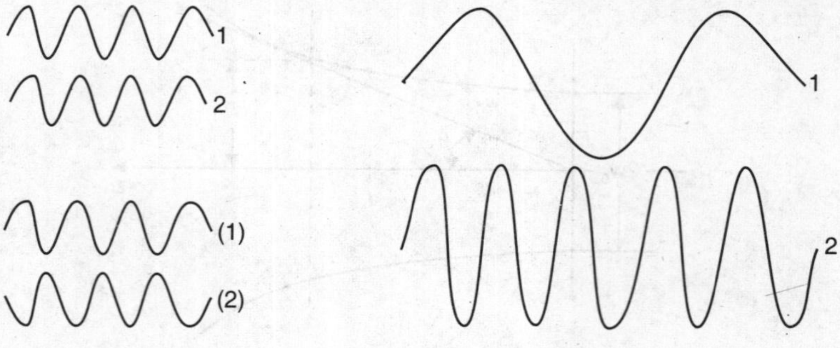

(*a*) Coherent waves (*b*) Incoherent waves

Fig. 3.3.

The light that emerges from a conventional light source is a jumble of short waves which combine with each other in a random manner. The resultant light is incoherent and the wave front varies from point to point and changes from instant to instant. On the other hand, the light from a laser is a resultant of a large number of identical photons which are in phase and is therefore exhibits a high degree of coherence.

Coherence requires that there is a connection between the amplitude and phase of the light at one point and time, and the amplitude and phase of the light at another point and time. Accordingly, we distinguish two classes of coherence, namely *temporal coherence* and *spatial coherence.* They are also known as *longitudinal coherence* and *transverse coherence* respectively. Temporal coherence refers to the constancy and predictability of phase as a function of time when the waves travel along the same path at slightly different times. Spatial coherence refers to the phase relationship between waves travelling side by side at the same time but at some distance from one another.

3.4.1 Temporal Coherence

The concept of temporal coherence can be easily understood with the help of the following example. Let us consider a single wave propagating along *x*-direction. Let us note the electric field at one point in space at two different times t_1 and t_2, as in Fig. 3.4. Let the phase difference between the field E_1 at t_1 and the field E_2 at t_2 be ϕ_1. Let us again note the electric fields at later times t_3 and t_4, where $(t_4 - t_3) = (t_2 - t_1)$. Let the phase difference be now ϕ_2.

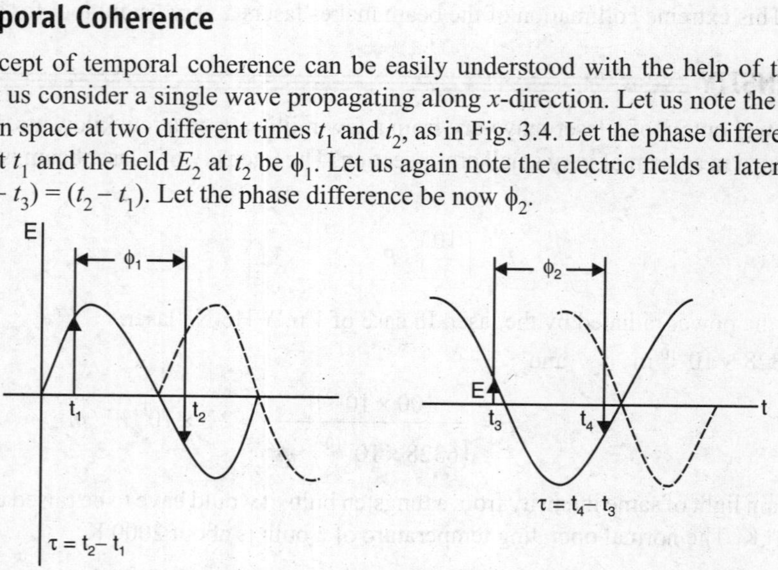

Fig. 3.4. Illustration of Temporal Coherence

If $\phi_2 = \phi_1$ and it is true for any time interval of same duration, then the wave is said to be *temporally coherent.* If the phase difference $\phi_2 \neq \phi_1$ and changes from interval to interval and in an irregular fashion, then the wave is said to the incoherent.

Temporal coherence or incoherence is a characteristic, essentially, of a single beam of light. We know that light is emitted by excited atoms of the source. An excited atom in the process of passing to a lower energy state, gives up the excess energy. The transition lasts for a short time of 10^{-8} sec. A light wave is emitted during this duration, which is not a continuous sine wave of infinite extension but is a wave packet, such as the one shown in Fig 3.5. The light from a source is a jumble of such wave packets emanating from different atoms. Each wave packet has a sustained phase for only about 10^{-8} s after which there is a random phase. On an average, a light beam undergoes random phase changes about 10^8 times per second. Temporal coherence may be expected only for a short distance in space and for a brief time. In view of this, temporal coherence is characterised by two parameters, namely coherence length, l_{coh} and coherencetime, t_{coh}. Both the coherence length and coherence time measure how long light waves remain in phase as they travel in space. The coherence length depends on the central wavelength λ and the bandwidth, Δλ, of the wave packet. It is given by

$$l_{coh} = \frac{\lambda^2}{2\Delta\lambda} \quad ...(3.4)$$

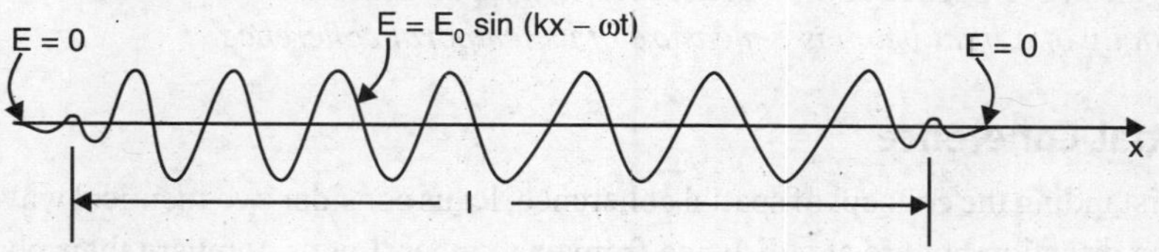

Fig. 3.5. A wave train

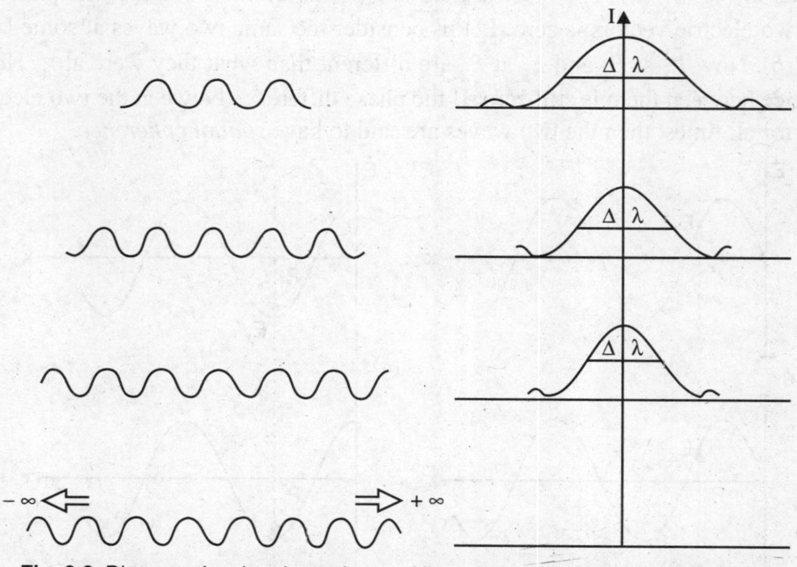

Fig. 3.6. Diagram showing dependence of line width on the length of wave train.

It can be expressed in terms of frequency as

$$l_{coh} = \frac{c}{\Delta\nu} \quad ...(3.5)$$

The coherence length is a very useful measure of temporal coherence because it tells us how far apart two points along the light beam can be, and remain coherent with each other.

Fluorescent tube lights emit visible light of wavelengths in the range, say 4000 Å to 7000 Å, with an average wavelength of 5500 Å. The coherence length

$$l_{coh} = \frac{\lambda^2}{2\Delta\lambda} = \frac{(5000 Å)^2}{2 \times 3000 Å} = 5040 Å$$

Light from a sodium lamp which is the traditional monochomatic source has a coherence length of about 0.3 mm.

$$l_{coh} = \frac{(5893 \text{ Å})^2}{2 \times 6 \text{ Å}} = 0.29 \text{ mm}$$

A He-Ne laser having a bandwidth of about 2×10^{-5} Å has a coherence length of about 100 m.

$$l_{coh} = \frac{(6328 \text{ Å})^2}{2 \times 10^{-5} \text{ Å}} = 100\text{m}$$

It is readily seen from Eqs. (3.4) and (3.5) that temporal coherence is an effect due to the finite line width $\Delta\nu$ of the source. A strictly monochromatic wave ($\Delta\nu = 0$) is an ideal harmonic wave of infinite extension, and of infinite coherence length. A real light wave consists of a train of wave packets because of which its temporal coherence decreases. The broader the line width, $\Delta\nu$, the shorter is the wave packet and its coherence length, as illustrated in Fig. 3.6. In other words, *monochromaticity of a light beam is a measure of its temporal coherence.*

3.4.2. Spatial Coherence

For understanding the concept of spatial coherence, let us consider two identical waves travelling along the same direction but are at a distance from each other. Let us compare their phases at some time, say t_1, as in Fig. 3.7(a). The electric field at P_1 is E_1 and at P_2 is E_2. The phase difference between the two electric vectors is zero. Let us consider the same two waves at some later time t_2, as in Fig. 3.7(b). Now E_1 at P_1 and E_2 at P_2 are different than what they were at t_1. However, the phase difference between them is still zero. If the phase difference between the two electric vectors remains zero for all times, then the two waves are said to have *spatial coherence.*

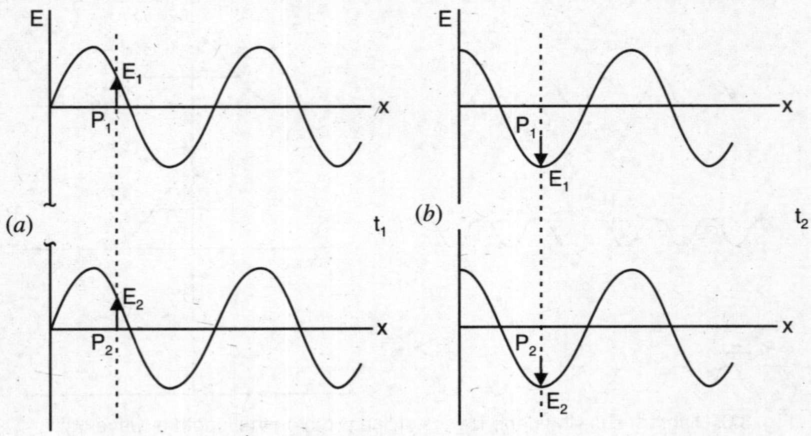

Fig. 3.7. Illustration of spatial coherence.

It implies that spatial coherence measures the area over which light is coherent. If we consider a point light source and look at two points, P_1 and P_2 as in Fig. 3.8 (a) that are at equal optical path distances from the source, the waves reaching these points will be exactly in the same phase.

As the source is made larger, it can no longer be claimed that the points P_1 and P_2 are in the same phase (Fig. 3.8(b)). Thus, spatial incoherence arises due to the size of the light source. As it characterizes the spatial variation in coherence across the wave front in the direction perpendicular to the light beam, spatial coherence is also known as *transverse coherence* of the beam.

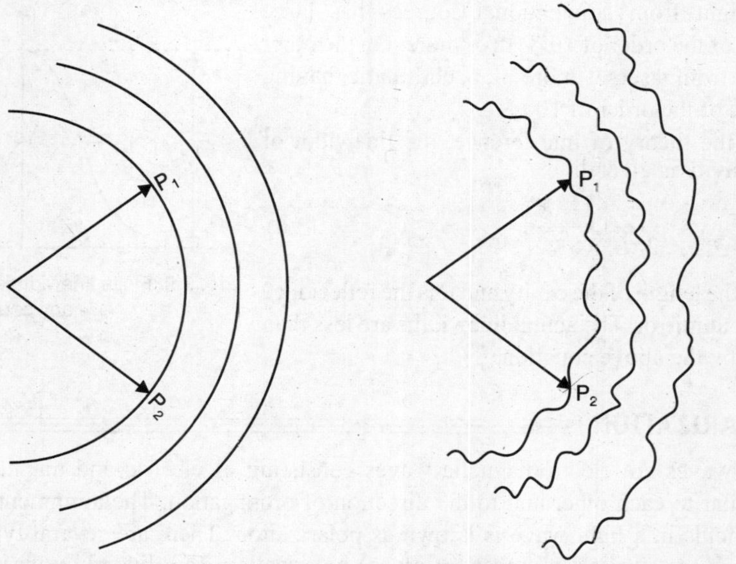

(*a*) Light waves from a point source, producing ideally spherical wave fronts, maintain spatial coherence. Points P_1 and P_2 have the same phase.

(*b*) Extended source produces wave fronts far from ideal. Points P_1 and P_2 do not have the same phase, leading to spatial incoherence.

Fig. 3.8.

The phenomenon of interference is a manifestation of coherence. It can be shown that the *number* of interference fringes will be larger, the smaller the line width or the longer the temporal coherence of the interfering waves. *The degree of contrast of the interference fringes obtained is a measure of the degree of spatial coherence.* It is also deduced that the higher the degree of spatial coherence of a source, the smaller is the divergence of the light beam and consequently, accounts for its high directionality.

Light emerging from a laser is both temporally and spatially coherent to a high degree. The reason is that each stimulated transition is a sort of forced oscillation of the excited atomic system and the atom radiates a photon in phase with the incoming photon. These two photons eventually interact with two other excited atoms, each causing the emission of a photon in phase with it. The process goes on and as all the photons are produced in phase with the original photon, the laser beam exhibits a high degree of coherence. Theoretically, laser light would be completely coherent. In practice, coherence is not ideal because all photons do not originate from the same original photon. So they do not all start out in phase. A laser that oscillates in only one longitudinal mode is more coherent than that oscillates in multiple longitudinal modes, because the different modes have slightly different wavelengths. CW laser beams are more coherent than the pulsed beams. Lasers with low gain are more coherent than high gain lasers. Thus, CW lasers of low gain operating in TEM_{00} mode give the most coherent light.

3.5 MONOCHROMATICITY

If light coming from a source has only one frequency of oscillation, the light is said to be *monochromatic* and the source a *monochromatic source.* In practice, it is not possible to produce light with only one frequency. Light coming out of any source consists of a band of frequencies closely spaced around a central frequency, ν_0 (Fig. 3.9). The band of frequencies, $\Delta\nu$,is called the *linewidth* or *bandwidth.*

The light from conventional sources has large linewidths of the order of 10^{10} Hz or more. On the other hand, light from lasers is more monochromatic, having line-widths of the order of 100 Hz.

From the theory of interference, the linewidth of mirrored cavity is given by

$$\Delta\nu = \frac{c}{2\pi L}\left(\frac{1-R}{\sqrt{R}}\right) \quad ...(3.6)$$

where L is the length of the cavity and R is the reflectance of the out put mirror. The actual linewidths are less than that given by the above equation.

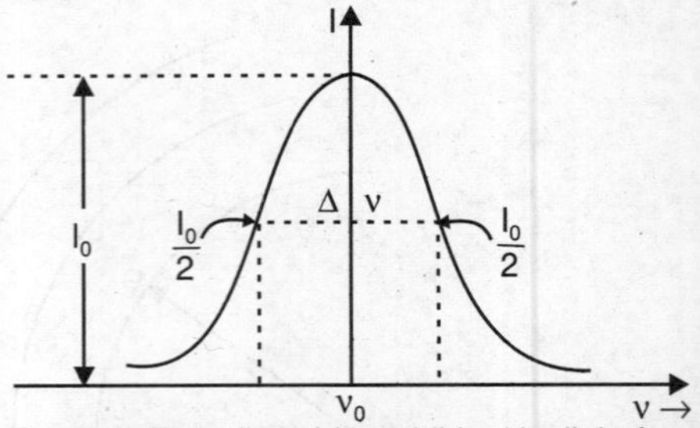

Fig. 3.9. Finite linewidth exhibited by light from any source.

3.6 POLARIZATION

Light waves are electromagnetic waves consisting of electric and magnetic fields vibrating perpendicular to each other and to the direction of propagation. The alignment of the electric and magnetic fields in a light wave is known as polarization. There are several types of polarization, out of which the simplest is linear (or plane) polarization. In a linearly polarized light beam, the orientation of the electric field remains in one plane while its magnitude changes with time. Any other type of polarized light can be viewed as a result of superposition of two linearly polarized waves that are having electric fields perpendicular to each other. For example, unpolarized light can be divided into two components with linear polarization, one with a vertical field and one with a horizontal field.

The conventional light sources produce unpolarized light. External devices such as polarizers and crystal plates are used to convert the unpolarized light into light of desired polarization. In case of lasers, most of them also emit unpolarized light. Laser output can be made linearly polarized light by adding a suitable device. Normally, Brewster window is used to obtain linearly polarized light from a laser. It is a surface arranged at an angle, called Brewster angle, with the resonator axis. Light incident on a surface at the Brewster angle will be reflected or transmitted depending on its polarization. Light polarized parallel to the plane of incidence is transmitted through the Brewster window while light polaized *perpendicular* to the plane of incidence is partly transmitted and partly reflected. The component that undergoes reflection will be lost to the resonator and will not contribute to the oscillation of the laser. The result is that the output beam obtained is plane polarized having polarization parallel to the plane of incidence.

3.7 SPECKLES

If a surface, such as of a wall or of a metal, is illuminated by a laser beam it appears to be granular. The granules are randomly distributed over the surface and appear to be shifting. Such grainy patterns are called laser *speckles.* The shifting speckles is an interference pattern created by slight differences in the paths that light rays travel. Ordinary surfaces are not optically flat and have random in homogeneties which are larger than the wavelength of the laser light. The light reflected diffusely from the inhomogeneties is coherent. The interference of these coherent reflected waves at the retina gives the appearance of alternate dark and bright speckles. If laser radiation is reflected from an optically flat surface, speckles are not observed. Thus, the speckles provide information on the quality of the surface. However, for practical purposes it is undesirable and is merely a background noise.

Example 3.1. *Find the coherent length for white light, the wavelength of white light ranges from 400 nm to 700 nm.*

Solution : Here $\delta\lambda = 700 \text{ nm} - 400 \text{ nm} = 300 \text{ nm}$

$$\text{Average wavelength } \lambda = \frac{\lambda_1 + \lambda_2}{2} = \frac{700 + 400}{2} = 550 \text{ nm}$$

$$\therefore \quad \text{Coherent length } L = \frac{\lambda^2}{\delta\lambda} = \frac{(550 \times 10^{-9})^2}{300 \times 10^{-9}} = \mathbf{10^{-6}\ nm.}$$

Example 3.2. *If the light of wavelength 660 nm has wave trains 13.2 × 10^{-6} m long, calculate coherent time.*

Solution : Coherent length L = Length of one wave train

$= 13.2 \times 10^{-6}$ m

Let T_c be the coherent time.

Now, $L = cT_c$

$$\therefore \quad T_c = \frac{L}{c} = \frac{13.2 \times 10^{-6}}{3 \times 10^{8}} = 4.4 \times 10^{-14}\ s = \mathbf{44 \times 10^{-15}\ s.}$$

Example 3.3. *The Doppler width for orange line of krypton is 550 × 10^{-15} m. If the wavelength of light is 605.8 nm, calculate the coherent length.*

Solution : Doppler width = line width $\delta\lambda = 550 \times 10^{-15}$ m

$\lambda = 605.8 \text{ nm} = 605.8 \times 10^{-9}$ m

$$\therefore \quad \text{Coherent length } L = \frac{\lambda^2}{\delta\lambda} = \frac{605.8 \times 605.8 \times 10^{-18}}{550 \times 10^{-15}} = 667.3 \times 10^{-3}\ m = \mathbf{66.73\ cm.}$$

Example 3.4. *Find the coherence length of a laser source of monochromatic light with frequency width 10,000 Hz.* **(Nagpur U. 2011)**

Solution : Frequency width = band width $\delta\nu$ = 10,000 Hz.

$$\therefore \quad \text{Coherence time } T_c = \frac{1}{\delta\nu} = \frac{1}{10{,}000} \text{ sec.}$$

$$\text{and} \quad \text{Coherence length } L = cT_c = 3 \times 10^8 \times \frac{1}{10{,}000} = 3 \times 10^4 \text{ m} = \mathbf{30\ km.}$$

Example 3.5. *For an ordinary source coherence time is nearly 10^{-10} sec. Calculate the degree of non-chromocity for λ = 5400 Å.*

Solution : Degree of non-chromacity = Purity factor = frequency stability = $\dfrac{\delta\nu}{\nu}$

Again, $$\Delta\nu = \frac{1}{T_c} \text{ and } \nu = \frac{c}{\lambda}$$

$$\therefore \quad \frac{d\nu}{\nu} = \frac{1}{T_c} = \frac{5400 \times 10^{-10}}{10^{-10} \times 3 \times 10^8} = \mathbf{18 \times 10^{-6}}$$

Example 3.6. *The coherence length of sodium D_2 line is 2.5 cm. Deduce the (i) coherence time (ii) spectral width of line and purity factor. Give λ = 5890 Å.*

Solution : Coherence time $T_c = \dfrac{L}{c} = \dfrac{2.5 \times 10^{-2}}{3 \times 10^8} = 0.83 \times 10^{-10}$ sec.

$$\therefore \text{ Spectral width} = \text{line width } \delta\lambda = \frac{\lambda^2}{L} = \frac{5890 \times 5890 \times 10^{-20}}{2.5 \times 10^{-2}} = 0.1387 \times 10^{-10}\text{m}$$

Spectral purity factor = frequency stability

$$= \frac{\delta\nu}{\nu} = \frac{c}{L} \times \frac{\lambda}{c} = \frac{\lambda}{L} = \frac{5890 \times 10^{-10}}{2.5 \times 10^{-2}} = \mathbf{2.356 \times 10^{-5}}.$$

Example 3.7. *A laser has an organic dye solution with a refractive index 1.4 as the active medium. The length of the resonator is 10 mm. The laser operating at 630 mm with a gain of 100. Then calculate : (i) The variation in the refractive index that can produce the required oscillation (ii) The line width of oscillation.*

Solution : (*i*) Refractive index, $\mu_t = \frac{\lambda}{L} \times \frac{ln(G)}{\sqrt{\pi.(G)}}$

$$= \frac{633 \times 10^{-9}}{10^{-2}} m \times \frac{ln\,(100)}{\sqrt{\pi}\,(100)} = 9.3 \times 10^{-6} = \mathbf{10^{-5}}$$

(*ii*) Line width of the output in the resonator.

$$\frac{\Delta\lambda}{\lambda_0} = \frac{\lambda_0}{4\pi\mu L} ln\,(G)$$

$$\therefore \quad \Delta\lambda = \frac{\lambda^2}{4\pi\mu l} ln\,(G) = \frac{(633 \times 10^{-9})^2}{4\pi \times 1.4 \times 0.01} ln\,(100)$$

$$= 2.27 \times 10^{-12} \text{ m} = \mathbf{2.27 \times 10^{-3}\ nm}$$

Example 3.8. *A laser beam has a wavelength of 7.2 × 10⁻⁷ m and aperture 5 × 10⁻³ m. The laser beam is focused towards moon the distance of which from earth is 4 × 10⁸ m. Calculate (i) the angular spread and (ii) axial spread when the beam reaches to moon.*

Solution : Here $\lambda = 7.2 \times 10^{-7}$ m; $d = 5 \times 10^{-3}$ m, Angular spread $d\theta$ = ?

As the aperture is circular, the angular spread

(*i*) $$d\theta = \frac{1.22\,\lambda}{d} = \frac{1.22 \times 7.2 \times 10^{-7}}{5 \times 10^{-3}} = \mathbf{1.75 \times 10^{-4}\ radian}$$

(*ii*) Distance of moon from the earth $D = 4 \times 10^8$ m

$\therefore$ Axial (aerial) spread $= (D.d\theta)^2$

$$= (4 \times 10^8 \times 1.757 \times 10^{-4})^2 = \mathbf{49.39 \times 10^8\ m^2}$$

Example 3.9. *A laser beam has a power of 100 mW. It has an apertrure of 5 × 10⁻³ m and emits a light of wavelength 6943 Å. The beam is focused with a lens of focal length 0.1 m. Calculate the area and the intensity of the image.*

Solution : Diameter of aperture $d = 5 \times 10^{-3}$ m

Wavelength $\lambda = 6943 \times 10^{-10}$ m; $f = 0.1$ m

Power of laser beam = 100 mW = 10^{-1} watt

As the aperture is circular,

$\therefore$ $$\text{Angular spread } d\theta = \frac{1.22\lambda}{d} = \frac{1.22 \times 6943 \times 10^{-10}}{5 \times 10^{-3}} = 1.694 \times 10^{-4} \text{ radian}$$

(*i*) Aerial spread $= (d\theta.f)^2 = (1.694 \times 10^{-4} \times 0.1)^2 = \mathbf{2.89 \times 10^{-10}\ m^2}$

(*ii*) $$\text{Intensity} = \frac{\text{Power}}{\text{Area}} = \frac{10^{-1}}{2.89 \times 10^{-10}} = \mathbf{3.46 \times 10^8\ watt / m^2}$$

Example 3.10. *A laser beam has a power of 0.2 watt and has an aperture of 1 mm. It emits light of wavelength 6000 Å. If it is focused by lens of F.L. 80 cm, Calculate the area and intensity of the image.*

Solution. Angular spread $d\theta = \frac{1.22\,\lambda}{d} = \frac{1.22 \times 6000 \times 10^{-10}}{10^{-3}} = 7.32 \times 10^{-4}$ radian

Area (Areal spread) $= (d\theta.f)^2 = (7.32 \times 10^{-4} \times 0.8)^2 = \mathbf{34.3 \times 10^{-8}\ m^2}$

$$\therefore \quad \text{Intensity} = \frac{\text{Power}}{\text{Area}} = \frac{0.2}{34.3 \times 10^{-8}} = \mathbf{5.83 \times 10^5\ watt/m^2}$$

Example 3.11. *Find the intensity of a laser beam 20 mW power and having a diameter of 1.3 mm. Assume uniform intensity across the beam.* **(B.Tech. Fourth sem. G.B.T.U. Luck 2010)**

Solution. Given : Power $P = 20$ mW, $d = 1.3$ mm and $r = 1.3/2$ mm

$$\therefore \quad \text{Area, } A = \pi r^2 = 3.14 \times \left(\frac{1.3}{2}\right)^2 = \frac{5.30}{4}\ (\text{mm})^2 = 1.32\ \text{mm}^2$$

$$\therefore \quad \text{Intensity } I = P/A = \frac{20}{1.32}\ \text{mW/mm}^2 = \mathbf{15.15\ mW/mm^2.}$$

Example 3.12. *A gas laser is generating a laser beam of 4 mW power. Calculate the number of photons emitted by the laser. The wavelength of the emitted radiation is 680 nm.* **(GBTU 2010)**

Solution. Given : $P = 4\ \text{mW} = 4 \times 10^{-3}$ J/sec.

$\lambda = 680\ \text{nm} = 680 \times 10^{-9}$ m

In one second, the energy rerlased $= 4 \times 10^{-3}$ J

$$\therefore \quad \text{Frequency } \nu = \frac{c}{\lambda} = \frac{3 \times 10^8\ \text{m/s}}{680 \times 10^{-9}\ \text{m}} = 0.0044 \times 10^{17}\ \text{Hz.}$$

Energy, $E = nh\nu$ or $n = E/h\nu$

$$\therefore \quad n = \frac{4 \times 10^{-3}}{6.6 \times 10^{-34} \times 0.0044 \times 10^{17}} = \mathbf{137 \times 10^{14}}$$

EXERCISE-3

1. What are the basic properties of laser light.
2. (*a*) How does laser light differ from a filament lamp ?

   (*b*) Discuss in detail the concept of directionality, monochromacity, intensity and coherence of laser light. **(K.U. 2001, G.N.D.U. 2000, P.U. 2007, 2006, 2005, 2004; Pbi. U. 2006, 2000; M.D.U. 2001, 2000)**
3. Explain the term 'Lateral coherence length'.
4. Sunlight has mean wavelength of 5400AU. Considensing sun as a source of light in a double slit experiment with $\theta = 32'$; find the lateral coherence ? **(Pune U. 2010)**

   What are the characteristics of laser light ? **(Pune U. 2010)**
5. What is the difference between coherence and Monochromatism of light. **(Pune U. 2010)**
6. Differentiate between spatial and temporal coherence. **(Pune U. 2010)**
7. A laser emits at 671 nm at 5mW. How many photons per second does it emit ? **(Pune U. 2010)**
8. A pinhole of diameter 1mm is placed from two pinholes at a diatance of 1m. Find the lateral coherence width of the pinhole which is illuminated with light of wavelength 5000 Å. **(Ans.** 0.06 cm)
9. A laser beam has a power of 50 mW. It has a aperture of 5×10^{-3} m and it emits light of wavelength 7200 Å. The beam is focused with a lens of focal length 0.1m. Calculate the area and the intensity of the image. **(Pune U. 2004)**

   (Ans. 2.074×10^{-10}m; 2.411×10^3 watt/m^2)

4 CHAPTER Techniques for Control of Laser Output

4.1 INTRODUCTION

Many practical applications of lasers require modification or control of their frequency characteristics or power output. It is carried out using optical accessories which modify the laser output. The laser beam may be modified either inside the resonant cavity by affecting the process of laser oscillation or outside the cavity by passing through a device with variable transmission. Optical devices such as prisms and gratings are called *passive devices*, as they do the same thing to light always. They are employed for selecting a narrow range of frequencies or to tune the laser to operate at different specific frequencies. Rotating mirrors and optical modulators are called *active optical devices* which do different things to light at different times. They are used when it is required to concentrate light energy into short time intervals.

4.2 SELECTING A NARROW FREQUENCY RANGE

Some lasers emit light at a set of frequencies but it is often desirable that a laser oscillates only at a single frequency. For example, in the He-Ne laser, neon atoms have three laser transitions at 3.39 mm, 1.15 mm and 0.63 mm. The transitions at 3.39 mm and 1.15 mm lie in the far IR while the one at 0.63 mm is in the visible range. For many practical applications, it is required that the He-Ne laser emits red light of 0.63 mm wavelength. In such cases, a simple method of obtaining the desired single frequency consists in suppressing the other possible oscillations. It may be achieved by coating the reflecting mirrors of resonant cavity with suitable materials such that they are highly reflective at the desired frequency and are absorptive at other frequencies. The result is that the desired frequency is sustained and amplified through repeated reflections whereas the other frequencies are strongly attenuated. Thus, in case of He-Ne laser, the mirrors are coated with materials which have high reflectivity at 0.63 mm and low reflectivity for 3.39 mm and 1.15 mm wavelengths.

Some of the lasers have broad emission lines consisting of a wide range of frequencies. In such cases a single frequency can be selected using a prism or a diffraction grating, which is inserted in the resonant cavity. If a prism is used, it disperses the incident laser beam into components of different frequencies which are reflected back by the cavity mirror (Fig.4.1). For a given setting of the prism, only one of the frequencies retraces its path into the cavity and gets amplified while other frequencies are lost. When a diffraction grating is used, the component which satisfies the grating equation, namely

$$\lambda = 2d \sin \theta \qquad ...(4.1)$$

is reflected back into the cavity and the laser oscillates at the wavelength λ (Fig. 4.1b). The laser can be tuned to different frequencies within the laser emission line by rotating the grating.

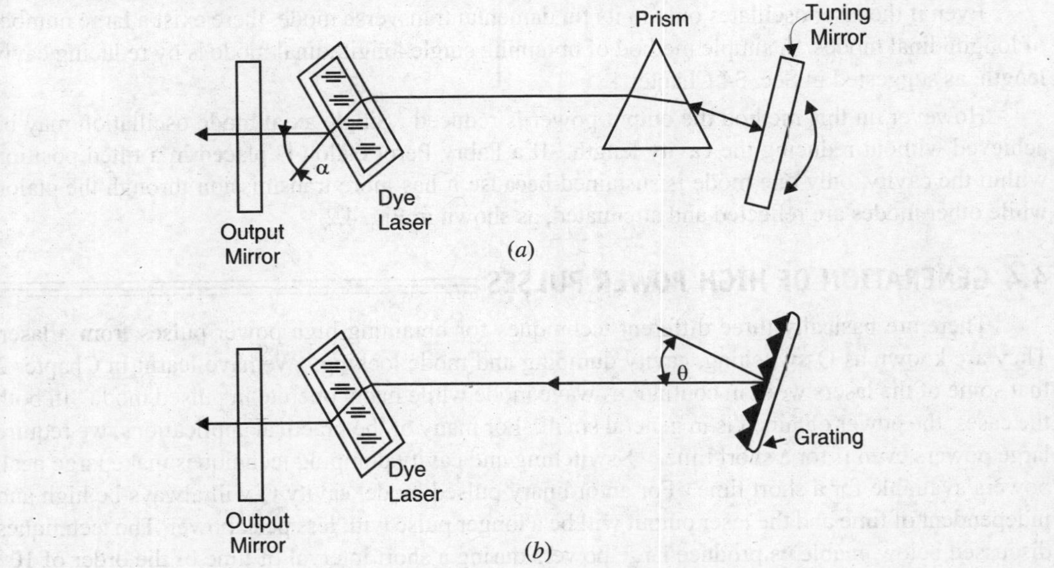

Fig. 4.1. Frequency tuning using (a) a prism (b) a diffraction grating.

4.3 SELECTION OF TEM_{00} MODE AND SINGLE LONGITUDINAL MODE

Out of the large number of transverse modes, it is often desired to select only the fundamental mode, TEM_{00}. It can be selected by keeping a circular aperture in the cavity. It can also be done by using an unstable optical resonator. An example of the unstable resonator is a confocal telescopic resonator.

The unstable resonators yield highly coherent beams. They exhibit high diffraction losses leading to suppression of high-order transverse modes. Thus, a laser with an unstable cavity supports the fundamental mode only.

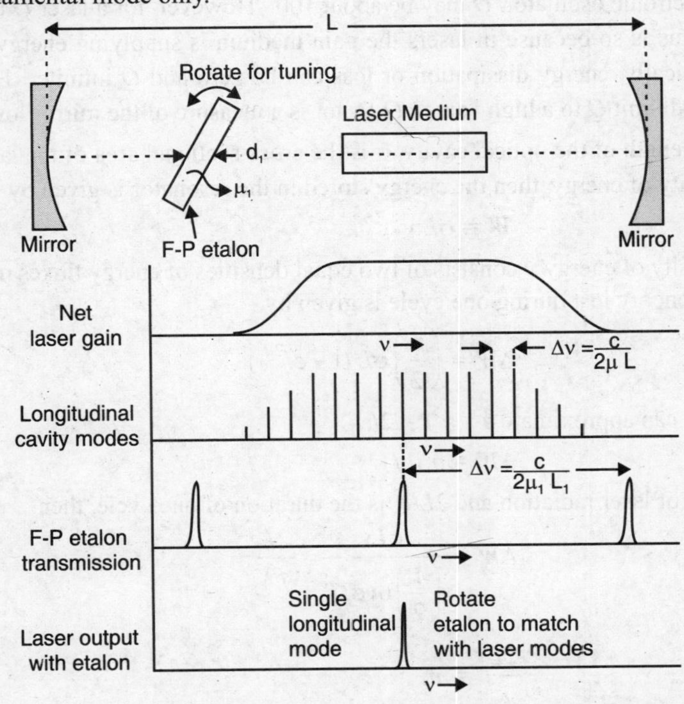

Fig. 4.2. Fixed spacing Fabry-Perot etalon for single longitudinal mode selection.

Even if the laser oscillates only in its fundamental transverse mode, there exist a large number of longitudinal modes. A simple method of obtaining single longitudinal mode is by reducing caviy length, as suggested in Sec. 34 Chapter 1.

However, in this method the output power is reduced. Single axial mode oscillation may be achieved without reducing the cavity length. If a Fabry-Perot etalon is placed in a tilted position within the cavity, only one mode is sustained because it has more transmission through the etalon while other modes are reflected and attenuated, as shown in Fig. 4.2.

4.4 GENERATION OF HIGH POWER PULSES

There are basically three different techniques for obtaining high power pulses from a laser. They are known as Q-switching, cavity dumping and mode locking. We have learnt in Chapter 2 that some of the lasers work in continuous wave mode while many operate in pulsed mode. In both the cases, the power obtained is in general small. For many of the practical applications, we require large powers even if for a short time. Q-switching and cavity dumping techniques make large peak powers available for a short time. For an ordinary pulsed mode, cavity Q will always be high and independent of time and the laser output will be a longer pulse with less peak power. The techniques discussed below enable us produce large powers during a short interval of time of the order of 10^{-9} sec.

4.5 Q-FACTOR

Since a laser is an optical oscillator, its resonator cavity may be characterized by the quality factor Q, as is done in case of an electronic oscillator. Q is defined as the ratio of the energy W stored in the system to the energy losses ΔW per cycle. Thus,

$$Q = 2\pi \frac{\textit{Energy stored in the resonator}}{\textit{Energy lost in a cycle}} = \frac{W}{\Delta W} \qquad \text{...(4.2)}$$

In case of an electronic oscillator, Q may be about 100. However, for a laser Q is very high, of the order of 10^8. This is so because in lasers the gain medium is supplying energy to the oscillating modes. Theoretically, energy dissipation or loss can be zero and Q infinite. However, there are always losses and limit Q to a high value. Q-factor is a measure of the mirror losses.

If L is the length of the optical cavity, σ is the cross-sectional area of the laser beam, and v is the volume density of energy, then the energy stored in the resonator is given by

$$W = \sigma L v$$

The volume density of energy v consists of two equal densities of energy fluxes moving in opposite directions. The energy lost during one cycle is given by

$$\Delta W = \left(\frac{\upsilon}{2}\right) \sigma L\,(1 - e^{-2f})$$

When $f << 1$, we can approximate $e^{-2f} \approx 1 - 2f$.

$$\therefore \qquad \Delta W = v\sigma L f$$

If T is the period of laser radiation and $2L/v$ is the duration of one cycle, then

$$\Delta W = \frac{\sigma L v}{\left(\frac{1}{2}\right)\upsilon v \sigma f T} = \frac{2L}{f\lambda} \qquad \text{...(4.3)}$$

But $$\frac{2L}{\lambda} = m. \qquad \text{...(4.4)}$$

$$\therefore \qquad Q = \frac{m}{f} \qquad \text{...(4.5)}$$

We can express the condition necessary for lasing, using $\alpha_0 L = f$, lasing threshold condition into Eq. (4·5). We obtain $Q = \frac{m}{\alpha_0 L}$. Using Eq. (4.4), we get

$$\frac{\alpha_0 \lambda}{2} = \frac{1}{Q} \qquad \text{...(4.6)}$$

Thus, the threshold condition depends on the reciprocal of the quality factor. The higher the Q-factor, the lower the lasing threshold. Q is also related to the laser linewidth in a laser cavity and is shown that it is given by

$$\boxed{Q = \frac{\nu_0}{\Delta\nu}} \qquad \text{...(4.7)}$$

4.6 Q-SWITCHING FOR GIANT PULSES

The power of a laser beam can be drastically increased, provided a very large number of atoms in the active medium participates in stimulated emission. It can happen only when a very high population inversion density is established in the active medium. Normally, when the laser starts oscillations, the high population inversion drops back to the threshold value of the steady state condition. It becomes therefore necessary that the onset of oscillations is delayed until largest possible number of atoms accumulates at the upper lasing level. The laser can be prevented from oscillating if, for example, the parallelism of the resonator mirrors is disturbed. If one of the end mirrors is misaligned, it cannot reflect incident photons into the active medium and therefore, stimulated emission cannot take place. Consequently, the pumping process can build up the population inversion to a very high value in the medium. In effect, the Q-factor of resonator is spoiled and optical losses are increased to a high value. If now the end mirror is aligned suddenly, it reflects photons into the laser medium. The feed back of photons triggers a chain of stimulated emissions and builds up rapidly a photon avalanche. Thus, laser oscillations set in suddenly and the cavity Q is increased abruptly. All the energy stored in the cavity is emitted in a single giant pulse with peak power much higher than the laser could produce otherwise. The pulse

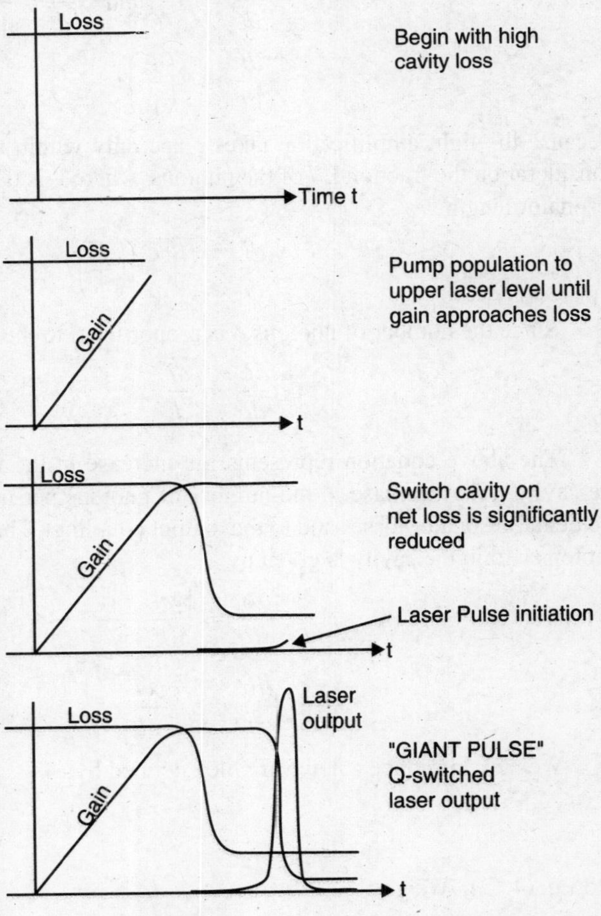

Fig. 4.3. Evolution of a giant laser pulse through Q-switching

lasts for a short time and depopulates the upper energy level quickly and the lasing action stops. This method of controlling the laser output power is called the *Q-switching method.* The schematic of evoluation of a giant laser pulse is shown in Fig. 4.3. Thus, by operating a laser in *Q*-swithced mode, the energy of the laser can be concentrated into a single high power pulse. The pulse duration depends on many factors but usually is of the order of 10^{-7} to 10^{-8} sec.

Q-switching is widely used, but not all lasers can be *Q*-switched.

We derive now an expression for the total number of photons ϕ in the laser cavity, and obtain expression for the peak power that can be generated.

Let the total population difference within the active medium be $n \equiv (N_2 - N_1)\,V$, and the decay time constant for photons in the cavity be

$$t_c = \frac{\mu l}{c(l - R)} \qquad \text{...(4.8)}$$

The change in light intensity per unit time within the amplifier is given by

$$\frac{dI}{dt} = \frac{dI}{dz} \bullet \frac{dz}{dt} \qquad \text{...(4.9)}$$

But $\quad I(z) = I.e^{\gamma z}$

$$\therefore \quad \frac{dI}{dt} = \gamma I \quad \text{and} \quad \frac{dz}{dt} = \frac{c}{\mu}$$

$$\therefore \quad \frac{dI}{dt} = \gamma\left(\frac{c}{\mu}\right) I \qquad \text{...(4.10)}$$

Because the light amplification takes place only within the active medium, we have to take into consideration the fraction L/*l* of the photons, where L is the length of the active medium and *l* is the resonator length

$$\therefore \quad \frac{dI}{dt} = \gamma\left(\frac{c}{\mu}\right)\left(\frac{L}{l}\right) I$$

Since the number of photons ϕ is proportional to the light intensity *I*, we can assume that

$$\frac{d\phi}{dt} \propto \frac{dI}{dt} \qquad \text{...(4.11)}$$

The above equation represents the increase in the number of photons per unit time within the cavity. The decrease in the number of photons per unit time is given by ϕ/t_c which is due to incidental resonator losses and to the output coupling. Therefore, the total change in the number of photons within the cavity is given by

$$\frac{d\phi}{dt} = \frac{\phi\gamma c L}{\mu\ell} - \frac{\phi}{t_c} \qquad \text{...(4.12)}$$

$$\therefore \quad \frac{d\phi}{dt} = \phi\left[\frac{rcL}{\mu L} - \frac{1}{t_c}\right]$$

We now introduce a new variable *t* defined by

$$\tau = \frac{1}{t_c} \qquad \text{...(4.13)}$$

into Eq. (4.12). We get

$$\frac{d\phi}{d\tau} = \phi\left[\frac{\gamma c L t_c}{\mu\ell} - 1\right]$$

$$\frac{d\phi}{d\tau} = \phi\left[\frac{\gamma}{\mu\ell / cLt_c} - 1\right] = \phi\left[\frac{\gamma}{\gamma_{th}} - 1\right] \quad ...(4.14)$$

where $\gamma_{th} = \dfrac{\mu\ell}{cLt_c}$ is the threshold gain at which laser oscillations will be sustained. Since there is no change in photon number in the cavity with time for a laser operating under steady state conditions,

$$\frac{d\phi}{d\tau} = 0 \quad \text{and} \quad \gamma = \gamma_{th}$$

As γ is proportional to the inversion n, eq.(4.14) may be written as

$$\frac{d\phi}{d\tau} = \phi\left[\frac{n}{n_{th}} - 1\right] \quad ...(4.15)$$

where $n_{th} = (N_2 - N_1)/V$ is the threshold inversion.

Every time a photon is emitted in the stimulated emission process, n changes by a factor of 2. Since in each transition, an atom leaves upper level and goes to the lower level, the total inversion decreases by 2. As the rate at which population difference is changing must be exactly equal to the rate at which photons are increasing, we have

$$\frac{dn}{d\tau} = -2\phi\frac{n}{n_{th}} \quad ...(4.16)$$

Dividing Eq. (4.15) by Eq. (4.16), we get

$$\frac{d\phi}{dn} = \frac{n_{th}}{2n} - \frac{1}{2} = \frac{1}{2}\left[\frac{n_{th}}{n} - 1\right] \quad ...(4.17)$$

Integrating the above equation (4.17), we get

$$\phi - \phi_0 = \frac{1}{2}\left[n_{th}\ell n\frac{n}{n_0} - (n - n_0)\right] \quad ...(4.18)$$

where ϕ_0 and n_0 are the initial values.

The variation of population inversion n with time is shown in Fig. 4.4. The variation of photon density ϕ is also shown in the same figure.

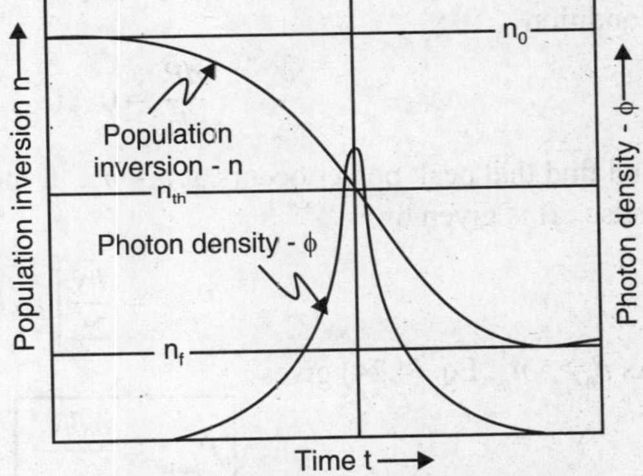

Fig. 4.4. The variation of population inversion and photon density with time associated with Q-switched pulse.

Eq. (4.18) describes the relationship between the number of photons in the cavity and the population inversion n at any particular time. If we assume that there are no photons initially in the cavity, $\phi_0 \cong 0$.

$$\therefore \quad \phi = \frac{1}{2}\left[n_{th}\ell n\frac{n}{n_0} - (n - n_0)\right] \quad ...(4.19)$$

For a time $t >> t_c$, there will also be no photons in the cavity.

$$\therefore \quad \phi = 0 \text{ at } t >> t_c$$

$$\therefore \quad ln\frac{n_f}{n_0} = \frac{n_f - n_0}{n_{th}}$$

where n_f is the final value of the population inversion.

$$\therefore \quad \frac{n_f}{n_0} = \exp\left[\frac{n_f - n_0}{n_{th}}\right] \quad ...(4.20)$$

The above transcendental equation can be solved numerically to obtain n_0 versus n_f. A plot of n_0/n_{th} is shown as a function of $(n_0 - n_f)/n_0$, in Fig. 4.5.

The ratio $(n_0 - n_f)/n_0$ is the fraction of energy stored in the population that gets converted to a laser pulse.

This energy utilization factor tends to unity as n_0/n_{th} increases. It suggests that the higher the threshold n is, before the cavity Q is switched, the more energy will be extracted from the cavity in the form of a giant pulse.

The instantaneous power output of the laser is given by

$$p = \frac{\phi h\nu}{t_c} \quad \text{...(4.21)}$$

Substituting the value of ϕ from Eq. (4.19) into Eq. (4.21), we get

$$P = \frac{h\nu}{2t_c}\left[n_{th}\ell n\frac{n}{n_0} - (n - n_0)\right] \quad \text{...(4.22)}$$

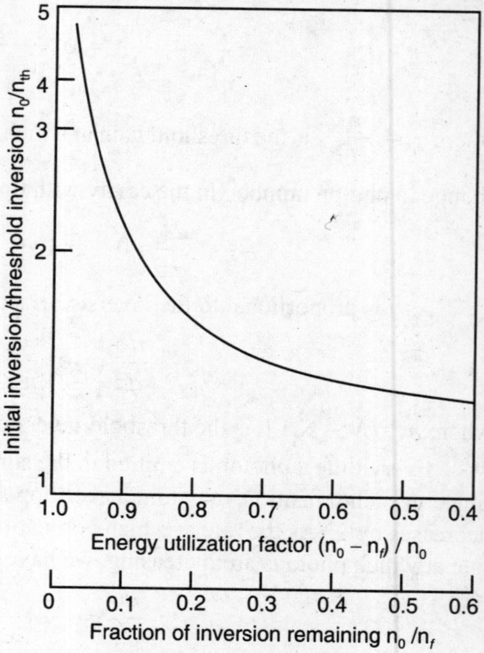

Fig. 4.5. Energy utilization factor during the generation of a pulse in a laser.

We are interested in knowing the peak power. To find the maximum power, we apply the condition

$$\frac{dP}{dn} = 0 \quad \text{...(4.23)}$$

and find that peak power occurs at $n = n_{th}$. It means that peak power occurs at the beginning of the pulse. It is given by

$$P_{peak} = \frac{h\nu}{2t_c}\left[n_{th}\ell n\frac{n_{th}}{n_0} - (n_{th} - n_0)\right] \quad \text{...(4.24)}$$

As $n_0 >> n_{th}$, Eq. (4.24) gives

$$\boxed{P_{peak} \cong \frac{n_0 h\nu}{2t_c}} \quad \text{...(4.25)}$$

Eq. (4.25) shows that the peak power of the Q-switched laser can be many times larger, since n_0 can be made very high compared to n_{th} when the laser is not oscillating, i.e. under the Q-spoiled situation. It should be noted that, although there is a manifold increase in the peak power of a Q-swithced laser, the total energy emitted will be less than the energy obtained in normal operation. The decrease in energy is owing to the losses associated with the Q-switching mechanism.

Total Energy :

The total energy contained in the pulse is

$$E_T = (P_{peak}) \times \textit{energy utilization factor}$$

or

$$E_T = \left(\frac{n_0 - n_f}{n_0}\right)\frac{n_0 h\nu}{2}$$

$\therefore$

$$E_T = \left[\frac{n_0 - n_f}{2}\right] h\nu \quad \text{...(4.26)}$$

Pulse duration

The width of the giant pulse, Δt can be estimated using the approximation

$$P_{peak}\,\Delta t \cong E_T$$

$$\therefore \qquad \boxed{\Delta t = \frac{E_T}{P_{peak}}} \qquad ...(4.27)$$

Q-switching is widely used, but not all lasers can be Q-switched.

The following two important requirements are to be met with for effective Q-switching:

(*i*) the rate of pumping must be faster than the spontaneous decay rate of the upper lasing level. Thus,

$$\nu_{pump} > A_{21}$$

Otherwise the upper level will be depleted faster than it can be populated and large population inversion cannot be attained.

(*ii*) The Q-switch must operate rapidly compared to the build up of laser oscillations. If this condition is not met, we obtain a longer pulse with less peak power.

4.7 METHODS OF Q-SWTCHING

The Q-switch can take several forms. Rotating mirrors, electro-optic modulators, acousto-optic modulators and saturable absorbers are some of the important devices used for Q-switching.

4.7.1 Rotating Mirrors

The rotating reflector method was the first method used for Q-switching of a laser. It involved a rapid rotation of one of the mirrors of the resonant cavity. The rotating mirror would serve as the rear mirror and an output mirror would be at the other end of the laser cavity. The mirrors come into parallel position for a short interval in each rotation. When the mirrors are not parallel, the cavity loss is very high and consequently cavity Q is very low. When the mirrors become parallel, the laser oscillations build up and Q-switching occurs. It causes the emission of a giant laser beam pulse. For a ruby laser with an upper laser level lifetime of 3 ms, the mirror would have to complete one full turn in 3 ms. It requires a rotation speed of 20,000 rpm which is very high. The required speed can be brought down if we use a hexagonal-shaped mirror assembly on a rotating shaft, such that for every 1/6th of a rotation of the shaft, a mirror would be aligned with the laser cavity and output mirror. In such a case, the shaft would have to rotate 1/6th of a turn in 3 ms, requiring a shaft rotation speed of 3,333 rpm. This method of Q-switching is cheap and reliable but is very slow.

4.7.2 Electro-optic Shutter

An electro-optic shutter can serve as a voltage-controlled gate which rapidly switches the cavity from a high loss to a low loss condition. It consists of a crystal that becomes double refracting (birefringent) when an electric field is applied across the crystal. The electro-optic shutter along with a polarizer is arranged inside the optical resonator, as shown in Fig 4.6.

A voltage is applied to the crystal during the pumping of the laser by the light from a flash lamp. The magnitude of the voltage is chosen such that it transforms the electro-optic crystal into a quarter-wave plate. Light emitted by the laser becomes linearly polarized light on passing through the polarizer. The linearly polarized light incident on the crystal splits into two mutually orthogonal components. As the two components travel through the electrooptic crystal, a relative phase retardation of 90^0 is produced between them. On emerging from the electrooptic crystal, the two components combine to produce circularly polarized light. The light beam reflects from

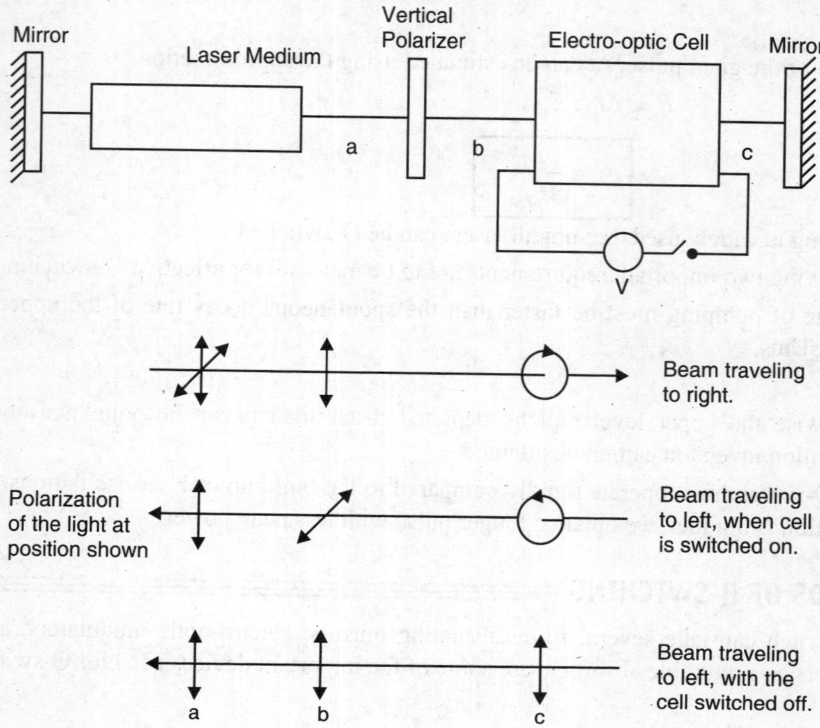

Fig. 4.6. Electro-optic cell used as a Q-switch.

the mirror and returns to the cavity travelling in the opposite direction. On reflection the sense of rotation of the circularly polarized light reverses and on re-passing through the electro-optic crystal, the two components of circularly polarized light experience a further retardation of 90^0 with respect to each other. Coming out of the crystal at the other end, the components recombine to produce linearly polarized light. The direction of polarization of the light is now at 90^0 with respect to its original direction of polarization and transmission axis of the polarizer. This light is not allowed to pass through the polarizer. Therefore, light does not come back to the laser rod and the cavity is switched off. Thus cavity Q is reduced to a low value. When the voltage applied to the electro-optic crystal is turned off, double refraction is absent in the crystal and the state of polarization of the light passing through the crystal is unaffected. Light, therefore, freely travels in both directions and returns to the laser rod to be further amplified. Now, the cavity is switched on and the Q regains its high value. The Q-switching is synchronized with pumping mechanism such that the volatge applied to the electro-optic crystal drops to zero value at the time when the population inversion in the laser medium attains its peak value.

Two types of electro-optic shutters, known as Kerr cell and Pockels cell are available for Q-switching. The Pockels cell is preferred over the Kerr cell because of the lower voltage needed to produce the desired effect.

4.7.3 Acousto-optic shutter

An acousto-optic shutter consists of a quartz crystal attached to a piezoelectric transducer. When an rf signal of 25-50 MHz frequency is applied to the transducer, acoustic waves are generated which propagate in the quartz crystal. The acoustic waves generate an effective diffraction grating in the quartz crystal. If an acousto-optic shutter is installed inside the resonator, and if it is activated during the pumping of the laser, the light beam is deflected out of the laser cavity by the phase grating established in the quartz crystal. Consequently, cavity Q goes to a low value. When the signal applied to the piezoelectric transducer is turned off, light passes through the crystal undeflected and re-enters the laser for further amplification. Thus, the Q is switched to a high value.

4.7.4 Saturable Absorber

Q-switching may be accomplished by using a saturable absorber. A saturable absorber is a suitable organic dye dissolved in an appropriate solvent. The organic dye selected for this purpose absorbs light and is opaque at low light intensity levels, but becomes transparent to light at high intensity levels, then the dye is said to be *bleached*. The saturable absorber is used as a Q-switching element by placing it within the laser cavity. At the beginning of the pumping. The light emitted is low and the saturable absorber blocks it and the cavity exhibits high loss. As the spontaneous emission increases, it causes the absorber to bleach through thereby creating low loss in the absorber. It causes the beam to traverse back and forth and oscillations build up within the cavity. The transmission coefficient of such a bleachable absorber increases from about 10^{-6} to 0.3 in a few nanoseconds. The effect is reversible so that the material recovers in time and be ready for the next pulse. A liquid dye solution such as DODCI is used for such a *Q*-switching element. Saturable absorbers are a simple and inexpensive method required for the operation, this method is known as a *passive Q*-switching.

4.8 CAVITY DUMPING

A cavity dumped laser employs two full reflective mirrors, one on each end of the cavity. When pumping of the lasing medium is done, power inside the cavity builds up due to stimulated emissions. The continuously operating pump maintains the population inversion above the threshold. Oscillations commence but the radiation is confined within the cavity, because the mirrors prevent the light from escaping. Thus, the cavity operates at high-Q. The cavity contains a beam deflector or other component that briefly deflects the beam outside of the laser. Thus, the energy within the cavity is dumped off the cavity in a single pulse.

In *Q*-switching mode, the energy within the cavity is low and the *Q* is low. Oscillations do not start in the initial state. When the cavity Q is swithced to the high value, oscillations start and a short intense pulse passes through the semitransparent output mirror. As the cavity-Q rises from a low value to a high value, it takes some time though very short, for the formation of the pulse. In the case of cavity dumping mode, the cavity is already within the high-Q condition and therefore the output pulse forms very quickly. The length of the cavity dumped pulse is $2\mu L/c$. For a 30 cm. long gas laser, it would be 2 ns, significantly shorter than a Q-switched pulse.

Cavity dumping generates lower energy pulses than the Q-switching. Lasers that cannot be Q-switched are operated in cavity dumped mode.

4.9 MODE LOCKING FOR ULTRASHORT PULSES

Q-switching produces giant pulses of duration of about 10^{-7} to 10^{-8}s. The technique of mode locking allows the generation of pulses of lesser duration of about 10^{-11} to 10^{-12} s. The matching of phases of different modes is called mode locking. An essential requirement for mode locking is existence of multimode operation in the active medium. In fact, the inhomogeneously broadened laser cavity in general supports oscillations in many axial modes simultaneously. The axial modes are uniformly spread on the frequency scale and are separated by $\Delta\nu$ given by

$$\Delta\nu = \frac{c}{2\mu L} \quad ...(4.28)$$

The resultant output of an ordinary laser (laser without mode locking)depends on the relative frequencies, phases and amplitudes of these modes. All these parameters are time-varying, because each of the modes oscillates independently of the other modes. Let the amplitude of the n_{th} mode be given by

$$E(t) = E_n \exp \{i(\omega_n t + \delta_n)\} \qquad ...(4.29)$$

where ω_n is the frequency and δ_n is the phase of the n_{th} mode. If there are N modes of equal amplitude oscillating simultaneously in the cavity, then as the modes are not in phase, *i.e.* incoherent, the total amplitude of all of the modes can be written as

$$E(t) = E_0 \sum_{n=0}^{N-1} \exp \{i(\omega_n t + \delta_n)\} \qquad ...(4.30)$$

The total intensity is given by the absolute square of the total amplitude. Thus,

$$I(t) = |E(t)|^2 = E_0^2 \sum_{n=0}^{N-1} \exp\{i(\omega_n t + \delta n_n\} \exp\{-i(\omega_n t + \delta_n\}$$

or

$$I(t)_{ave} = NE_0^2 = NI_0 \qquad ...(4.31)$$

It means that when the modes are incoherent, the total intensity due to the N modes is equal to N times the intensity of the individual modes. The intensity may exhibit small fluctuations if two or three modes happen to be in phase at any given time, but for large N, it more or less maintains constant.

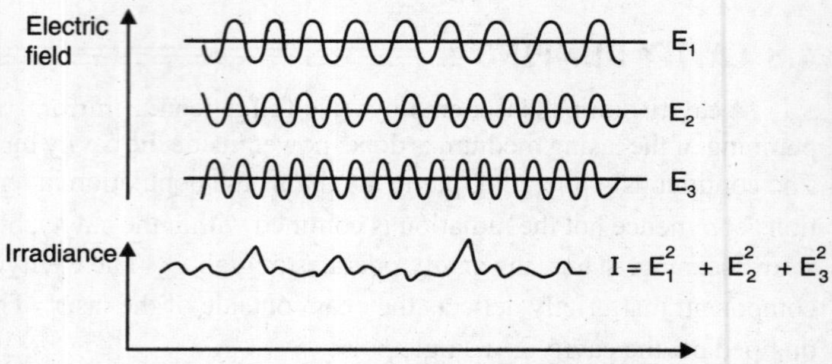

Fig. 4.7. Output of a non-mode locked laser, consisting of random fluctuations in intensity

Let us now suppose that the modes are forced to oscillate in phase, such that $\delta_n = \delta$. In other words, the laser is *mode locked.* The resultant electric field can now be expressed using Eq. (4.29), as

$$E(t) = E_0 \exp(i\delta) \sum_{n=0}^{N-1} \exp(i\omega_n t) \qquad ...(4.32)$$

For convenience, we write the angular frequency ω_n as $\omega_n = \omega - n\delta\omega$, where ω is the angular frequency of the highest frequency mode, and $\delta\omega$ is the angular frequency separation between modes. We can then rewrite Eq. (4.32) as

$$E(t) = E_0 \exp(i\delta) \sum_{n=0}^{N-1} \exp\{i(\omega - n\delta\omega)t\} \qquad ...(4.33)$$

which upon expansion becomes

$$E(t) = E_0 \exp\{i(\omega t + \delta)\} \sum_{n=0}^{N-1} \exp(-i\delta\omega t)$$

$$= E_0 \exp\{i(\omega t + \delta)\}\{1 + e^{-i\delta\omega t} + e^{-2i\delta\omega t} + ... + e^{-i(n-1)\delta\omega t}\}$$

$$= E_0 e^{i(\omega t + \delta)} \left[\frac{1 - e^{-iN\delta\omega t}}{1 - e^{-i\delta\omega t}}\right]$$

$$\therefore \qquad E(t) = E_0 e^{i(\omega t + \delta)} \frac{\sin (N\delta\omega t/2)}{\sin (\delta\omega t/2)} \qquad \text{...(4.34)}$$

$$\therefore \qquad I(t) = |E(t)|^2 = E_0^2 \frac{\sin^2 (N\delta\omega t/2)}{\sin^2 (\delta\omega\, t/2)} \qquad \text{...(4.35)}$$

The intensity $I(t)$ is periodic and exhibits maxima at $\frac{\delta\omega t}{2} = 0, \pi, 2\pi, ..., n\pi$

$$\therefore \qquad \frac{\sin^2 (N\delta\omega\, t/2)}{\sin^2 (\delta\omega\, t/2)} = N^2 \qquad \text{at } \frac{\delta\omega\, t}{2} = 0, \pi, \qquad \text{...(4.36)}$$

$$\therefore \qquad I(t)_{\max} = N^2 E_0{}^2 = N(NE_0{}^2) = NI(t)_{ave} \qquad \text{...(4.37)}$$

The difference between two successive maxima at t_{n+1} and t_n is given by

$$\Delta t_{sep} = t_{n+1} - t_n = \frac{2(n+1)\pi}{\delta\omega} - \frac{2n\pi}{\delta\omega} = \frac{2\pi}{\delta\omega} \qquad \text{...(4.38)}$$

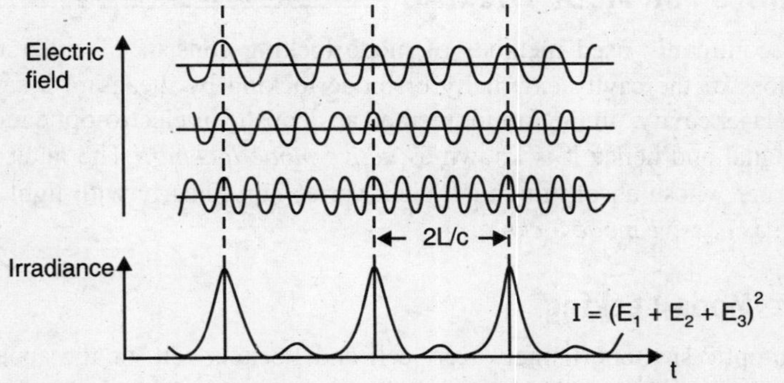

Fig. 4.8. Output of a mode locked laser, consisting of a narrow intense pulses.

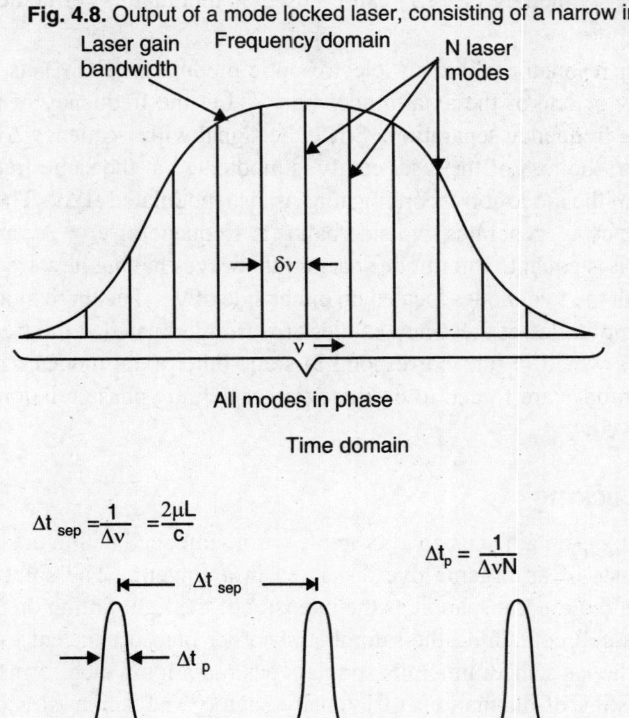

Fig. 4.9. Time spacing and duration of pulses produced by a mode-locked laser

Using Eq. (4.28), we can write

$$\delta\omega = \frac{2\pi c}{2\mu L}$$

$$\Delta t_{sep} = \frac{2\pi(2\mu L)}{2\pi c} = \frac{2\mu L}{c} \quad \text{...(4.39)}$$

This is the interval between pulses of $I(t)_{max}$.

We thus find that the output of a mode locked laser consists of a sequence of short pulses at time intervals of $2\mu L/c$, which is equal to the round trip transit time for light within the cavity. Each pulse has a peak power equal to N times the average power. The mode locked condition can be viewed as a condition in which a pulse of light is bouncing back and forth inside the cavity and every time it hits the mirror, a fraction of it is transmitted as the output pulse.

4.10 TECHNIQUES FOR MODE LOCKING

The most commonly used methods of mode locking consists of either mode locking by modulating the loss of the cavity externally or mode locking by means of a satrurable absorber placed inside the laser cavity. In the former method an acoustic or electro-optic modulator is driven by an external signal and hence it is known as *active mode locking*. The latter method employs a saturable absorber whose absorption coefficient varies non-linearly with light intensity and the method is known as *passive mode locking*..

4.10.1 Active Mode Locking

The electro-optic shutters, namely Kerr cell and Pockels cell, or the acousto-optic shutter described in the context of Q-swithching (sec4.7) can be used as modulators for achieving mode locking.

Lect us consider a laser resonator where an electro-optic modulator (EOM) is kept inside the cavity. Let the frequency of one of the axial modes be ν_m. Let the frequency of modulation be equal to $\Delta\nu$, the intermode frequency separation. Since the signal with frequency $\Delta\nu$ is applied to the electro-optic modulator, the loss of the laser cavity is modulated at the same frequency $\Delta\nu$. Consequently, the amplitude of the mode corresponding to ν_m is also modulated at $\Delta\nu$. The amplitude modulated mode at a frequency ν_m generates two side bands at frequencies $\nu_m + \Delta\nu$ and $\nu_m - \Delta\nu$. Since $\Delta\nu$ is chosen such that it is equal to intermode spacing, the waves having new frequencies $\nu_m + \Delta\nu$ and $\nu_m - \Delta\nu$ coincide with the two modes located on either side of ν_m. The three modes become locked in phase. Since the amplitudes of these new modes are also modulated at the frequency $\Delta\nu$, they generate new side bands which in turn correspond to some other axial modes. The process continues and at the end all modes are forced to oscillate with a definite phase relationship. As a result, mode locking is achieved.

4.10.2 Passive Mode Locking

The passive mode locking is obtained using a saturable absorber placed within the laser cavity. The saturable absorber consists of an organic dye dissolved in a solvent. The solution has the property of becoming more and more transparent as the intensity of the light falling on it increases. Let us consider the case of a cell containing the saturable absorber placed adjacent to one of the resonant mirrors. Initially, the laser medium emits spontaneous radiation which forms within the cavity a radiation field consisting of fluctuations of intensity having a noiselike structure. Some of the intensity peaks within the fluctuations bleach the saturable absorber more than the other components and pass through. They get amplified in its round trip through the active medium. If the cell recovery time is very short, the cell becomes opaque (absorbing) immediately after this pulse passed through. As other fluctuations are weak, the are attenuated in the cell. Thus, the strongest pulse will grow faster while the other less intense fluctuations are suppressed. At regular intervals of time T, this pulse arrives at the output mirror and is partially emitted. Therefore, the laser output consists of a regular sequence of pulses with a pulse repetition time of T.

4.11 FREQUENCY CONVERSIONS USING NONLINEAR CRYSTALS

Although a large number of active materials are available and lasers can be built using them, it is not possible to produce light covering all wavelengths of interest. It becomes therefore, necessary to transform the frequency of light generated by lasers into light of desired frequency. The frequency conversion can be conveniently affected with the help nonlinear optical media. Harmonic generation, sum frequency generation and parametric oscillation are three important effects exhibited by nonlinear crystals which are utilized in laser light frequency transformations.

4.11.1 Nonlinear medium

Dielectric materials or insulators are nonconductors. They do not allow electric current to pass through when an electric field is applied across them. Instead, they get polarized. A dielectric is said to be *polarized* when the electric dipoles in it orient in the direction of the externally applied electric field. The extent to which the dielectric is polarized is denoted by the polarization vector, P, which is given by

$$\boldsymbol{P} = \chi \boldsymbol{E} \qquad \ldots(4.40)$$

where χ is called the *dielectric susceptibility*. The stronger the electric field, the greater will be the polarization. Materials, in which this kind of linear relationship holds, are known as *linear dielectrics*. Dielectrics are polarized, for example, when they are placed between the charged plates of a capacitor. In a similar way, they are polarized when electromagnetic wave propagates through them. Light waves are electromagnetic waves and the electric field of the waves polarizes the dielectric. Hence the optical parameters are closely related to the dielectric polarization. The refractive index of the medium is given by

$$\mu = \sqrt{\varepsilon} = \sqrt{1 + 4\pi\chi} \qquad \ldots(4.41)$$

As long as the intensity of the light propagating in a dielectric medium is small, the parameters χ and μ are constant quantities and are independent of the intensity of light. Ordinary light sources generate light of field strengths of the order of 10^5 V/m which are very small compared to atomic fields, and therefore cannot affect the optical parameters of the medium. Lasers produce light of field strengths of the order of 10^7 to 10^{11} V/m, which are of the order of atomic field strengths. Therefore, the intense light of lasers is in a position to influence the optical parameters of a medium. When the electric field E in the light wave is very large, the parameters χ, ε and μ become functions of E. thus we can express

$$\chi(E) = \chi_0 + \chi_1 E + \chi_2 E^2 + \ldots \qquad \ldots(4.42)$$

and

$$P = \chi\,(E)E$$

$$\therefore \quad P = \chi_0 E + \chi_1 E^2 + \chi_2 E^3 + \ldots. \qquad \ldots(4.43)$$

where $\chi_0, \chi_1, \chi_2, \ldots$ are parameters of the medium characterizing its polarizability. The second and higher order terms represent the *nonlinear* part of the polarization. Materials, in which polarization exhibits nonlinear dependence on the strength of the electric field, are called *nonlinear medium.* Fig. 4.10 shows the variation of electric polarization with the electric field strength in a nonlinear medium.

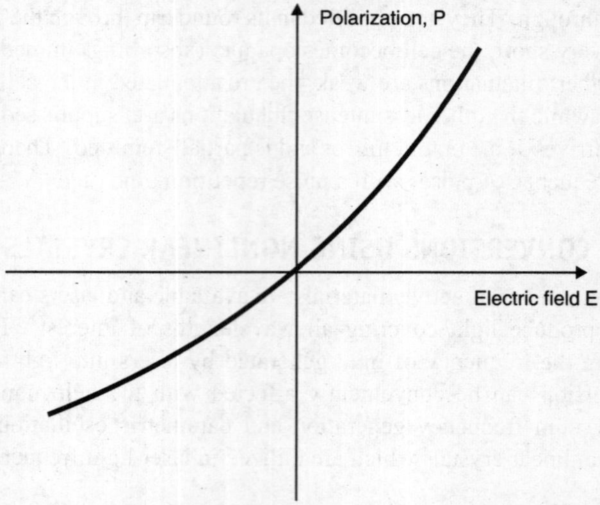

Fig. 4.10. A typical plot of variation of electric polarization with electric field in a nonlinear medium.

4.11.2 Second harmonic generation

Strictly speaking, any medium becomes "nonlinear" provided the electric field of the incident radiation is very high. In case of optically isotropic materials, the coefficients of even powers of E in Eq. (4.3) are zero. But in case of an isotropic materials, coefficients of both even and odd powers of E exist. However, the third order term is substantially smaller than the second order term. Hence, Eq. (4.43) can be rewritten in case of anisotropic crystal as

$$P = \chi_0 E + \chi_1 E^2 \qquad \text{...(4.44)}$$

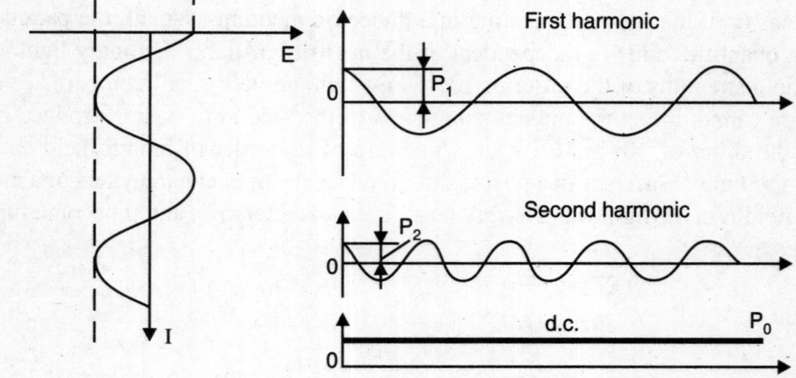

Fig. 4.11. Fourier analysis of the asymmetrical polarization wave into fundamental, second harmonic and d.c. components.

Fig. 4.11 shows the response of a nonlinear medium, in which a symmetrical electric field produces an asymmetrical polarization. A Fourier analysis of the polarization shows that it consists of components having frequencies ν, and 2ν as well as d.c. component. Consider a plane monochromatic light wave of frequency travelling at the velocity v_1 in a nonlinear medium. The electric field of the wave may be of the form

$$E = E_0 \sin [2\pi\nu (t - x/v_1) + \chi_1 E_0^2 \sin^2 [2\pi\nu (t - x/v_1)]$$

$$= E_0 \sin [2\pi\nu (t - x/v_1) + \frac{1}{2} \chi_1 E_0^2 - \frac{1}{2} \chi_1 E_0^2 \cos [4\pi\nu (t - x/v_1)] \quad ...(4.45)$$

Equation (4.45) contains a term in $4\pi\nu$ which corresponds to a wave of polarization propagating in the same direction and at the same speed of light wave having twice higher frequency. The magnitude of the term in $4\pi\nu\{= 2(2\pi\nu)\}$ approaches the magnitude of the first term $\chi_0 E_0$ at high values of E_0. Secondly, under certain conditions, this polarization wave may cause emission of a new light wave. It means that a strong light wave of frequency ν propagating in a nonlinear medium can generate a new light wave at frequency 2ν. It is a second harmonic and hence the phenomenon is known as *second harmonic generation.*

While the polarization wave propagates at the velocity v_1, the second harmonic propagates at the velocity v_2 in the medium. The velocities are given by

$$v_1 = \frac{c}{\mu(\nu)} \quad \text{and} \quad v_2 = \frac{c}{\mu(2\nu)}$$

The velocity v_2 of the generated wave differs from that of the incident wave,v_1, because the refractive index is a function of frequency. Efficient transfer of energy from the polarization wave to the second harmonic requires that the two waves must be matched in velocity, *i.e.* they should have identical refractive indices. Thus, the condition is that

$$\mu (\nu) = \mu (2\nu) \quad ...(4.46)$$

The above condition is known as the *phase-matching condition.*

This condition is not satisfied in general cases. It is satisfied only in anisotropic crystals known as *uniaxial crystals*. When a light ray is incident on uniaxial crystals, it splits into two rays namely, an *ordinary ray* (o-ray) and an *extra ordinary* ray (e-ray). The refractive index of e-ray depends on the direction of propagation. The refractive index surface is a sphere in case of o-ray and an ellipsoid in case of e-ray. The refractive index surfaces in case of a negative uniaxial crystal is shown in Fig. 4.12(a). Fig 4.12(b) shows the refractive index surfaces for frequency ν and for frequency 2ν. It is seen that the refractive index surface of the o-ray with frequency ν and that of the e-ray with frequency 2ν intersect; one of the points of intersection is point *P*. It is easy to see that the phase-matching condition is satisfied for light waves propagating in the direction *PQ*. Thus, the direction *PQ* is the *direction of phase –matching. Along PQ.*

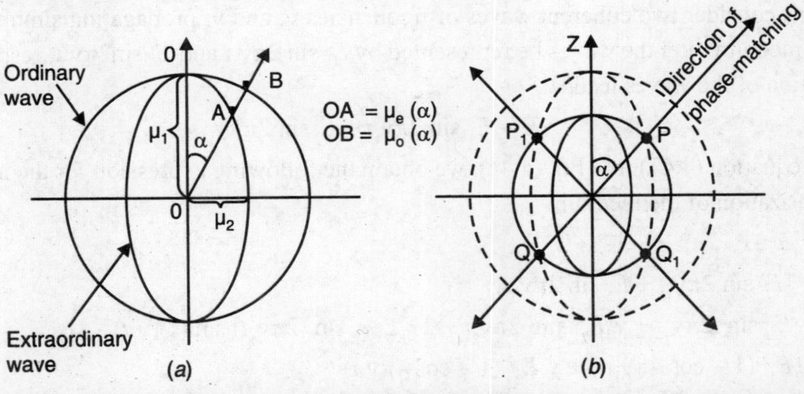

Fig. 4.12. (a) Index ellipsoid and sphere corresponding to a e-ray and 0-ray in a negative uniaxial crystal. (b) Index matching in a negative crystal.

$$\boxed{\mu_0(\nu) = \mu_e(2\nu)} \qquad ...(4.47)$$

The process of second harmonic generation requires a uniaxial crystal exhibiting high nonlinear susceptibility χ_2. It must be cut in the shape of a rectangular perallelepiped with the axis long the direction of phase matching. The incident light must be plane-polarized, with the plane of polarization perpendicular to the plane of principal section of the crystal. When these conditions are satisfied, a wave of frequency ν produces another wave of frequency 2ν, which is the second harmonic. The direction of propagation of the second harmonic coincides with that of the fundamental wave. The plane of polarization of second harmonic will be perpendicular to that of the fundamental wave, as shown in Fig. 4.13.

There are crystals like KDP which offer a conversion efficiency of 20 to 50%. Second harmonic generation helps us in extending the range of laser wavelengths into the blue and ultraviolet part of the spectrum, which are not rich in naturally occurring laser lines.

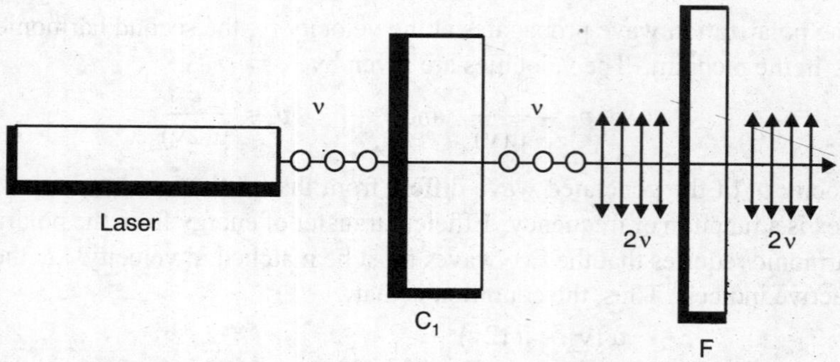

Fig. 4.13. Arrangement for second harmonic generation

Second harmonic generation was first demonstrated by Franken and his coworkers in 1961. They focussed a 3 kW ruby laser pulse (λ = 6943Å) onto a quartz crystal and obtained a very low intensity output at a wavelength of 3471.5 Å.

4.11.3 Sum frequency conversion

Besides second harmonic generation, nonlinear crystals make it also possible to generate beams of sum and difference frequencies when two light waves of different frequencies interact with them. Thus, interaction of waves with frequencies ν_1 and ν_2 generate light waves with frequencies $\nu_1 + \nu_2$ and $\nu_1 - \nu_2$.

Let us consider two coherent waves of frequencies ν_1 and ν_2 propagating simultaneously in a nonlinear medium. Let the waves be represented by $E_1 \sin 2\pi\nu_1 t$ and $E_2 \sin 2\pi\nu_2 t$, respectively. The superposition of the waves leads to

$$E = E_1 \sin 2\pi\nu_1 t + E_2 \sin 2\pi\nu_2 t \qquad ...(4.48)$$

Using the equation (4.47) into Eq. (4.44), we obtain the following expression for the nonlinear term in the polarization of the medium.

$$\begin{aligned} P_{nl} &= \chi_1 E^2 \\ &= \chi_1[E_1 \sin 2\pi\nu_1 t + E_2 \sin 2\pi\nu_2 t]^2 \\ &= \chi_1 E_1^2 \sin^2 2\pi\nu_1 t + \chi_1 E_2^2 \sin^2 2\pi\nu_2 t + 2\chi_1 E_1 E_2 \sin(2\pi\nu_1 t)\sin(2\pi\nu_2 t) \\ &= \chi_1 E_1^2 (1 - \cos 4\pi\nu_1 t) + \chi_1 E_2^2 (1 - \cos 4\pi\nu_2 t) \\ &\quad + \chi_1 E_1 E_2 \{\cos 2\pi(\nu_1 - \nu_2)t - \cos 2\pi\nu(\nu_1 + \nu_2)t\} \end{aligned}$$

or $P_{nl} = \chi_1 (E_1^2 + E_2^2) - \chi_1 E_1^2 \cos 2\pi(2\nu_1)t - \chi_1 E_2^2 \cos 2\pi(2\nu_2)t$

$$+ \chi_1 E_1 E_2 \{\cos 2\pi (\nu_1 - \nu_2)t - \cos 2\pi (\nu_1 + \nu_2)t\} \quad ...(4.48)$$

The eq (4.48) shows that the interaction of light waves of frequencies ν_1 and ν_2 propagating in a nonlinear medium generates new light waves of sum $(\nu_1 + \nu_2)$, difference $(\nu_1 - \nu_2)$ and double $2\nu_1$ and $2\nu_2$ frequencies.

In sum frequency generation, ν_1 is taken as a weak infrared wave that sums with a strong pump wave ν_2 to generate a high frequency wave at $\nu_3 = (\nu_1 + \nu_2)$. Thus, the light of wavelength of 10.6 μm produced by CO_2 laser is made to interact with a laser beam of 1.06 μm wavelength produced by Nd : YAG laser to yield a light beam of wavelength of 0.964 μm. Such upconversion is done because it is difficult to detect the radiation at 10 μm while the radiation at 0.964 μm can be easily detected.

4.11.4 Parametric oscillation

The production of coherent radiation at frequencies in the ultraviolet region is very much essential. Apart from the technique of harmonic generation, tuning of frequency by parametric oscillation is another technique available for this purpose. The method of parametric excitation of oscillations is widely used in electronics, where nonlinear capacitors are made use of. The parametric generation of light involves use of nonlinear medium and thus bears close similarity with the electronic process.

In sum frequency generation, light at frequencies ν_1 and ν_2 add together to produce the frequency $(\nu_1 + \nu_2)$. On the other hand, if a strong beam at frequency $(\nu_1 + \nu_2)$ alone is applied to a suitable nonlinear material such as lithium niobate, two beams at lower frequencies ν_1 and ν_2 can be generated. For exciting parametric oscillations, the nonlinear medium is placed between two mirrors, which form a resonant cavity comparable to a laser cavity. The wave of frequency $(\nu_1 + \nu_2)$ is known as *pump wave* and it should be highly intense such that it can induce nonlinear behavior of the crystal. Secondly, the process requires that the phase-matching condition is satisfied. The wave of higher frequency, say ν_1, is called the *signal wave* and the lower frequency wave, say at ν_2, is called the *idler wave*. Both the idler and signal waves are weak and are always present in the crystal in the form of *noise* which arises due to spontaneous photons. If the pump frequency $(\nu_1 + \nu_2)$ is fixed, then the two frequencies ν_1 and ν_2 are free to spread over a wide range of values.

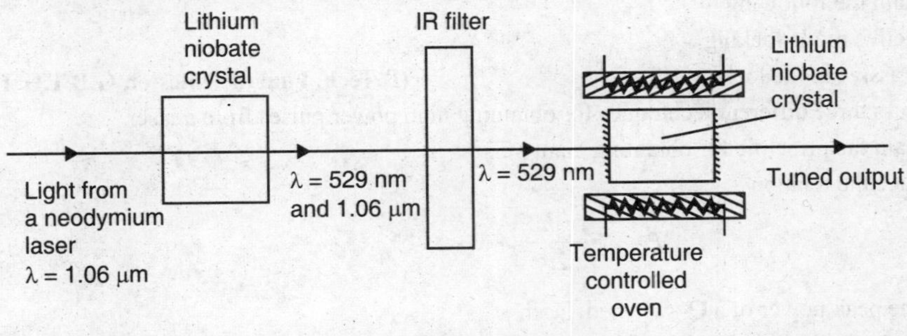

Fig. 4.14. A typical arrangement for observing parametric oscillation in a nonlinear medium

This effect is known as *parametric amplification*. By varying the orientation of the crystal, the values of ν_1 and ν_2 can be varied. If the mirrors of the resonant cavity are reflective at ν_1 and ν_2 and not at the $(\nu_1 + \nu_2)$ frequency , resonance of these waves in the cavity enhances the interaction in the nonlinear crystal and builds up the waves. Thus, the system acts as a parametric oscillator. Tuning can be achieved by varying the phase-matching conditions through mechanical or temperature control of the cavity.

Parametric oscillation was first observed by Giordmaine and Miller in 1965. The schematic of set up used by them is shown in Fig. 4.14. The output was tuned by changing the temperature of lithium niobate crystal. A temperature range of about 11°C produced wavelengths in the range 9680 Å to 11540 Å. The conversion efficiency was as low as 1%. Currently, conversion efficiency is improved to 40% to 50%.

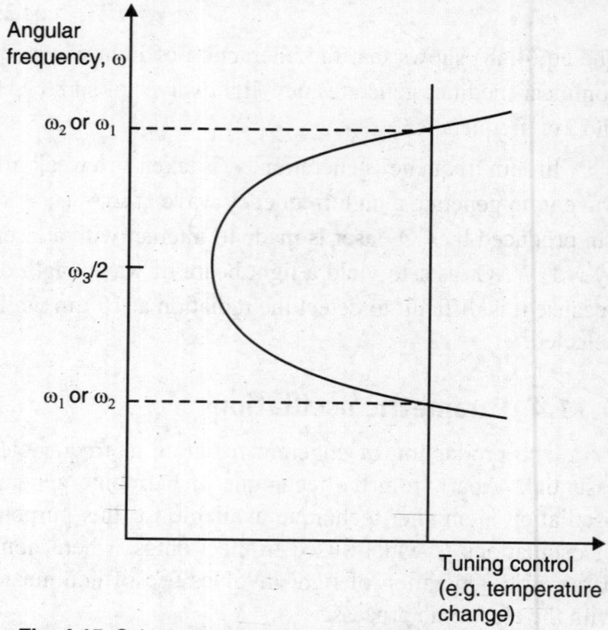

Fig. 4.15. Schematic tuning curve for a parametric oscillator.

EXERCISE-4

1. What is optical cavity ? Derive relation for quality factor in cavity.
2. Explain in brief Q-factor of a cavity. **(G.N.D.U. 2007, 2006)**
3. What do you mean by Q-switching ? Describe different types of Q-switches. **(G.N.D.U. 2007, Pbi. U. 2007)**
4. Write a short note on mode locking. **(G.N.D.U. 2008, 2007, P.U. 2006)**
5. Describe various methods of Q-switching. How it is helpful in generating laser pulse. **(B.Tech. Fourth semester, G.B.T.U. Luck 2010)**
6. Explain the following
   (*i*) Active mode locking
   (*ii*) Passive mode locking **(B.Tech. Fourth semester, G.B.T.U. Luck 2010)**
7. Discuss three different techniques for obtaining high power pulses from a laser.
8. Explain the principle for obtaining giant pulses.
9. Deduce the relation

$$P_{\text{peak}} = \frac{n_0 h\upsilon}{2t_c}$$

for the peak power of a Q-switched laser.

10. What do you understand by cavity dumping ?
11. What are parametric oscillations ? Draw a schematic disgram for observing parametric oscillations in a non-linear medium.
12. Explain how frequency conversions are established using non-linear crystals.

5 CHAPTER
Applications of Lasers

5.1 INTRODUCTION

It is not an exaggeration if we say that next to computers it is the laser that is bringing revolutionary changes in our lives. Lasers are profitably used in almost every field. They are used in fundamental research. Many scientific, military, medical and commercial laser applications have been developed since the invention of the laser in the 1958. The coherency, high monochromaticity, and ability to reach extremely high powers are all properties which allow for these specialized applications.

In **science**, lasers are used in many ways, including: A wide variety of interferometric techniques, Raman spectroscopy, Laser induced breakdown spectroscopy, Atmospheric remote sensing, Investigating nonlinear optics phenomena, Holographic techniques employing lasers also contribute to a number of measurement techniques. Laser based Light Detection And Ranging (LIDAR) technology has application in geology, seismology, remote sensing and atmospheric physics. Lasers have been used aboard spacecraft such as in the Cassini-Huygens mission. In astronomy, lasers have been used to create artificial *laser guide stars*, used as reference objects for adaptive optics telescopes. Lasers may also be indirectly used in spectroscopy as a micro-sampling system, a technique termed Laser Ablation (LA). They are applied in entertainment electronics, industrial electronics, communications, mechanical working, metrology, surveying, surgery and related medical fields, information processing and even in warfare. The list is unending and continues to grow. The numerous fields of laser applications can be broadly divided into two groups, one involving laser beams of high power and the other involving beams of low power. High power gas and solid state lasers are used in materials processing, nuclear fusion, medical field, defence etc. applications. The low power semiconductor lasers are used in CD players, laser printers, optical floppy discs, optical memory cards, data processing and information processing devices, hologram preparation and viewing, range finders, optical communications etc applications.

It is not possible for going into the details of each and every application. We therefore confine our study only to some of the important and well-established applications of lasers. Some of the main features and applications of lasers are listed in Table 5.1.

Table 5.1. Characteristics and applications of some common lasers

Laser	*Principal output wavelength μm*	*Output*	*Maximum power W*	*Normal application*
Ruby	0·6943	Pulsed	10^8	Drilling, opthalmology,
Nd^{3+} : YAG/glass	1·06	Pulsed	10^9	Drilling, welding,
		CW	10^3	rangefinding, holography

Helium-neon	0·6328	CW	10^{-1}	Surveying, metrology, holography, velocity measurement
	1·1152	CW	10^{-3}	
	3·39	CW	10^{-3}	
Argon	0·458 0·515	CW	10	Opthamology, holography
Carbon dioxide	9·2-10·8	Pulsed	10^{7}	Cutting, welding
		CW	10^{4}	
Nitrogen	0·3371	Pulsed	10^{-2}	Pumping dye lasers
Dye	0·265-0·960	Pulsed	10^{8}	Spectroscopy, pollution detection
Gallium arsenide	0·85-0·90	Pulsed	10	Communications, rangefinding, pollution detection
		CW	0·01	

5.2 LASER: 'A TOOLBOX'

Laser light is highly directional, or collimated. This means that all the photons travel in the same direction. Laser light is also coherent, meaning that all the waves in a laser beam have the same wave pattern. These propertiesmake laser light more intense than ordinary light and allow it to travel long distances. Ordinary light waves scatter in all directions from their source. As the beam of ordinary light travels, the more it spreads out and gets dimmer. It tends interfere with one another. On the other hand, collimated laser light can travel a great distance without losing very much of its energy and brightness. Ordinary light could never have made it to the moon mirror and back, whereas laser light did quite easily. There is one more important quality of laser light. It is coherent, or very organized. This means that the light waves are lined up with each other and moving along in step, almost like a regiment of soldiers marching in a parade. By contrast, ordinary light is incoherent. Its waves become mixed up as they move along, like the crowd of people watching the parade. So, the laser light is special. It is concentrated, directional, and organized.

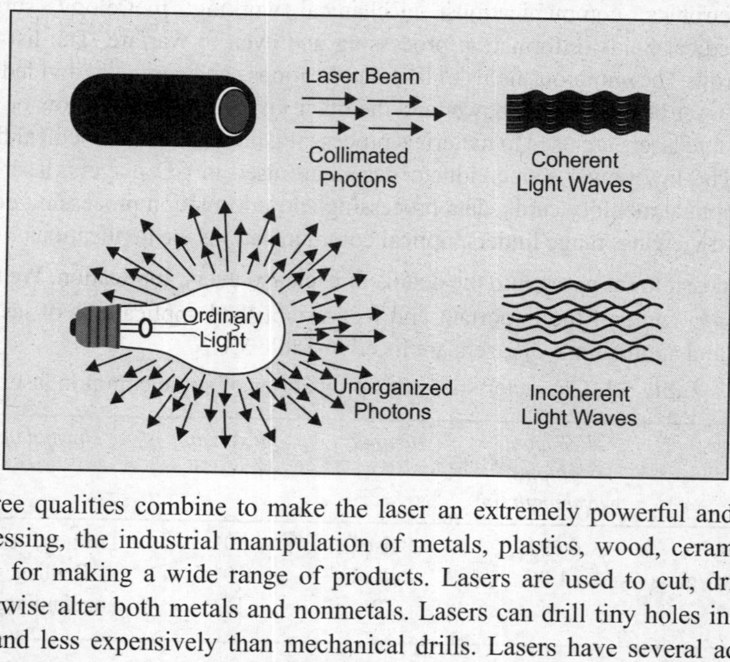

These three qualities combine to make the laser an extremely powerful and useful tool in materials processing, the industrial manipulation of metals, plastics, wood, ceramics, cloth, and other materials for making a wide range of products. Lasers are used to cut, drill, weld, heat-treat, and otherwise alter both metals and nonmetals. Lasers can drill tiny holes in turbine blades more quickly and less expensively than mechanical drills. Lasers have several advantages over conventional techniques of cutting materials. For one thing, unlike saw blades or knife blades, lasers never get dull. For another, lasers make cuts with better edge quality than most mechanical cutters. The edges of metal parts cut by a laser rarely need to be filed or polished because the laser makes such a clean cut. A laser beam excels as an industrial drill because it can be focused into a tiny bright point. Of course, ordinary light can be focused in a similar way. For instance, a magnifying glass held up to the sun will focus the sun's rays, into a tiny, very bright point, a point that is also hot enough to burn a leaf or ignite a piece of paper.

5.3 LASERS IN MATERIAL PROCESSING

Materials working is most important and fundamental in mechanical industry. It consists of cutting, drilling, welding etc. jobs to be done on both metals and nonmetals. These processes require transfer of energy from the laser beam to the workpiece. It can happen only if the material has high absorption at the wavelength corresponding to the laser beam. Once the surface of the materials absorbs energy, the material starts to melt and then vapourize. At high intensity of radiation, the vapour will be ionized to produce plasma. The plasma layer formed between the laser and the workpiece absorbs light and prevents the laser beam from reaching the workpiece. Therefore, it is essential that the plasma should be removed to increase energy coupling.

Material working requires large amounts of energy to be localized at specific areas in order to cause heating there. Therefore, the lasers should deliver large amount of power. The intensity of the laser beam can be enhanced with a suitable optical system that can focus the beam into a spot of about 10 to 100 μm diameter. The lasers that are widely used in material working are the CO_2 laser and Nd : YAG laser. CO_2 lasers operate at 10.6 μm and metals have high reflectivity at this wavelength. Instead of a cw CO_2 laser, a pulsed mode CO_2 laser produces high peak powers and makes possible to work on metals. Nd : YAG lasers operate at 1.06 μm where metals are less reflective and are better candidates for working on metals. However, as CO_2 lasers are cheaper compared to Nd : YAG lasers, CO_2 lasers are more generally preferred. Nd: YAG lasers offer the advantage of compactness. Different materials and working processes rerquire laser power outputs of different magnitudes. For example, relatively low peak power pulses of 10^{-3} to 10^{-2} s width are suitable for welding while more intense and short pulses of 10^{-4} to 10^{-3} s are required for drilling.

5.3.1 The Surface Treatments

The surface treatments are given to the given piece of material as per the particular properties required for putting it into work or its use. For this, laser surface treatment is employed. The particular type of laser is indicated in the field of the surface treatments depending upon how much it allows to obtain high energetic concentrations, and the attainment of elevated skin temperatures in fractions of second, with small thermal alteration of the Bulk properties of the material with negligible residual deformations.

According to the energy density of the incident laser beam, the inter-action time, the spatial distribution of the beam and its wavelength and the thermo-physical properties of the material, it is possible to obtain various effects on the surface of the piece to be treated, like heat treating process.

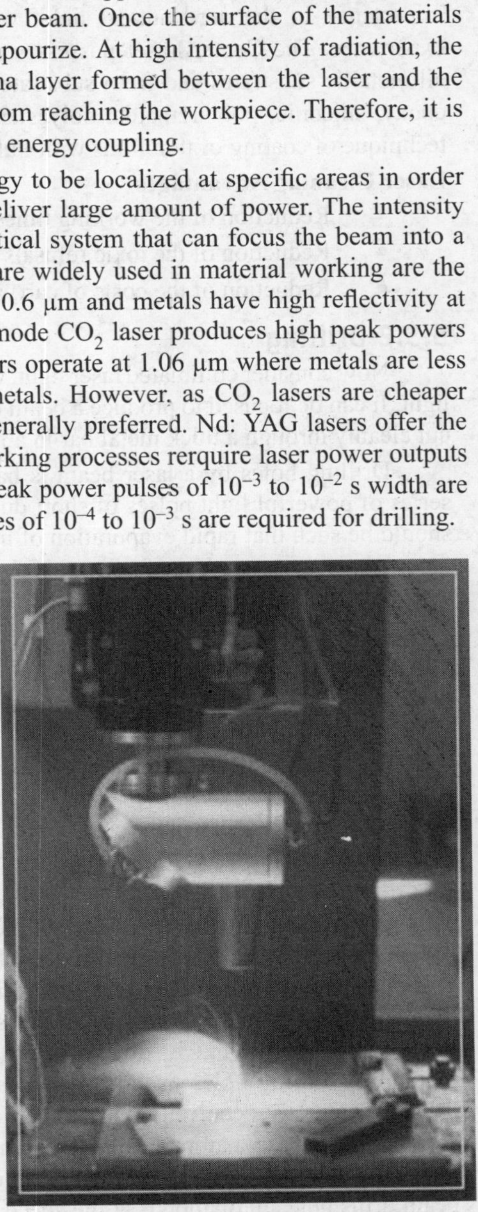

Fig. 5.1 CO_2 laser surface treatment

So it is possible to obtain as a thermal process type (the quick heating and consequent the fast cooling of the material interested) from the passage of the laser beam can provoke, second of the caught up values temperature on the surface of a material, an heat treatment or a fusion or a vaporization, as a photochemical process type, that removes thin superficial layers of material or coating or other material because of the molecular bond breaking caused by the absorption of the incident laser radiation.

De-Rusting De-Oxidizing De-Painting Laser Cleaning

The removal of the painting is obtained by means of a process of vaporization produced from the impact of the beam with material to remove. The products that are obtained are vapors and solid particles ofs dimensions that by means of particular-exhauat fans and suitable filters are removed from the workspace. The possibility to transport the laser beam in the space by means of mirrors or fiber glasses allows to catch up diffcultly accessible points. The applications are of particular importance in the aeronautical field, where the Center Laser has developed one remarkable experience. The Center Laser possesses the technology in a position to removing paintings from all metallic substrates and from substrates composed in thermoplastic material and materials like epoxy, siliconic, polyurethanic, polyesters, acrylic. Plasma assisted Laser deposition utilizes the technique of coating of thin films with Pulsed Laser Deposition (P.L.D.).

Laser Cleaning Advantages

- Reduction of the working times (following removal and re-painting)
- Reduction of the toxic refusals of working
- Reduction of the costs of carrying outs and transports of the refusals.

5.3.2 Drilling

Now consider collimated laser light, which is hundreds of times morej directional than ordinary light. It can be focused to produce a beam of light, much hotter than the surface of the surp, that can cut cleanly through a thick metal bar in a few millionths of a second.

Drilling holes by a laser beam is based on the intense evaporation of material heated by a series of powerful light pulses of short duration of 10^{-4} to 10^{-3} s. The energy supplied for drilling should be such that rapid evaporation of material takes place before radial distribution of heat into the workpiece occurs. Use of a series of short pulses minimises the energy diffused laterally into the workpiece and assists in controlling the size and shape of the hole.

Laser drilling is a non contact process and does not require a physical drill bit. The problems of wear and broken drills do not arise. As there is no physical contact between drill and workpiece, the process becomes faster, and is also highly reproducible. The drilling operation can be done with extremely high precision and in any desired direction

CO_2 laser and Nd : YAG laser are used widely. Metals have high reflectivity at 10.6 μm which is the wavelength of CO_2 laser. However, using high peak power pulses metals can be drilled with CO_2 laser. On the other hand Nd : YAG laser wavelength is 1.06 μm where metals have more absorption and Nd : YAG laser is used for drilling holes in metals only. CO_2 laser is equally suitable for drilling in metallic and non-metallic materials.

One of the more important uses of the laser drill in industry is in the production of metal (viz. copper) wire. The wire is formed by forcibly pulling copper metal through tiny holes in diamond "dies". However, diamond is a very hard material and it is difficult to drill holes in it. Normally, diamond drill bits are used to drill and these drill bits get dull as they drill. The hard diamond acts like a mold, and the much softer copper squeezes out the other end in the form of wire. The old method of drilling holes in industrial diamonds was very time-consuming and expensive. Since the only naturally occurring material hard enough to cut through a diamond is another diamond, workers had to use diamond drills. But diamonds are expensive. Furthermore, the drilling process took several hours, so a worker could drill only two or three holes in a workday. In contrast, a laser beam drills holes in diamonds at the speed of light. One worker using one laser can bore hundreds or even thousands of holes in a single hour. And the same method is used for drilling holes in other gems that are used as moving parts in watches.

Lasers have advantage where high speeds and small holes in deformable materials are required, for example perforation in plastic nozzles and nylon buttons. An interesting example of laser perforation is the piercing of three holes in a baby bottle nipple – one at the top for the liquid and the other two for air inlets. The rubber used for nipples is very soft and flexible. Earlier holes are made in the nipple mechanically with fine wire pins but the rubber bends and the pins can be caught and broken by the rubber. If a laser beam is used instead, tiny holes are burnt through the rubber. Using a beam splitter, the laser output is split into three beams and the job is done in a fast and efficient manner.

Similarly, lasers are used to drill tiny holes in the soft plastic valves of spray cans (such as those of hair spray or glass cleaner). One such laser can punch over a thousand valve holes in one minute.

Lasers are routinely used for drilling in ceramic materials. Ceramic materials become brittle after they arc burnt and therefore conventional drilling has to be performed prior to firing. The size and position of the hole may ; change after firing. Laser drilling is carried out after firing and therefore, size variation does not take place.

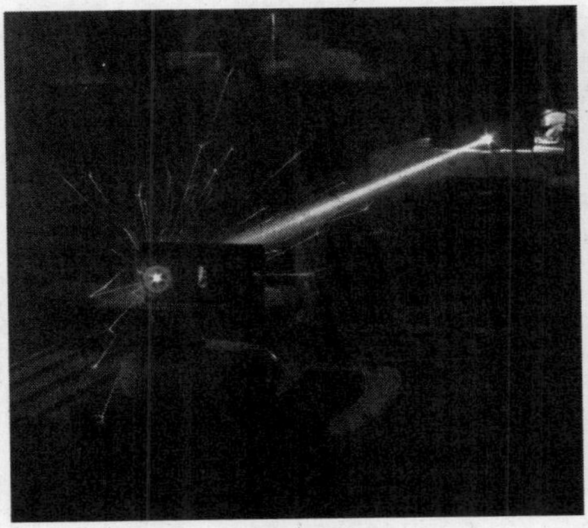

Fig. 5.2 : Laser Drilling

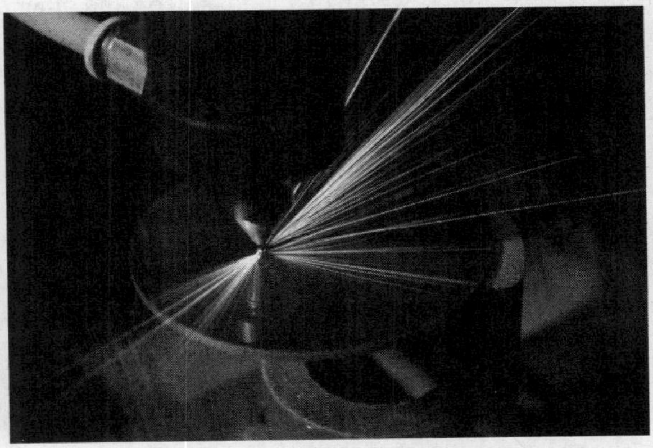

Fig. 5.3 Drilling hole in carbon steel

The laser technology, has improved the industrial field in the microdrilling. The elevated ability to focalization of a laser beam concurs to concentrate the energy in reduced areas a lot. In this way it is possible to locally raise the temperature until that one of fusion and vaporization of the material so as to realize of the holes or microholes. By means of focusing lenses or mirrors the beam laser can be transformed and be focused in shapes and different geometries so as to obtain the holes of the *intentional shape*, i.e., either circular (Fig. 5.2.) or rectangular, moreover it can be addressed in unattainable hidden points with job tools that operate to contact. They are moreover eliminates the problems to you deriving from the vibrations generated from the traditionally used tools that they produce breaches in the fragile materials. The automation possibility concurs a high reliability and repeatability, is possible to realize matrices of holes with the highest precision, tied to the systems to numerical control uses you. The altered zone that it encircles the edge of the microhole is practically absent.

Laser drilling is a process used to drill alloys and composites that usually cannot be drilled by conventional methods. In many cases, laser drilling is the only solution for many drilling challenges. By laser drilling very high position and diameter tolerances, less than lrm (micron) can be achieved. Some of the benefits of utilizing laser hole drilling technique include:

- Low dimensional variability
- Low distortion
- Minimal mechanical force on workpiece
- Highly repeatable
- Faster than any conventional method
- 5-or 6-axis laser control
- Creating entrance angles not possible with contact machining
- Low tool cost, no tool wear
- Drills angled holes and difficult access holes
- Economic setup time

Materials:

Laser drilling can be used to produce micro-holes in nozzles, orifices, vias, photovoltaic cells, fuel injection components, vertical probe cards, metered dose inhaler products, pinholes and slits for scientific instrumentation, inkjet printer nozzles, detectors, sensors, high resolution circuitry, fuel cells, fibre optic interconnects and medical devices etc. Micro-drilling micro-holes using laser in metals, ceramics, wood, diamond, silicon and other semiconductors, polymers, plastics, glass and in almost any material is very common today.

5.3.3 Cutting

Every good toolbox has a hacksaw and a pair of scissors: the saw to cut metal, the scissors to cut cloth. The toolbox laser can do the jobs of both, hence does the job of "laser scissors". Making saw blades themselves is an excellent example of using lasers to cut metal. The old methods of producing saw blades involved many steps, each of which required a person to handle the blades with his or her hands; not surprisingly, injuries were common. In contrast, a laser cuts the blade out of the sheet metal in only one step. Only the beam touches the metal, so as long as the operator is wearing protective glasses there is no chance for injury. In addition, reflective substances like glass can be cut by a laser if their surfaces are first coated with a dark substance. That way the laser light is absorbed rather than reflected.

Laser cutting is a technology that uses a laser to cut materials, and is typically used for industrial manufacturing applications. Laser cutting works by directing the output of a high-power laser, by computer, at the material to be cut. The material then either melts, burns, vaporizes away, or is blown away by a jet of gas, leaving an edge with a high-quality surface finish. An example of the use of "laser scissors" is to cut patterns for clothes. A laser cloth-cutting system was designed by Hughes Aircraft, the company that employed Theodore Maiman, the inventor of the ruby laser. The system works in the following way: Pieces of cloth are laid out on a large table while the patterns are entered into a computer, which decides the best way to trace them out on the cloth. Next, the computer directs the laser beam to cut out the traced patterns very precisely. Cloth for hundreds of suits can be cut in an hour, and as an added advantage the heat of the beam keeps the edges of the cloth from fraying.

The laser cutting of the materials is the common and diffused application of the laser systems. During the cutting process the energy, absorbed from the material in the zone in which the laser beam has focused (power densities 10^5 - 10^8 W/cm^2) comes transformed in heat as shown in Fig. 5.4. This heat locally provokes to a quick increase of the temperature of the piece; the fusion and/or the vaporization of the interaction zone determine the formation of a hole. In the first case the removal of the material happens through aspiration of particles, smoke and vapors produced, in second one generally the expuls. Industrial laser cutters are used to cut flat-sheet material as well as structural and piping materials.

The idea to extend such applications also in the field of the plastics too. For that it concerns the cut of such materials, considering as an example most common plastic rolled sheets (PC, PE and PP). The vaporization for thermal effect of a smallest volume of material and the very reduced dimensions "of the tool laser" (0.2 mm - 200 microns approximately), concur to obtain of the cut kerfs of a hundred of micron respect to the various millimeters obtained with conventional cutting tools. This reduces remarkably the amount of removed material and remarkably increases the precision in finishes of the object to realize. To the ends of the quality, an object is obtained that, even if of complex geometry, it does not need of ulterior workings of cleaning and polishing of the cut edge. But since it enunciates strongly regarding the most common techniques than cut, it is in a generalized manner the elevated speed of cut (cutting speeds). Such speed values depend clearly on the type of material and its thickness and not for last from the employed laser power and other laser working parameters.

Ion of the melted material has helped from the action of a gas pressure, called cutting gas that impinges on the piece surface together the laser beam. If the passing hole has moved along the piece in working, it generates the separation of the two cutting pieces.

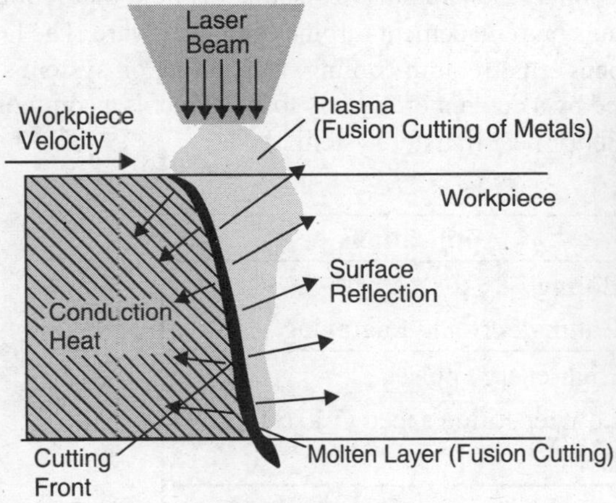

Fig. 5.4 Outline of the laser cutting

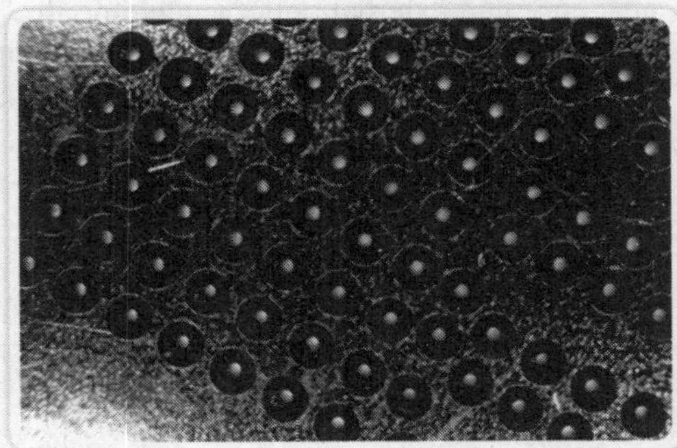

Fig. 5.5 Laser Cut Edge of Kevlar sheet viewed by an electron microscope (SEM)

A wide range of materials can be cut by CO_2 lasers. The materials include paper, wood, cloth, glass, quartz, ceramics, composites, steel etc. The cutting process essentially consists of removing material. Laser cutting is done with the assistance of air, oxygen or dry nitrogen gas jet as shown in Fig. 5.6. The role of the jet is to cool the surface of the material and blow away the debris from the cutting zone.

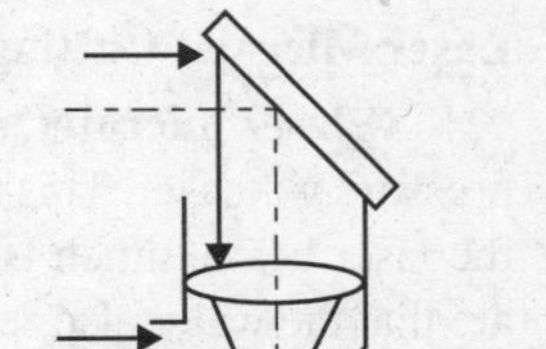

Fig. 5.6 An arrangement for a gas-laser cutter

The advantage of laser cutting is that it is fine and precise. It introduces a minimum mechanical distortion and thermal shock in the material being cut. The process does not introduce any contamination. It is easily automatised and high production rates can be achieved.

The power required for cutting depends on the material to be cut. Plywood can be cut with a 8 kW CO_2 while metals can be cut with 100 to 500 W output power CO_2 laser. 1 cm thick glass requires 20 kW output power. A 100 W CO_2 laser prepares cut pieces for 50 suits during an hour.

Brittle materials such as ceramics and glasses are cut using pulsed lasers in order to minimize micro cracking.

In metal cutting, the laser beam heats the metal to a high temperature where it burns as oxygen passes over it. It is therefore actually the oxygen that does the job of cutting. It is widely used in fabrication of aircraft engine parts.

CO_2 lasers are used for selective removal of insulation from electrical wires and cables. While the insulation is stripped off, the electrical wire inside is not affected because of its high reflectivity.

There are three main types of lasers used in laser cutting. The CO_2 laser is suited for cutting, boring, and engraving. The neodymium (Nd) and neodymium yttrium-aluminum-garnet (Nd-YAG) lasers are identical in style and differ only in application. Nd is used for boring and where high energy but low repetitions are required. The Nd-YAG laser is used where very high power is needed and for boring and engraving. CO_2 lasers are used for industrial cutting of many materials including mild steel, aluminum, stainless steel, titanium, paper, wax, plastics, wood, and fabrics. YAG lasers are primarily used for cutting and scribing metals and ceramics. CO_2 lasers are used for selective removal of insulation from electrical wires and cables. While the insulation is stripped off, the electrical wire inside is not affected because of its high reflectivity.

In addition to the power source, the type of gas flow can affect performance as well. In a fast axial flow resonator, the mixture of carbon dioxide, helium and nitrogen is circulated at high velocity by a turbine or blower. Transverse flow lasers circulate the gas mix at a lower velocity, requiring a simpler blower. Slab or diffusion cooled resonators have a static gas field that requires no pressurization or glassware, leading to savings on replacement turbines and glassware. The laser generator and external optics (including the focus lens) require cooling. Depending on system size and configuration, waste heat may be transferred by a coolant or directly to air. Water is a commonly used coolant, usually circulated through a chiller or heat transfer system.

Lasing Materials and their Applications:

Lasing Material	Applications
CO_2	Boring Cutting/Scribing Engraving
Nd	High-energy pulses Low repetition speed (1 kHz) Boring
Nd-YAG	Very high energy pulses Boring Engraving Trimming

Laser Microjet Cutting:

A **laser microjet** is a water-jet guided laser in which a pulsed laser beam is coupled into a low-pressure water jet. This is used to perform laser cutting functions while using the water jet to guide the laser beam, much like an optical fiber, through total internal reflection. The advantages of this are that the water also removes debris and cools the material. Additional advantages over traditional "dry" laser cutting are high dicing speeds, parallel kerf and omnidirectional cutting.

Generation of the laser beam involves stimulating a lasing material by electrical discharges or lamps within a closed container. As the lasing material is stimulated, the beam is reflected internally by means of a partial mirror, until it achieves sufficient energy to escape as a stream of monochromatic coherent light. Mirrors or fiber optics are typically used to direct the coherent light to a lens, which focuses the light at the work zone. The narrowest part of the focused beam is generally less than 0.01.25 in; (0.3175 mm) in diameter. Depending upon material thickness, kerf widths as small as 0.004 in (0.1016 mm) are possible. In order to be able to start cutting from somewhere else than the edge, a pierce is done before every cut. Piercing usually involves a high-power pulsed laser beam which slowly makes a hole in the material, taking around 5–15 seconds for ½-inch-thick (13 mm) stainless steel, for example.

The parallel rays of coherent light from the laser source often fall in the range between 1/16 inch to 1/2 inch (1.5875 mm to 12.7 mm) in diameter. This beam is normally focused and intensified by a lens or a mirror to a very small spot of about 0.001 inch (0.0254 mm) to create a very intense laser beam. In order to achieve the smoothest possible finish during contour cutting, the direction of beam polarization must be rotated as it goes around the periphery of a contoured workpiece. For sheet metal cutting, the focal length is usually between 1.5 inches and 3 inches (38.1 mm and 76.2 mm).

Different Methods of Cutting

There are many different methods in cutting using lasers, with different types used to cut different material. Some of the methods are vaporization, melt and blow, melt blow and burn, thermal stress cracking, scribing, cold cutting and burning stabilized laser cutting as explained below:

1. **Vaporization cutting:** In vaporization cutting the focused beam heats the surface of the material to boiling point and generates a keyhole. The keyhole leads to a sudden increase in absorptivity quickly deepening the hole. As the hole deepens and the material boils, vapor generated erodes the molten walls blowing eject out and further enlarging the hole. Non melting material such as wood, carbon and thermoset plastics are usually cut by this method.
2. **Melt and blow:** Melt and blow or fusion cutting uses high-pressure gas to blow molten material from the cutting area, greatly decreasing the power requirement. First the material is heated to melting point then a gas jet blows the molten material out of the kerf avoiding the need to raise the temperature of the material any further. Materials cut with this process are usually metals.
3. **Thermal stress cracking:** Brittle materials are particularly sensitive to thermal fracture, a feature exploited in thermal stress cracking. A beam is focused on the surface causing localized heating and thermal expansion. This results in a crack that can then be guided by moving the beam. The crack can be moved in order of m/s. It is usually used in cutting of glass.
4. **Stealth dicing of silicon wafers:** The separation of microelectronic chips as prepared in semiconductor device fabrication from silicon wafers may be performed by the so-called stealth dicing process, which operates with a pulsed Nd:YAG laser, the wavelength of which (1064 nm) is well adopted to the electronic band gap of silicon (1.11 eV or 1117 nm).
5. **Reactive cutting:** Also called "burning stabilized laser gas cutting" or "flame cutting". Reactive cutting is like oxygen torch cutting but with a laser beam as the ignition source. In metal cutting, the laser beam beam heats the metal to a high temperature where it burns as oxygen passes over it. It is therefore actually the oxygen that does the job of cutting. Mostly used for cutting carbon steel in thicknesses over 1 mm. This process can be used to cut very thick steel plates with relatively little laser power. It is widely used in fabrication of aircraft engine parts.
6. **Pulsing:** Brittle materials such as ceramics and glasses are cut using pulsed lasers in order to minimize micro cracking. Pulsed lasers which provide a high-power burst of energy for a short period are very effective in some laser cutting processes, particularly for piercing, or when very small holes or very low cutting speeds are required, since if a constant laser beam were used, the heat could reach the point of melting the whole piece being cut. Most industrial lasers have the ability, to pulse or cut CW (Continuous Wave) under NC (numerical control) program control. Double pulse lasers, use a series of pulse pairs to improve material removal rate and hole quality. Essentially, the first pulse removes material from the surface and the second prevents the ejecta from adhering to the side of the hole or cut.

Materials:

Organic Materials	Inorganic Materials	Metallic Materials
Acrylic	Glass	Steels
Polyethylene	Ceramics	Iron
Polypropylene	Quartz	Nickel
Polycarbonate	Alumina	Tin
Rubber	Asbestos	Lead
PVC	Mica	Aluminum
Wool	Rocks	Titanium
Cotton		Copper
Leather		Molybdenum
Wood		Brass
		Bronze
		Tungsten

Advantages and disadvantages:

The advantage of laser cutting is that it is fine and precise. Cutting edges are tight and parallels. Advantages of laser cutting over mechanical cutting include easier workholding and reduced contamination of workpiece (since there is no cutting edge which can become contaminated by the material or contaminate the material). Precision may be better, since the laser beam does not wear during the process. Reduced heat affected zone. There is almost no chance of distortion (or deformation) in the material that is being cut, as laser systems have a small heat-affected zone. Possibility to operate on complex profiles with reduced curving radius. Some materials are also very difficult or impossible to cut by more traditional means. No influence of the hardness of the material. No problems to cut materials previously coated.

Laser cutting for metals has the advantages over plasma cutting of being more precise and using less energy when cutting sheet metal, however, most industrial lasers cannot cut through the greater metal thickness that plasma can. Newer lasers machines operating at higher power (6000 watts, as contrasted with early laser cutting machines' 1500 watt ratings) are approaching plasma machines in their ability to cut through thick materials, but the capital cost of such machines is much higher than that of plasma cutting machines capable of cutting thick materials like steel plate.

The main disadvantage of laser cutting is the high power consumption. The power required for cutting depends on the material to be cut. Industrial laser efficiency may range from 5% to 15%. The power consumption and efficiency of any particular laser will vary depending on output power and operating parameters. This will depend on type of laser and how well the laser is matched to the work at hand. The amount of laser cutting power required, known as *heat input*, for a particular job depends on the material type, thickness, process (reactive/inert) used, and desired cutting rate. The amount of heat input (power) required for various materials at various thicknesses using a CO_2 laser (in watts) is given in the following table:

Material	Material thickness (in)				
	0.02	0.04	0.08	0.125	0.25
Stainless steel	1000	1000	1000	500	250
Aluminum	1000	1000	1000	3800	10000
Mild steel	–	400	–	500	–

Titanium	250	210	210	–	–
Plywood	–	–	–	–	650
Boron/epoxy	–	–	–	3000	–

5.2.3 Welding

Welding is the joining of two or more pieces into a single unit. Let us consider the welding of two metal plates. The metal plates are held in contact at their edges and a laser beam is made to move along the line of contact of the plates. The laser beam heats the edges of the two plates to their melting points and causes them to fuse together where they are in contact. The main advantage of laser welding is that it is a contact less process and hence there is no possibility of introduction of impurities into the joint. Unlike electron beam welding, it can be done in atmosphere. The workpieces do not get distorted, as the total amount of power input is very small compared to conventional welding processes. The heat affected zone is relatively small because of rapid cooling. Laser welding can be done even at difficult to reach places as illustrated in Fig. 5.7. The process is easily automated.

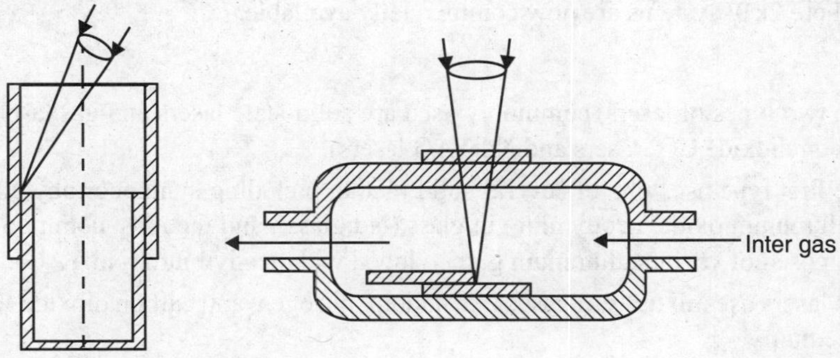

Fig.5.7 Inert gas laser welding for otherwise locations

Laser welding works on both large and small scales. On the large scale, the Indian Navy uses lasers to weld together huge metal parts in shipbuilding. Experts estimate that millions of Rupees are saved in the welding process and millions more in reduced need for later repairs. Such common items as automobile spark plugs, portable batteries, and metal braces for the teeth are also routinely welded by laser beams.

On a smaller scale, lasers weld the parts for tiny electrical circuits used in computers, calculators, and miniature television sets. In the past, welding these small parts was accomplished by soldering— melting a metallic substance called solder around them to ensure a proper electrical connection. But soldering tools cannot be made small enough to weld the very tiny electrical parts now being produced; and manipulating the smallest available soldering tools is very painstaking work, produces uneven results, and can damage the delicate parts. By contrast, such tiny welds, some of them even microscopic, are easily made by the hot, razor-thin beam of a carbon dioxide laser.

Like electron beam welding (EBW), laser beam welding has high power density (of the order of 1 megawatt/cm^2(MW)) resulting in small heat-affected zones and high heating and cooling rates. The spot size of the laser can vary between 0.2 mm and 13 mm, though only smaller sizes are used for welding. The depth of penetration is proportional to the amount of power supplied, but is also dependent on the location of the focal point: penetration is maximized when the focal point is slightly below the surface of the workpiece.

A continuous wave (CW) or pulsed laser, beam may be used depending upon the application. Milliseconds long pulses are used to weld thin materials such as razor blades while continuous laser systems are employed for deep welds.

Laser beam welding (LBW) is a versatile process, capable of welding carbon steels, HSLA steels, stainless steel, aluminum, and titanium. Due to high cooling rates, cracking is a concern when welding high-carbon steels. The weld quality is high, similar to that of electron beam welding. The speed of welding is proportional to the amount of power supplied but also depends on the type and thickness of the workpieces. The high power capability of gas lasers make them especially suitable for high volume applications. LBW is particularly dominant in the automotive industry. Here, the welding of the curved contours under the body of automobiles is carried out. The automobile is made to move while computer -controlled beam deflectors control the laser beam which perform the welding. There are two main types of industrial laser of interest to structural fabrication; CO_2 and Nd:YAG lasers. High power CO_2 lasers are predominantly used for the welding of automotive components, such as gears and transmission components, which require circular and annular welds and in tailored blank applications. The majority of lasers have a power of 6kW or less. Within the laser industry, one of the main advances in the past two years has been in diode lasers (wavelength 0.8-0.9μm), where 2kW systems are now commercially available.

Equipment :

(*i*) The two types of lasers, commonly used are solid-state lasers and gas lasers (especially carbon dioxide CO_2 lasers and Nd:YAG lasers).

(*ii*) The first type uses one of several solid media, including synthetic ruby and chromium in aluminum oxide, neodymium in glass (Nd:glass), and the most common type, crystal composed of yttrium aluminum garnet doped with neodymium (Nd:YAG).

(*iii*) Gas lasers use mixtures of gases like helium, nitrogen, and carbon dioxide (CO_2 laser) as a medium.

(*iv*) Regardless of type, however, when the medium is excited, it emits photons and forms the laser beam.

Solid state laser

Solid-state lasers operate at wavelengths on the order of 1 micrometer, much shorter than gas lasers, and as a result require that operators wear special eyewear or use special screens to prevent retina damage. Nd:YAG lasers can operate in both pulsed and continuous wave (CW) mode, but the other types are limited to pulsed mode. The original and still popular solid-state design is a single crystal shaped as a rod approximately 20mm in diameter and 200mm long, and the ends are ground flat. This rod is surrounded by a flash tube containing xenon or krypton. When flashed, a pulse of light lasting about two milliseconds is emitted by the laser. Disk shaped crystals are growing in popularity in the industry, and flash-lamps are giving way to diodes due to their high efficiency. Typical power output for ruby lasers is 10-20 W, while the Nd:YAG laser outputs between 0.04-6,000 W. To deliver the laser beam to the weld area, fiber optics are usually employed.

Gas laser

Gas lasers use high-voltage, low-current power sources to supply the energy needed to excite the gas mixture used as a lasing medium. These lasers can operate in both continuous and pulsed mode, and the wavelength of the laser beam is 10.6 μm. A CO_2 laser can also weld sheets or films of plastic materials. Fiber optic cable absorbs and is destroyed by this wavelength, so a rigid lens and mirror delivery system is used. Power outputs for gas lasers can be much higher than solid-state lasers, reaching 25 kW.

Fiber laser

In fiber lasers, the gain medium is the optical fiber itself. They are capable of power up to 50 kW and are increasingly being used for robotic industrial welding.

Advantages

The laser is hotter, faster, more accurate, and also safer because the welder does not have to go near the hot metal. Some of the advantages of LBW are as follows:

(*i*) The laser beam can be transmitted through air rather than requiring a vacuum,

(*ii*) The process is easily automated with robotic machinery,

(*iii*) X-rays are not generated, and

(*iv*) LBW results in higher quality welds.

5.3.5. Heat Treating

Heat treating is a process which consists of heating metals and certain other materials for sometime to harden them. Heat treating converts the surface layer to a crystalline state that is harder and more resistant to wear. In general CO_2 lasers of about 1 kW output power operating in CW mode are used for heat treating. As metals are more reflecting at 10.6 mm, an absorptive coating such as graphite or zinc phosphate is applied on the surface of the workpiece to help it absorb the laser energy more efficiently. Laser heat treating requires a low amount of energy input to the workpiece. Laser processing is advantageous as it can provide selective treatment of the desirable areas. Heat treatment is used to strengthen cylinder blocks, gears, camshafts etc in the automobile industry.

5.4 LASERS IN ELECTRONICS INDUSTRY

Lasers are widely used for material processing in electronics industry also. They are mainly employed in wafer scribing, in repairing photomasks, in making electrical connections for thick film hybrid circuits etc areas. Investigations are being carried out on deposition of conducting metal films by decomposing metal-containing gases just above the surfaces of semiconductors and insulators.

5.4.1. Scribing

Scribing of silicon and ceramic substrates is very important in the electronics industry. Lasers are extensively used for breaking silicon chips having hundreds of ICs on them. Scribing involves drawing fine lines in brittle ceramic and semiconductor wafers. When the ceramic or silicon wafer is bent, it breaks along the weakened line scribed by the laser. Computer controlled scribing with a laser is a non-contact process and does not cause any damage to the surface. Low power single mode CO_2 lasers are generally used in these operations. Because of high absorption of nonmetallic materials at 10.6 μm, CO_2 laser is highly suitable. In order to obtain a sufficiently small diameter of the beam after focussing, single mode operation is preferred.

5.4.2. Soldering

Lasers are used in welding of leads of materials such as platinum, silver, palladium which are usually difficult to solder. Nd : YAG laser is used for this process of fluxless soldering. The main advantages of laser soldering are as follows:

(*i*) It is a non-contact method and areas visible but difficult-to-reach can also be soldered.

(*ii*) Leads as thin as 0.001 inch. diameter can be soldered without causing damage to them.

(*iii*) The process is highly cost effective.

5.4.3. Trimming

Existing methods of manufacturing the thick and thin film resistors are not capable of producing required values of resistance. The resistance depends on the surface area of the film and as such the desired resistance value can be obtained by removing some surface material. Initially, a greater amount of film is deposited on the surface and the initial resistance is of low value. Then, the resistor is trimmed by removing part of the film material using a laser beam while the value of the resistor is continuously monitored. Use of lasers gives about 50 % cost saving in comparison to other techniques. Pulsed Nd : YAG lasers are used, as most metals have high absorption at 1.06 μm. Pulsed operation is preferred since it minimizes the thermal damage to the substrate.

5.4.4. Photolithography

Photolithography is a process used to form patterns on the surface of semiconductor wafers. In this process, the surface of a wafer is coated with a "photoresist" which is a light sensitive material. Selective exposure of certain areas of the resist to light through a circuit pattern leaves the pattern in the resist material. Unexposed resist material is developed and washed. After etching the semiconductor surface by chemical methods, the desired circuit pattern is obtained on the semiconductor surface.

With growing miniaturization, smaller and smaller lines are to be made on the semiconductor wafer. The minimum width of the lines depends on the wavelength of light that exposes the photoresist. The shorter the wavelength, the finer are the features. Pulsed excimer lasers are used nowadays in making circuits with finer features.

Silicon chips could be made more quickly and cheaply using a new technique developed by physicists in the US. Stephen Chou and colleagues at Princeton University have successfully imprinted patterns onto silicon using quartz moulds instead of the usual combination of lithography and etching. With a resolution of just 10 nm and an 'imprint time' of 250 ns, the new process could revolutionize the semiconductor industry.

But such 'photolithography' is expensive and complex, and the resolution of the technique is fast approaching the diffraction limit. This means that it will not be able to make features much smaller than the current minimum size of about 130 nm - and that the semiconductor industry could soon violate one of its guiding principles, known as Moore's Law. Coined in 1965, this law described how the density of components on a chip doubled every 18 months, and was soon adopted by the semiconductor industry as a target.

Now Moore's Law could be back on track. Chou and co workers say that their technique known as laser-assisted direct imprint - can create features as smalsilicon wafers. The new process also eliminates the need for the resist and washing steps.

The team inscribed the pattern into a quartz mould using 'impact lithography', which is much cheaper and easier than the photolithography of silicon. The mould was then placed on top of a silicon wafer. A pulse of light from a helium-neon laser'-with a wavelength of 633 nm - was then fired through the mould to melt the top layer of silicon. The researchers then pressed the mould into this liquid silicon and removed it after the silicon had solidified, leaving the pattern.

Since silicon reflects more light as a liquid han it does as a solid, the team could tell when the mould needed to be removed by measuring how much of the laser pulse was reflected. The researchers also say that the mould can be used several times.

5.5 LASERS IN NUCLEAR ENERGY

Isotope Separation :

The natural uranium ore used to fuel nuclear reactors contains mainly two principal isotopes, U-238 and U-235. U-238 is the more abundant isotope but it cannot sustain the fission chain reaction needed to drive the nuclear reactor. It is the U-235 isotope that sustains a fission reaction. Natural uranium contains only 0.7% of U-235 and to be useful for nuclear power generation or for production of atomic bombs, it is essential that about 3% U-235 is present. The isotope enrichment of uranium is a very important problem and in general this enrichment is performed using gascon diffusion, which is very expensive and time consuming. Isotopes are chemically almost identical. The differences in the nuclear mass shift the electronic energy levels slightly and therefore each isotope absorbs light at different characteristic wavelength. The absorption bands are fairly narrow and lie close to each other. If the mixture of isotopes is irradiated by a source of narrow bandwidth it is possible to excite one isotope without disturbing the other. The desired energy may be obtained by tuning a dye laser to a precise wavelength with a very narrow linewidth. The selectively excited U-235 atoms can be ionized by applying another short wavelength light to the mixture. The ionized U-235 atoms can be separated from the neutral U-238 atoms using electrostatic fields.

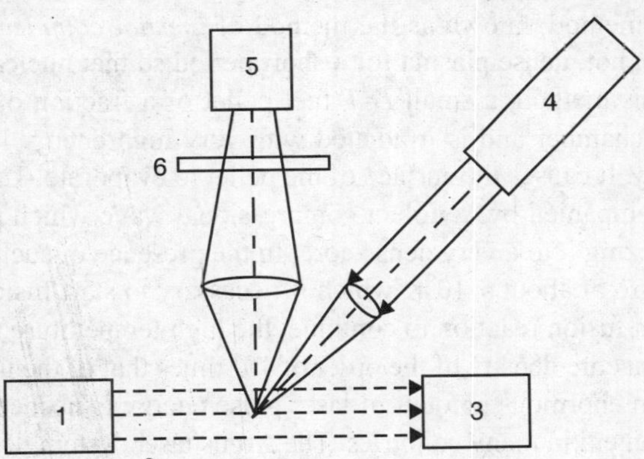

1 furnace, 2 uranium atoms beam, 3 mass spectrometer, 4 dye laser, 5 mercury lamp, 6 bandpass filter

Fig. 5.8 Schematic of a uranium isotope separation facility

Nuclear Fusion :

Fusion refers to releasing enormous amounts of energy through the fusion of nucleus of lighter elements to form nuclei of heavier elements. It is alternative source of energy and its chief attractions over nuclear fission are as follows :

(*i*) It offers a low cost and pollution free energy.

(*ii*) It produces less radioactive nuclear waste materials.

(*iii*) The light nuclei required for a fusion reaction are available in abundance on earth than the heavy elements needed for fission.

(*iv*) If the efforts succeed, fusion will be a practically inexhaustible source of energy.

The main disadvantage is that the right conditions for carrying out controlled fusion reactions are very hard to produce. Extremely high temperatures and pressures are required in order to make light nuclei to overcome their mutual repulsion and combine to release energy. It is concluded that fusion would be easier to achieve in a hot plasma containing heavy isotopes of hydrogen, namely deuterium D and tritium, *T*. The plasma must have a temperature of about 10^8K for the reaction to take place

$$D + T \rightarrow 4\,\text{He} + n + 17.6\ \text{MeV}.$$

where n stands for a neutron.

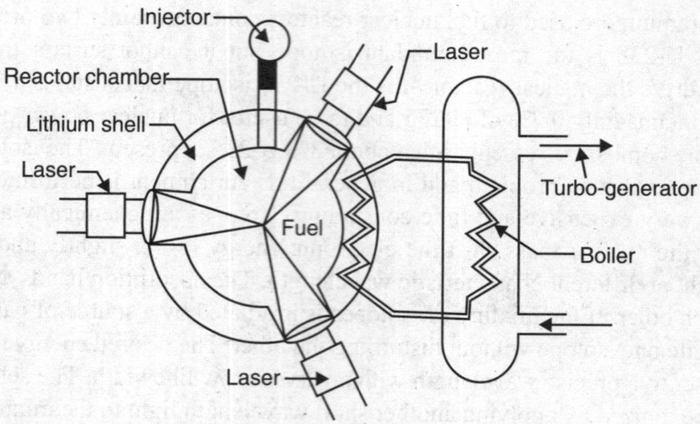

Fig. 5.9 Thermal energy conversion from nuclear fusion.

The laser assisted method, known as the method of *inertial confinement* seeks to compress the fuel pellet into a tiny, hot, dense plasma for a short period so that nuclear fusion can occur and energy is released. In this method, a small *D-T* fuel pellet of a fraction of a millimeter diameter is kept in an evacuated chamber and is irradiated with very high energy laser pulses from many directions simultaneously. It causes the surface of the pellet to evaporate. The rapid vaporization of the pellet surface is accompanied by a violent compressional wave which propagates towards the centre of the pellet squeezing it to a very dense core. In the presence of such high compression, the core reaches a temperature of about $\approx 10^8$K which is necessary to start fusion reaction. It has been estimated that for a chain fusion reaction to continue, the high temperature must be maintained for about 1 picosecond with a core density of the order of 10^4 times that of the uncompressed pellet. To fulfill these conditions an enormous amount of laser pulse energy is needed. Laser fusion projects have been under development in many countries. The Argus laser system, USA used Nd:YAG laser followed by Nd-glass disc amplifiers. The Shiva system used 20 amplifier chains which irradiated the pellet from 20 directions. The Delphin system in USSR used a Nd:glass system with 256 beams launched through 256 amplifiers. Inspite of the efforts, the quantities of fusion energy generated so far are very small.

Nuclear fission:

Many probes had been used to induce fission, particularly neutrons both fast and slow. Other nuclear probes had also been used e.g. protons, deuterons, particles and heavy ions as well as non nuclear beams e.g. γ-rays; electrons and muons. It is perhaps not surprising for a sufficiently intense light source to induce fission especially since a number of lasers had pulse powers of a terawatt (10^{12} W) or even a petawatt (10^{15} W). A terawatt is the total electrical power generated in the USA. This initiated number of groups around the world working on an exciting new area of physics named "laser-induced nuclear physics". Today laser systems are readily available with several Terawatts of power and some of the largest laser systems such as the VULCAN laser situated at the Rutherford Appleton Laboratory, near Oxford in the UK, will shortly be able to reach Petawatts. This laser is currently able to deliver over 100 Joules of energy in a picosecond (10^{-12} s) pulse. The VULCAN laser uses Nd:glass (optical glass doped with Neodymium) as its amplifying medium. This is an excellent medium for producing pulses of high energy and short duration, although at these enormous power densities, damage to the amplifying medium can occur. This problem is overcome by the use of an ingenious technique called Chirped Pulse Amplification (CPA).

Laser-induced fission of ^{238}U was observed at the VULCAN laser facility at Rutherford Appleton Laboratory (RAL) in 1999. Uranium nucleus has a rugby ball shape-the attractive short-range nuclear force balances the repulsive Coulomb force due to the 92 protons to give the nucleus its prolate deformation. Because uranium is such a large nucleus, it behaves like a drop of liquid. Supply of energy from the outside (from, e.g., an incident neutron or γ-ray) causes the nucleus to vibrate and rotate. The vibrations produce increased average distance between the nucleons resulting in a reduction of the binding effects of the short-range nuclear force, while the infinite ranged Coulomb force is little affected. At some stage in the vibrations, the attractive forces are no longer able to hold the nucleus together and it flies apart with two main fragments, a few evaporated neutrons and gamma rays. The fragments are more tightly bound than the original uranium nucleus and the excess energy is released as kinectic energy of motion.

Three 80 J pulses from the VULCAN laser, irradiated a 1.75mm Ta target backed with a 2 mm thick ^{238}U sample. The ^{238}U was shrink wrapped in plastic to contain any gaseous activity and enclosed in an aluminum container. The laser pulses produced a flux of high energy gamma rays which induced fission in uranium nuclei as described above. Unambiguous characteristic gamma rays from the fission fragments ^{134}I, ^{138}Cs and ^{92}Sr were observed.

From the measured intensities, it is concluded that about a million fission events are generated by a 10^{19} Wcm^{-2} laser shot in a 2 mm thick ^{238}U target.

5.6 LASER IN MEDICINE AND SURGERY

Lasers are playing nowadays a very important role both in diagnosis and surgery. Laser surgery is a surgery using a laser to cut ***tissue*** instead of a ***scalpel***. Examples include the use of a laser scalpel in otherwise conventional surgery, and soft tissue laser, surgery, in which the laser beam vaporizes soft tissue with high water content. Laser resurfacing is a technique in which molecular bonds of a material are dissolved by a laser. Lasers have become in association with optical fibres a powerful tool in diagnostics. Interiors of body which otherwise be difficult or impossible to see, now be seen very easily with the help of optical fibres. Fibre-optic sensors are, very useful both before and during surgery. The focused laser beam proved to be a new and unique 'scalpel' in the hands of surgeons. Laser surgery is a highly sterile process since contact does not occur between surgical tool and tissues being cut. Hence, laser surgery is much safer than normal surgery as it makes no physical contact so no infections are spread. Further advantage is that the laser does not cuts but also "welds" blood vessels being cut. Operations are done very fast and the patient does not feel pain and can be discharged very soon from the hospital and can resume his/her job in a normal way. Green laser surgery is used for the treatment/reduction of enlarged prostates.

Laser surgery has been extensively used in Cosmetic surgery particularly in removing tattoos, scars, stretch marks, sunspots, wrinkles, birthmarks, and hair removal. Laser types used in dermatology include ruby (694 nm), alexandrite (755 nm), pulsed diode array (810 nm), Nd:YAG (1064 nm), Ho:YAG (2090 nm), and ErYAG (2940 nm). Laser is also used in Eye surgery and refractive surgery; Soft tissue surgery: CO_2, ErYAG laser; Laser scalpel (General surgery, gynecological, urology, laparoscopic); Photobiomodulation (i.e. laser therapy); "No-Touch" removal of tumors, especially of the brain and spinal cord. In dentistry for caries removal, endodontic/periodontic procedures, tooth whitening, and oral surgery.

We discuss here a few, interesting applications of lasers in the medical field. Some of them are listed in table 5.2.

Table 5.2. Applications of Lasers in Medicine

Laser type	*Applications*
Argon	Eye surgery
	Brain surgery
	Removal of birthmarks
	Photocoagulation
	Angioplasty
	Photochemical treatments
ArF	Eye surgery
	Angioplasty
ArF excimer	Radial keratomy (eye surgery)
Carbon dioxide	Oviduct surgery
	Brain surgery
	Removal of precancerous lesions
	Spinal surgery
	Cancer removal
Gold vapor	Photochemical treatments (Hpd)
He-Ne	Doppler velocimeters
	Trophic ulcer treatment
	Holographs
Nitrogen	Genetic engineering
	DNA analysis
Nd : YAG	Brain surgery
	Spinal surgery
	Photocoagulation
	Cataract (eye) surgery
XeCl excimer	Angioplasty

5.7 EYE LASER SURGERY

The first big success of lasers in medicine was in the treatment of eye. Argon laser has been in use for several years to treat the detachment of the retina. The retina is the light sensitive layer at the back of the eye. Some times it becomes torn and gets detached from the back of the eye ball. The damage can spread and lead to total blindness. Conventional open-eye surgery was very risky. Laser treatment, on the other hand, is very simple and fast. Argon laser beam is focussed on the desired point of the retina. The green beam of the laser is strongly absorbed by the red blood cells of the retina and causes thermal effects which weld the retina back to the eyeball.

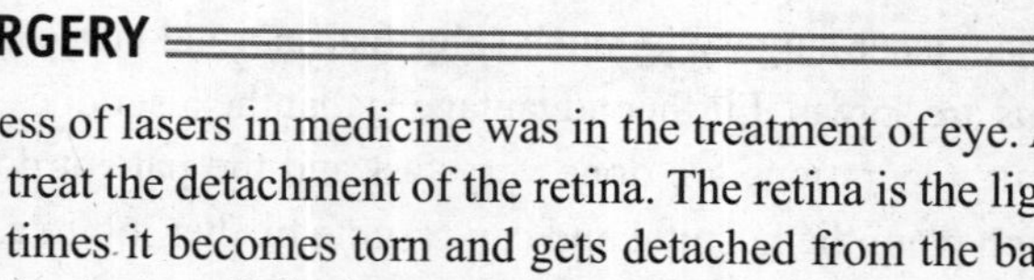

Fig. 5.10 Lasers are used to "weld" detached retina in eyes.

Cataract occurs when the natural lens of the eye becomes cloudy and obstructs vision, leading to blindness. It is treated by removing the natural lens and implanting a plastic lens there. In some of the cases, the natural membrane behind the lens becomes cloudy after cataract operation and again obstructs vision. Open-eye surgery is performed to remove the clouded membrane. Nd:YAG laser can be performed to remove the clouded membrane. Nd:YAG laser can be used to treat the complication. A series of laser pulses of high intensity focussed to a small spot (10-50 μm) on the back of lens produce high electric field sparks and ruptures the membrane. As the light spreads out behind the membrane, the laser pulses do not damage the retina.

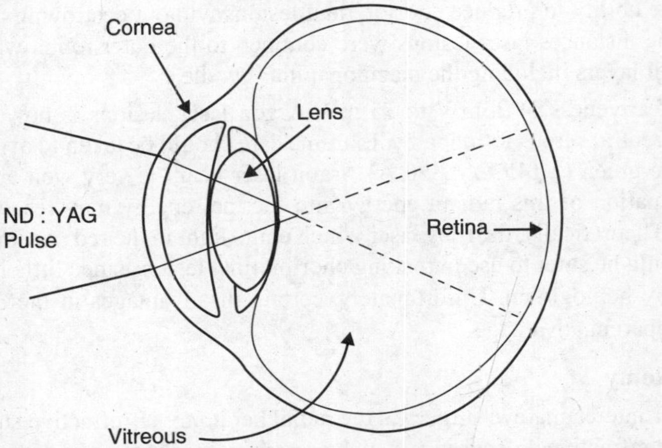

Fig. 5.11 Lasers are used to rupture the cloudy membrane behind the eye lens.

Nowdays, laser eye surgery has become very common. Techniques used include LASIK, which is used to correct near and far sightedness in version, and photorefractive keratectomy, a procedure which permanently reshapes the cornea using an excimer laser to remove a smalll amount of tissue.

5.8 PHOTOCOAGULATIONS

Laser photocoagulations of retinopathies is a common treatment. In people suffering from diabetes, for some unknown reasons, abnormal blood vessels (streamers) spread across the surface of the retina. These blood vessels are very fragile and leave blood into the clear liquid of the eye. This process causes gradual dimming of vision. If these blood streamers are not removed, the patient can become blind. Laser photocoagulation is used to destroy areas of new blood vessels and areas of hypoxic retina which is believed to further increase blood vessel proliferation.

Diabetic retinopathy is retinopathy (damage to the retina) caused by complications of diabetes mellitus, which can eventually lead to blindness. It is an ocular manifestation of systemic disease which affects up to 80% of all patients who have had diabetes for 10 years or more. Despite these intimidating statistics, research indicates that at least 90%, of these new, cases could be reduced if there was proper and vigilant treatment and monitoring of the eyes.

Signs and symptoms

Diabetic retinopathy often has no early warning signs. Even macular edema, which may cause vision loss more rapidly, may not have any warning signs for some time. In general, however, a person with macular edema is likely to have blurred vision, making it hard to do things like read or drive. In some cases, the vision will get better or worse during the day.

As new blood vessels form at the back of the eye as a part of Proliferative Diabetic Retinopathy (PDR), they can bleed (ocular hemorrhage) and blur vision. The first time this happens, it may not be very severe. In most cases, it will leave just a few specks of blood, or spots, floating in a person's visual field, though the spots often go away after a few hours.

Laser treatment

Argon laser photocoagulation has been chiefly used for the treatment of neovascular diseases such as diabetic retinopathy. Recently, photocoagulation therapy with the argon laser has also been advocated for the treatment of macular disease; such as central serous retinopathy, ocular histoplasmosis, and senile macular degeneration. The effectiveness of such treatnent is dependent, in part, on the ability to produce chorioretinal lesions without destroying the overlying nerve fiber layer. In some instances laser lesions were confined to the outer retina while in others alterations occurred in all layers including the maculopapillar bundle.

The effectiveness of 25mW to 75 mW Argon laser, lesions is now being evaluated for the treatment of central serous retinopathy. Extreme care should be taken to avoid retinal blood vessels, since the blue-green (5,145 to 4,880 Å) Argon laser beam is very well absorbed by hemoglobin. The transformation of this radiant energy into thermal energy may result in damage to adjacent nerve fibers. Theoretically, the ruby laser which emits light in the red portion of the visible spectrum (6,943 Å) would be safer to use in making chorioretinal lesions since little if any of this wavelength is absorbed by hemoglobin. Unfortunately, certain disadvantages in the design of the ruby laser negate this apparent advantage.

Radial Keratomy

Another interesting eye surgery is the radial keratomy or refractive surgery, which is done for correcting the refractive defects of eye such as near sightedness and astigmatism. The cornea is the outer transparent layer of the eye and is responsible for most of the eye's refractive power. Lasers are used to change the shape of the cornea by making small incisions on it. ArF excimer laser (λ =193 nm) with a pulse energy of 250 mJ and pulse length of 14 ns is typically used to alter shape of the cornea.

5.9 LIGHT INDUCED BIOLOGICAL HAZARDS : EYE AND SKIN

Light causes biological damage through both temperature effects due to absorbed energy and through photochemical reactions. The chief mode of damage depends on the wavelength of the light and on the tissue being exposed. For control of hazards from lasers, the damage is believed to be due principally to temperature effects, and the, critical organs are the **eye** and the **skin**. Let us first study in detail the detrimental effects (hazards) induced by laser beam (which includes wavelengths ranging from far-infrared, red, visible, near UV and far UV) on eye and then on skin. Before studying the detrimental effects on eye, let us get acquainted with the structure of eye.

5.9.1 Structure of the Eye

The structure of the eye is shown below. The optical components of the eye - those components that act together to focus an image of an object on the retina - are the cornea, aqueous humor, lens, and vitreous humor. The components of the eye most susceptible to laser damage are the cornea, retina, and lens. The active components of the eye are described in more detail below:

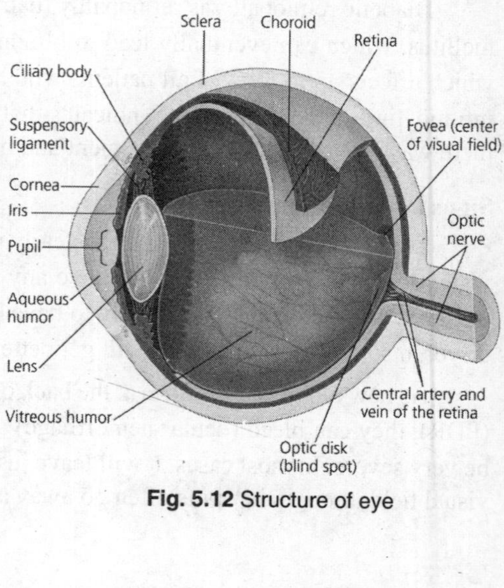

Fig. 5.12 Structure of eye

Cornea

The Living tissue exposed directly to the environmental elements. It is protected by a thin tear film. The corneal epithelium has one of the highest metabolic rates in the entire body. The tear layer of 6-10 μm thickness that protects this cell layer is fairly well balanced. The outer most surface of the tear layer is a superficial lipid mono-multilayer less than 0.5 μm in thickness, then beneath this are mucin layers with gradually increasing concentrations of mucin. The result is the cornea has a mean index of refraction of 1.376. This provides approximately 70% of the refractive power of the eye. The cornea has a high metabolic rate - rejuvenating itself in 24 to 48 hours.

Pupil - Iris - Sclera

Pupil forms the aperture of the eye. Normal range is of 2 - 7 mm. Range decreases with age. Iris adjusts the pupil of the eye. Iris is circular, pigmented membrane. It lies behind the cornea. Sclera is adense fibrous shell. It maintains the roughly spherical shape of the eye along with the internal pressure of the eye.

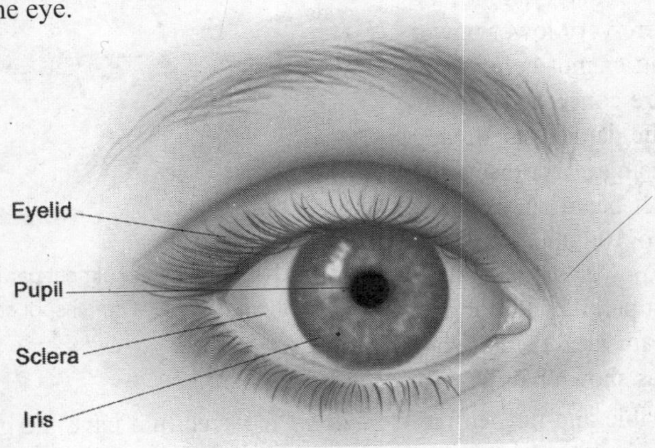

Fig. 5.13

Retina

The retina is an extension of the brain and consists of several complex layers of nerve cells. It is made up of rods and cones - rods for night and peripheral vision, cones for color and resolution. The macula is where the highest resolution takes place. The cones have a yellowish pigment to filter out blue light. Sharp vision is dependent on the formation of a real image on the macula. The fovea is in the center of the macula and is where the cones are concentrated.

Lens

The crystalline lens (containing α -crystallin protien) is supported in its location by fine ligaments which are connected to the ciliary muscles. The ciliary muscles control the eye's focusing ability. The lens is constructed of layers of cells, similar to the make-up of an onion. The lens provides fine tuning for the eye. It provides approximately 30% of the refractive power of the eye. The lens has a slow metabolism. Effects are delayed (cataracts). The lens hardens and becomes yellow with age.

The part of the eye that provides the most acute vision is the Fovea Centralis (also called the Macula Lutea). This is a relatively small area of the retina (3 to 4%) that provides the most detailed and acute vision as well as your colour perception. This is why you move your eyes when you read or when you look as something; the image has to be focused on the fovea for detailed perception. The balance of the retina can perceive light, and movement but not detailed images (peripheral vision).

5.9.1.1 Eye Damage : Wavelength Dependence

Laser irradiation of the eye may cause damage to the cornea, lens, or retina, depending on the wavelength of the light and the energy absorption characteristics of the ocular tissues. The major danger of laser light shows is hazards from beams entering the eye since this is the organ most sensitive to light. The simplest way to explain this is to say, "*just as a magnifying glass can be used to focus the sun and burn wood, the lens in the human eye focuses the laser beam into a tiny spot than can burn the retina*".

If a laser burn occurs on the Fovea, you can loose most of you fine (reading and working) vision in an instant. If a laser burn occurs in the peripheral vision it may produce little or no effect on fine vision. Repeated retinal burns can lead to blindness. Fortunately the eye has a self defense mechanism, the blink or aversion response. When a bright light hits the eye it tends to blink or turn away from the light source (aversion). This may defend the eye from damage where very lower power lasers are involved but cannot help where higher power lasers are concerned. By the time the eye reacts, the damage is already done. Due to the focusing effect discussed above, a one watt laser beam entering the eye can be focused to 100,000 watts per square centimeter of power at the retina. Laser beam are almost parallel and the eye lens will focus the beam into a small spot causing retinal burns as shown below:

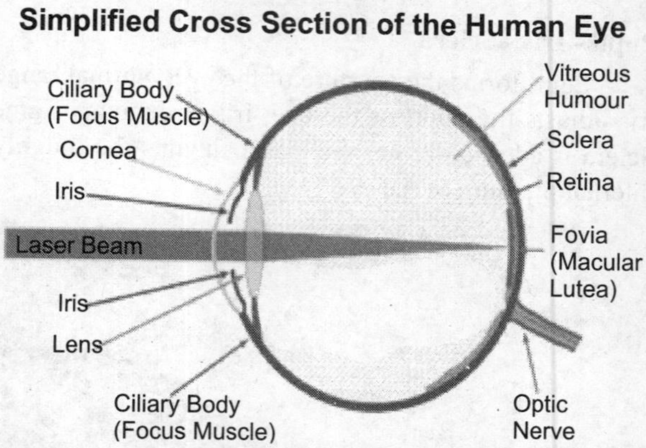

Fig. 5.14 Laser beams are almost parallel thus the eye's lens will focus them down to a small spot causing retinal burns.

Lasers in the visible and near infrared range of the spectrum have the greatest potential for retinal injury, as the cornea and lens are transparent to these wavelengths and the lens can thus focus the laser energy onto the retina. The maximum transmission by the cornea and lens, and the maximum absorption of laser energy in the retina occurs in the range from 550 nm to 400 nm as shown below. ***Argon and YAG lasers operate in this range clearly making them the most hazardous lasers.*** Wavelengths of less than 550 nm can cause a photochemical injury similar to sunburn. Photochemical effects are cumulative and result from long exposures (over 10 seconds) to diffuse or scattered light.

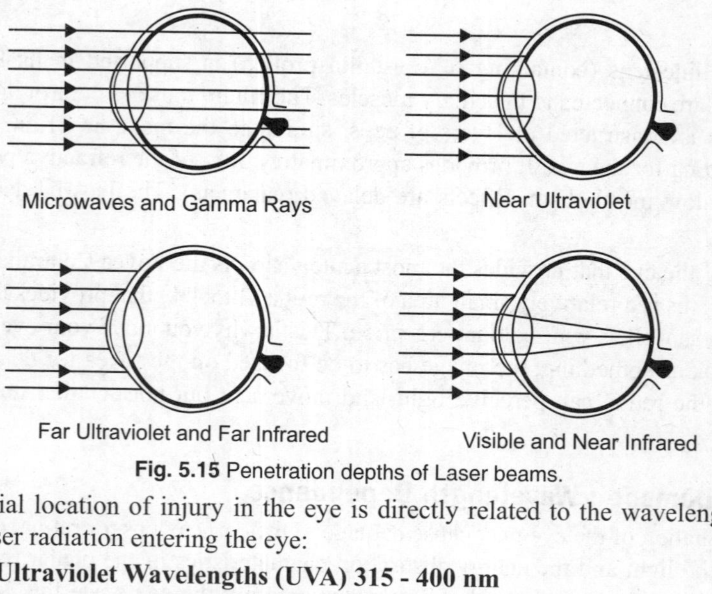

Fig. 5.15 Penetration depths of Laser beams.

The potential location of injury in the eye is directly related to the wavelength of the laser radiation. For laser radiation entering the eye:

1. **Near Ultraviolet Wavelengths (UVA) 315 - 400 nm**
   - Most of the radiation is absorbed in the lens of the eye.
   - The effects are delayed and do not occur for many years (e.g.; cataracts).

2. **Far Ultraviolet (UVB) 280 - 315 nm and (UVC) 100 - 280 nm**
   - Most of the radiation is absorbed in the cornea.
   - Keratocojunctivitis (snow blindness/welder's flash) will result if sufficiently high doses are absorbed.
3. **Visible (400 -760 nm) and Near Infrared (760 -1400 nm)**
   - Most of the radiation is transmitted to the retina*.
   - Overexposure may cause flash blindness or retinal burns and lesions.
4. **Far Infrared (1400 nm - 1 mm)**
   - Most of the radiation is transmitted to the cornea.
   - Overexposure to these wavelengths will cause corneal burns.

Laser damage and retinal burns can occur when a laser beam enters the human eye. Lasers are a 'point source' of light much like the sun thus the eye focuses on infinity when viewing lasers especially in a darkened space where no other light sources are present. Laser beams are almost parallel thus the lens of the human eye will focus them down to a small spot. A laser beam with low divergence entering the eye can be focused down to a spot 10 to 20 microns in diameter.

Eye damage can also occur when laser beams are scanned across the eye even for very brief periods. The amount of exposure is difficult to estimate as 'dwell' or 'transit' time must be taken into account in your calculations. For example a 1 mW laser illuminating a 7 mm aperture (the average size of the dark adapted iris) for one second is the equivalent of a 10 mW laser illuminating the same 7 mm aperture for 1/10 of a second.

The laws of thermodynamics do not limit the power of lasers. The second law states that the temperature of a surface heated by a beam from a thermal source of radiation cannot exceed the temperature of the source beam. The laser is a non-thermal source and is able to generate temperatures far greater than it's own. A 30 mW laser operating at room temperature is thus capable of producing enough energy (when focused) to instantly burn through paper!

Due to the law of the conservation of energy, the energy density (measure of energy per unit of area) of the laser beam increases as the spot size decreases. This means that the energy of a laser beam can be intensified up to 100,000 times by the focusing action of the eye. A one watt laser beam when focused down to a small spot can produce temperatures higher than the surface temperature of the sun! Thus even a low power laser in the milliwatt range can cause a burn if focused directly onto the retina.

5.9.1.2 Ocular Damage Mechansim

1. Thermal burns (lesions) in the eye are caused when the choroid layer blood flow cannot regulate the heat loading of the retina. Secondary bleeding into the vitreous humor may occur as a result of burns which damage blood vessels. This bleeding can obscure vision well beyond the area of the lesion.
2. Although the retina can repair minor damage, major injury to the macular region of the retina may result in temporary or permanent loss of visual acuity or blindness. Photochemical injury to the cornea by ultraviolet exposure may result in photokeratoconjunctivitis (often called welders flash or snow blindness). This painful condition may last for several days and is very debilitating. Long term UV exposure can cause cataract formation in the lens.
3. The duration of exposure also plays a role in eye injury. For example, if the laser is a visible wavelength (400 to 700 nm), the beam power is less than 1.0 mW and the

Laser retinal injury can be severe because of the focal magnification (optical gain) of the eye which is approximately 100,000 times. This means that an irradiance of 1 mW/cm^2 entering the eye will be effectively increased to 100 W/cm^2 when it reaches the retina.

exposure time is less than 0.25 second (the human aversion response time), no injury to the retina would be expected to result from an intrabeam exposure. Class 1, 2a and 2 lasers fall into this category and do not normally present a retinal hazard. Unfortunately, intrabeam or specular reflection viewing of Class 3a, 3b, or 4 lasers.

Not all lasers are used for eye, surgery. It should be noted here:that laser emitting radiation in ultra-violet (UV) region (200 nm - 400 nm) are detrimental for human eye. The biological molecule α-crystallin is the major protein of the mammalian eye lens. Studies show that the mutagenesis of α-crystallin (gradual loss of protein) may lead to cataract formation in human eye due to exposure to UV radiation, especially from the Sun and in the most severe cases, result in blindness.

Two predominant effects of radiation can occur in terms of photon energy. With the shorter wavelengths in the blue to UV range, single photons may have sufficient energy to induce photochemical changes in target molecules causing *photochemical damage*. In contrast, photochemical reactions induced by lower energy red to infrared photons depend on many photons hitting the molecules with multiphoton absorption causing vibrational modes and *damage by heat*. Therefore, selection of laser of a particular wavelength is of prime importance.

The following table 5.3 summarizes the various pathological effects, particularly that of eye caused by lasers at different wavelengths.

Table 5.3 : Pathological effects with wavelength

Wavelength range	Pathological effects
180 - 315 nm (UVB, UVC)	Photokeratitis (inflation of cornea)
315 - 400 nm (UVA)	Photochemical cataract (clouling of eye lens)
400 - 750 nm (visible)	Photochemical damage of retina, ratinal burn
750 - 1400 nm (near IR)	Cataract, ratinal burn (protein in aqueous humour)
1.4 - 3.0 μm (IR)	Corneal burn
3.0 μm - 1 mm	Corneal burn

International Safety Standard

The international laser safety standard, IEC-825, defines *the Maximum Permissible Exposure (MPE) that, the eye can, receive a short exposure of 2.5 mW per square centimeter without causing any eye injury.* Each jurisdiction has it's own maximum exposure levels for laser radiation. You should consult with your local regulatory authorities and, get their official methodology and formula(s) for calculating the MPE. Symptoms of a laser burn in the eye include a headache shortly after exposure, excessive watering of the eyes, and sudden appearance of many 'floaters' in your vision. Floaters are those swirling distortions that occur randomly in normal vision most often after a blink or when you have had your eyes closed for a couple of seconds. Floaters are caused by dead cell tissues that detach from the retina and choroid and float in the Vitreous Humour. Ophthalmologists often dismiss minor laser injuries as floaters due to the very difficult task of detecting minor retinal injuries.

Precautionary Measures

Never point a laser, even a laser pointer, at someone's eyes no matter how low the power of the laser. If you can prevent the laser beam and the beam reflections from entering the eye, you can prevent a painful and possibly binding injury.

Protect your eye by wearing goggles (sun glasses) which absorb these radiations as a complete shield or covering.

5.9.2. Human Skin and Damages

The skin is usually much less sensitive to laser than eye, but excessive exposure to ultraviolet (UV) light from any source causes short or long term effects like erythema (sunburn). The human skin consists of two major layers namely **epidermis** and **dermis.** The richly cellular, relatively thin outer layer is called the epidermis and the thicker and tougher inner part which is largely composed of fibrous connective tissue, is known as the dermis or cutis or corneum.

Biological molecules which absorb light radiation are called chromophores. :When irradiated at appropriate wavelengths, the chromophore is transformed into a more energetic electronic state, either the unstable singlet or the more persistent triplet state. Although the duration of the singlet state is infinitesimal, the triplet state may persist for up to a few seconds, allowing time for photochemical reactions and subsequent thermal chemistry with mediator production and observable biological responses. The radiation absorption spectrum of a substance is the plot of its relative absorption by the substance at different wavelengths, while the radiation action spectrum for a photobological response is the likelihood of induction of the response at different wavelengths, ideally corresponding to the absorption spectrum of the absorbing molecule. However, the action and absorption spectra related to the suspected skin chromophore do not usually correspond exactly since scattering and absorption within skin modify the radiation before it reaches the chromophore.

Skin Hazards

Exposure of the skin to high power laser beams (1 or more, watts) can cause burns. At the under five watt level, the heat from the laser beam will, causes a flinch reaction before any serious damage occurs. The sensation is similar to touching any hot object you tend to pull your hand away or drop it before any major damage occurs. With higher power lasers, a burn can occur even though the flinch reaction may rapidly pull the affected skin out of the beam. These burns can be quite painful as the affected skin can be cooked and forms a hard lesion which takes ages to heal.Ultraviolet light has multiple effects on living tissue. Continuous bombardment of UV radiation from the sun, due to unchecked depletion of ozone in the stratosphere is hampering the immunity and tampering the DNA causing mutation of human race on the earth. Potential molecular targets of ultraviolet light include DNA, RNA, proteins and lipids. UV irradiation not only alters DNA, it causes cytoskeletal reorganization in keratinocytes. Although UVA induced erythema in normal skin requires 1000 times more energy than from UVB, daily exposure to UVA is much greater than to UVB, and at, the level of the dermal capillaries, the UVA effect due to greater penetrance is much stronger that UVB. Thus, UV light of varying wavelengths can provoke abnormal skin responses and development of pre-malignant and malignant skin lesions non-melanoma (basal cell and squamous cell) skin cancer (NMSC) and malignant melanoma (MM). Another detrimental effect on skin is erythema (severe sunburn) and photoageing.

5.9.2.1 Application of 308-nm Excimer in psoriasis and vitiligo

The excimer laser proved the next logical step for the treatment of localized resistant psoriasis and vitiligo, the severe skin diseases. The 308-nm excimer laser treatment was well tolerated by all patients suffering from above skin conditions. The machine was easy to operate and quick. Repigmentation was from the periphery of most lesions on the hands and feet, and also from around the hair follicles in subjects with pigmented hair on the back of the hands.

The 308 -nm excimer laser is manufactured by Photomedex, Carlsbad, CA. It is a excimer laser based on a self-contained gas system of Xe Cl. The output is initiated by a foot switch and consists of a train of short pulses with a pulse -width of 30 nanoseconds delivered through a fiberoptic hand piece and is operated at 3 mJ per pulse with pulse repetition of up to 200 Hz. The beam diameter is

0.7 in square. The laser allows fixed fluences to be delivered starting from 100 mJ/cm^2 and with 50 mJ/cm^2 increments to a maximum dose of 2100 mJ/cm^2. Additional fluences can be delivered by pressing the foot switch a second time.

Photobiologically, the wavelengths of NBUVB at, 311 nm and the, excimer laser at 308 nm are very close to each other, and presumably the therapeutic effects may well be similar. The mechanisms for improvement of vitiligo with UV light remain speculative. Narrow-band UVB was first used for psoriasis and subsequently was successfully used for polymorphic light eruption (PLE), atopic dermatitis, and vitiligo.

Precautionary Measures

Protect your skin as minimum laser exposure as possible.

Protect your skin, by sun screens which absorb, these radiations as a complete shield or covering.

5.9.2.2 Skin- Conditioning using Erb-YAG Laser

The Erb-YAG Laser is the first clinically proven laser for skin resurfacing and management of wrinkles and scars. It's a gentle treatment that ensures minimal discomfort and rapid recovery. It works by removing skin layer by layer. Once the damaged skin is removed it is replaced by the new skin which grows quickly. Conditions treatable with this laser are acne, chicken pox and, surgical scars, birth marks, wrinkles around the mouth and eyes. It can also be used for Vitiligo, skin grafting & hair transplantation.

5.10 HOMEOSTASIS

Homeostasis is a medical term which refers to the natural tendency of attaining balance in a body of mammals by activating the auto/self-regulatory system. It's a gift of nature to us.

In dermatology, laser is used to cause homeostasis, the cessation of bleeding and removal of warts, freckles, acne and various other growths both malignant and benign. One typical example is bleaching birthmarks. Dark-red birthmarks called "portwine stains" appear on the face or neck due to abnormal blood vessel networks under the surface of the skin. Such marks are treated by illuminating with the blue-green light from an argon laser. The radiation is absorbed by the blood and heats it up. The area thus receives burns and in the process closes the blood vessels. Gradually the burns heal and the birthmarks are bleached.

An advantage of homeostatic regulation is that it allows an organism to function effectively in a broad range of environmental conditions. Most homeostatic regulation is controlled by the release of hormones into the bloodstream. However, other regulatory processes rely on simple diffusion to maintain a balance. Homeostasis includes the control of body temperature, regulation of the pH of the Blood at 7.365 (a measure of alkalinity and acid). All animals also regulate their blood glucose, as well as the concentration of their blood. Mammals regulate their blood glucose with insulin and glucagon. The human body maintains glucose levels constant most of the day, even after a 24-hour fast. Even during long periods of fasting, glucose levels are reduced only very slightly. Insulin, secreted by the beta cells of the pancreas, effectively transports glucose to the body's cells by instructing those cells to keep more of the glucose for their own use. The kidneys are used to remove excess water and ions from the blood. These are then expelled as urine. The kidneys perform a vital role in homeostatic regulation in mammals, removing excess water, salt, and urea from the blood. These are the body's main waste products. Many diseases are a result of disturbance of homeostasis, a condition known as homeostatic imbalance. As it ages, every organism will lose efficiency in its control systems. The inefficiencies gradually result in an unstable internal environment that increases the risk for illness. In addition, homeostatic imbalance is also responsible for the physical changes associated with aging. Even more serious than illness and other characteristics of aging is death. Heart failure has been seen where nominal negative feedback mechanisms become overwhelmed, and destructive positive feedback mechanisms then take over. Diseases that result from a homeostatic imbalance include diabetes, dehydration, hypoglycemia, hyperglycemia, gout, and any disease caused by a toxin present in the bloodstream.

5.11 LASER APPLICATIONS IN DENTISTRY

Dentistry has entered an exciting era of high technology. The laser technique has become popular in most branches of dentistry because of its ability to vaporize tissue, coagulate and seal small blood vessels. Moreover, the treated wound surface would be sterilized. Since the golden years of the 1960s, Lasers were developed and approved for soft tissue procedures. In 1964 Goldman, Ruben and Sherman reported an issue on laser for spectrographic analysis of the inorganic components or supragingival and subgingival calculus. Researchers after that continued to prove the uses of laser in dentistry. There are many different types of lasers such as Gas laser, Chemical laser, Dye laser, Metal-vapor laser, Solid-state laser and Semiconductor laser used in dentistry and their applications vary. Each laser type has its own unique features, advantages and disadvantages. The most common use of lasers is for removing dental caries, old composite fillings, sterilizing infected root canals, bone shaping, aphthous ulcer treatment, crown lengthening, apicectomy, periodontal treatment and oral surgery. In soft tissue procedures (gignivectomy, frenectomy, curettage, crown lengthening, lesion removal) CO_2 laser, Nd: YAG laser, Ar (Argon) laser, Er: YAG (Erbium doped Yattrium Aluminum Garnet) laser and diode laser are used. CO_2 laser was a valuable aid in direct pulp capping in human patients. While in hard tissue procedures (cutting of enamel) Er: YAG laser is preferable. In composite curing: Argon laser is used.

In 1989, the Er: YAG laser was first described as being able of removing dental hard tissues without causing thermal undesirable effects, such as cracking or charring to the remaining dental tissues.

Table 5.4: Laser applications in density

Laser type	Wavelength (nm)	Mode of operation	Power (W)	Applications
Argon	488,514	CW	1-20	• Curing composite resin • Excition and photocoagulation of the vascular lesion-carries detection
Diode	650,980	CW Pulsed	1-10	• Biostimulation • Excision of soft tissue pathology. • Gingivectomy
Nd:YAG	1064	CW Pulsed	50-100	Coagulation • Cutting (frenectomy, gingivectomy) • Subgingival curettage
Ho:YAG	2100	Pulsed	25	• Arthroscopic surgery for TMJ • Gingivectomy, frenectomy, implant exposure
Er:YAG	2940	Pulsed	3	Cravity preparation, Periodontal treatment, excision and incision of soft tissue (for dental use)
CO_2	9600, 10600	CW	10-100	For excision of benign and malignant lesion, cutting, drilling, vaporization of tissue and caries prevention

5.12 LASER ANGIOPLASTY

An important area of application of lasers is the laser angioplasty. Angioplasty (Greek word) is the technique of mechanically widening a narrowed or obstructed blood vessel, typically as a result of atherosclerosis*. The principal method of dealing with coronary artery blockage is **bypass surgery**, in which clogged blood vessels are replaced by high cost procedure with some risk. A non-surgical alternative is available to some patients is balloon angioplasty. An empty and collapsed balloon on a guide wire, known as a balloon catheter, is passed into the narrowed locations and then inflated to a fixed size using water pressures some 75 to 500 times normal blood pressure (6 to 20 atmospheres). The balloon crushes the fatty deposits, opening up the blood vessel for improved flow, and the balloon is then deflated and withdrawn. For several years, medical researchers have been exploring another alternative that would help a wider circle of patients than the balloon treatment and entail less risk than bypass surgery: **laser angioplasty**, in which a laser inserted in a catheter vaporizes the plaque and thus opens the artery. Several types of lasers have been employed in tests on human cadaver and animal arteries. In laser angioplasty, an Argon laser delivers energy to the blockage through a fibre-optic catheter and a disposal balloon. The inflated balloon and laser pulses assist each other in cleaning a path through the blockage. This technique opens even totally blocked arteries. Angioplasties are safer than bypass surgery and according to .statistics less than 1% of people die from complications after this procedure.

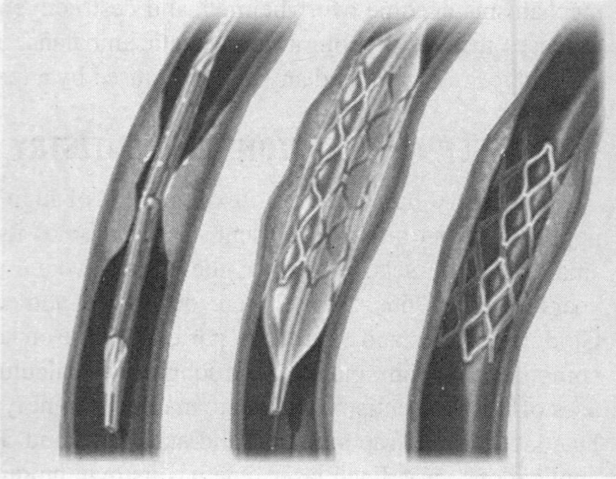

Fig. 5.16 Balloon catheter in Angioplasty

The problem has been that the lasers are too hot. Human tissue can withstand heat up to 154 degrees Fahrenheit; most laser heat tissue to much higher temperatures, which can cause damage to arterial walls. Now, however, a research group is on the verge of an exciting development: laser angioplasty with a "cool" type of laser called an excimer laser that does not damage blood vessel walls and offers non-surgical cleansing of clogged arteries with extraordinary precision. The system is the Dymer 200$^+$ Excimer Laser Angioplasty System, developed and produced by Advanced Interventional Systems, Irvine, California. The Dymer 200+ is based on NASA-patented technology developed at Jet Propulsion Laboratory (JPL) for satellite-based atmospheric studies.

Lasers are used in destroying kidney stones and gallstones. An optical fibre is threaded through until it faces the stone directly. Laser pulses launched through the fibre shatters the stone into small pieces that can pass through the uterer without pain. Tunable dye lasers are used in such "laser lithotriptor".

A leading cause of sterility in women is the blockage of one or both oviducts. Conventional surgery is risky in view of the delicate nature of the organs. However, treatment using laser is much simpler and more successful. The technique is similar to that of laser angioplasty.

---

* Atherosclerosis is the buildup of fatty deposits called plaque in human arteries. In time the plaque partially or totally obstructs the flow of blood through the clogged artery and threatens life. For example, extreme blockage of a coronary artery, which carries blood to and from the heart, can result in a heart attack and death. Blockages in the arteries may be caused by hypertension, diabetes, sedentary lifestyle, smoking, high cholesterol levels, diets high in saturated fats, and cardiovascular disease.

At times tumors develop in brain and spinal cord. They cannot be operated in conventional way because of their delicate nature. Laser surgery in such areas is possible and much safer.

Laser welding of human blood vessels is more advantageous than conventional suturing. Argon laser with 1 mm spot size is used for this purpose.

Argon pumped dye lasers are used to destroy viruses found in donor blood.

In dentistry, Q-switched ruby, CO_2 and Nd lasers are found to arrest decay in teeth by reducing the rate of dimineralisation of tooth enamel. The lasers are also used for drilling tooth cavities.

Laser canes and spectacles assist the blind. The laser cane is equipped with three diode lasers. One aims its beam downward, one upward and one straight ahead. Distance is sensed by vibration of the handle. The obstacle level is indicated by the pitch of vibration. In case of spectacles, the temple pieces house IR LEDs.

It is possible with lasers to selectively destroy cancer. In the treatment, a dye called hemotopophyrin derivative (HpD) is injected into the patient's body. The healthy cells flush out the dye while the dye is concentrated in cancer cells. The suspect areas are illuminated with red light (628-632 nm). Cells having HpD absorb light strongly. As a result, the dye reacts chemically with oxygen in cancer cells and produces a highly reactive form of oxygen that kills the cancer cells. However this treatment is effective for small tumors since light penetrates living tissue upto about 0.5 cm only.

5.13 DIFFERENT LASER THERAPIES

Large number of different laser therapies are popular for different conditions. Laser therapy is often given through a flexible endoscope (a thin, lighted tube used to look at tissues inside the body). The endoscope is fitted with optical fibers (thin fibers that transmit light). It is inserted through an opening in the body, such as the mouth, nose, anus, or vagina. Laser light is then precisely aimed to cut or destroy a tumor.

Laser-induced interstitial thermotherapy (LITT) (or interstitial laser photocoagulation) also uses lasers to treat some cancers. LITT is similar to a cancer treatment called hyperthermia, which uses heat to shrink tumors by damaging or killing cancer cells. During LITT, an optical fiber is inserted into a tumor. Laser light at the tip of the fiber raises the temperature of the tumor cells and damages or destroys them. LITT is sometimes used to shrink tumors in the liver.

Photodynamic therapy (PDT) is another type of cancer treatment that uses lasers. In PDT, a certain drug, called a photosensitizer or photosensitizing agent, is injected into a patient and absorbed by cells all over the patient's body. After a couple of days, the agent is found mostly in cancer cells. Laser light is then used to activate the agent and destroy cancer cells. Because the photosensitizer makes the skin and eyes sensitive to light for approximately 6 weeks, patients are advised to avoid direct sunlight and bright indoor light during that time.

Lasers Therapy

Laser therapy uses high-intensity light to treat cancer and other illnesses. Lasers can be used to shrink or destroy tumors. Lasers are most commonly used to treat superficial cancers (cancers on the surface of the body or the lining of internal organs) such as basal cell skin cancer and the very early stages of some cancers, such as cervical, penile, vaginal, vulvar, and non-small cell lung cancer. Lasers also may be used to relieve certain symptoms of cancer, such as bleeding or obstruction. For example, lasers can be used to shrink or destroy a tumor that is blocking a patient's trachea (windpipe) or esophagus. Lasers also can be used to remove colon polyps or tumors that are blocking the colon or stomach. Laser therapy can be used alone, but most often it is combined with other treatments, such as surgery, chemotherapy, or radiation therapy. In addition, lasers can seal nerve endings to reduce pain after surgery and seal lymph vessels to reduce swelling and limit the spread of tumor cells. In clinical trials (research studies), doctors are using lasers to treat cancers of the brain and prostate, among others.

Three types of lasers are used to treat cancer: carbon dioxide (CO_2) lasers, argon lasers, and Neodymium:Yttrium-Aluminum-gamet (Nd:YAG) lasers. Each of these can shrink or destroy tumors and can be used with endoscopes. CO_2 and argon lasers can cut the skin's surface without going into deeper layers. Thus, they can be used to remove superficial cancers, such as skin cancer. In contrast, the Nd:YAG laser is more commonly applied through an endoscope to treat internal organs, such as the uterus, esophagus, and colon. Nd:YAG laser light can also travel through optical fibers into specific areas of the body during LITT. Argon lasers are often used to activate the drugs used in PDT.

Advantages of laser therapy

Lasers are more precise than standard surgical tools (scalpels), so they do less damage to normal tissues. As a result, patients usually have less pain, bleeding, swelling, and scarring. With laser therapy, operations are usually shorter. In fact, laser therapy can often be done on an outpatient basis. It takes less time for patients to heal after laser surgery, and they are less likely to get infections. Patients should consult with their health care provider about whether laser therapy is appropriate for them.

Disadvantages of laser therapy

Laser therapy also has several limitations. Surgeons must have specialized training before they can do laser therapy, and strict safety precautions must be followed. Also, laser therapy is expensive and requires bulky equipment. In addition, the effects of laser therapy may not last long, so doctors may have to repeat the treatment for a patient to get the full benefit.

5.14 LASER ENDOSCOPY

Now a days, endoscopy has been become an essential tool for a doctor to diagnosis a patient. For this, the instruiment mainly consists of optic fibra catheter illuminated by laser beam and a computer monitoring system. With this doctor can view each and every part in 3-D of the internal organs of the body like intestine etc.

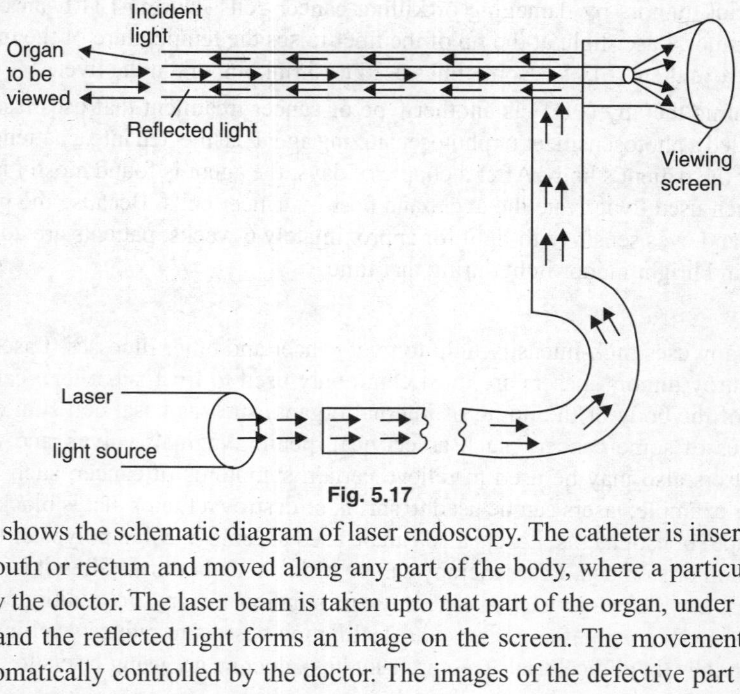

Fig. 5.17

Fig. 5.17 shows the schematic diagram of laser endoscopy. The catheter is inserted in the body through the mouth or rectum and moved along any part of the body, where a particular defects has been studies by the doctor. The laser beam is taken upto that part of the organ, under study, the light gets reflected and the reflected light forms an image on the screen. The movement of endoscope catheter is automatically controlled by the doctor. The images of the defective part can be viewed and stored on video CD in the form of a moving film.

5.15 LASER IN DEFENCE

Military uses of lasers include applications such as *target designation and ranging, defensive countermeasures, communications and directed energy weapons*. This involves mainly ranging, guiding weapons to the intended target and laser beam itself acting as a weapon. Directed energy weapons are also in use, such as Boeing's Airborne Laser which was constructed inside a Boeing 747. It disrupts the trajectory of shoulder-fired missiles.

Lasers are used in measurement of distances in the same way as microwaves do in radars. The optical radars, using laser beams, are employed in detecting distant objects and collecting information about them. This is known as *range finding*. In the optical range finding system, a laser transmits a signal which is typically a pulse or a sequence of pulses. A portion of the signal reflected from the object returns and is received by the receiver. By measuring the round-trip delay time, the range of the object is ascertained. If the object is moving, the Doppler shift in signal frequency is determined and from that the speed of the moving object is deduced. Q-switched ruby, Nd: YAG or Nd: glass lasers with peak powers in the range 1 to 10 MW are used.

Nowadays, low flying aircrafts are used in ground attacks. They are fitted with laser instrumentation for measuring the range of the target, and guiding the bomb onto it thereby reducing the human element to the minimum. The principle of laser guidance is as follows: The target is selected by a viewing telescope. The bomb is released from the aircraft at a certain height from where it falls freely under gravity. It carries a direction sensing seeker head and a servo loop. The head houses a laser which illuminates the target and the radiation scattered from the target is received by the head, whereby it determines the angle of line of slight. This angle is compared with the glide angle of the trajectory of bomb at that particular time. The difference between these two angles is fed to the servo loop as the error signal. The error correcting signal controls the direction of motion of the bomb and brings it onto the target. CO_2 lasers are used in this application as its beam is less affected by the conditions of fog, smoke or haze in the atmosphere.

It is required from the laser weapons that their beam may either disable the enemy weapons or destroy them. It is easier to achieve the goal of damaging the infrared sensors on guided missiles or the sensitive electronic eyes of spy satellites. Laser powers of laser beam suffice for such type of operation. However, if it is intended to destroy the enemy weapons, high intensity laser beams having powers of megawatts are needed. The major problems involved in this is that as the laser beam travels through air, it is absorbed though to a small extent. The absorption of energy leads to heating up of the air. Therefore, random turbulence is created there around and the laser beam is bent away from the target. It makes it hard both to track the target and to focus the laser beam onto it effectively.

Laser-based military weapons

On March 18, 2009 Northrop Grumman announced that its engineers in Redondo Beach had successfully built and tested an electric laser capable of producing a 100-kilowatt ray of light, powerful enough to destroy cruise missiles, artillery, rockets and mortar rounds. An electric laser is theoretically capable, according to Brian Strickland, manager for the United States Army's Joint High Power Solid State Laser program, of being mounted in an aircraft, ship, or vehicle because it requires much less space for its supporting equipment than a chemical laser.

On 19 July 2010 an anti-aircraft laser was unveiled at the Farnborough Airshow. It was described as the Laser Close-in Weapon System.

On April 6, 2011, the U.S. Navy successfully tested a laser gun, manufactured by Northrop Grumman, that was mounted on the former USS Paul Foster, which is currently used as the navy's test ship. When engaged during the test that occurred off the coast of Central California in the Pacific Ocean test range, the laser gun was documented as having a destructive effect on a high-speed cruising target', said Chief of Naval Research Admiral Nevin Carr. While classified, the range of the laser gun is attributed to miles, not yards.

Defensive countermeasures

Defensive countermeasure applications can range from compact, low power infrared countermeasures to high power, airborne laser systems. IR countermeasure systems use lasers to confuse the seeker heads on heat-seeking anti-aircraft missiles. High power boost-phase intercept laser systems use a complex system of lasers to find, track and destroy intercontinental ballistic missiles (ICBM). In this type of system a chemical laser, one in which the laser operation is powered by an energetic chemical reaction, is used as the main weapon beam. The Mobile Tactical High-Energy Laser (MTHEL) is another defensive laser system under development; this is envisioned as a field-deployable weapon system able to track incoming artillery projectiles and cruise missiles by radar and destroy them with a powerful deuterium fluoride laser.

Another example of direct use of a laser as a defensive weapon was researched for the Strategic Defense Initiative (SDI, nicknamed "Star Wars"), and its successor programs. This project would use ground-based or space-based laser systems to destroy incoming intercontinental ballistic missiles (ICBMs). The practical problems of using and aiming these systems were many; .particularly the problem of destroying ICBMs at the most opportune moment, the *boost phase* just after launch. This would involve directing a laser through a large distance in the atmosphere, which, due to optical scattering and refraction, would bend and distort the laser beam, complicating the aiming of the laser and reducing its efficiency.

One of the strategic defence systems chalked out by USA is Star War where a possible nuclear attack by Intercontinental Ballistic Missiles from USSR is proposed to be thwarted. When a missile is launched, it can hit the target within a few minutes. It is required that the missile should be detected and destroyed before it can reach the target. It is only laser weapons that can do the job of intercepting the missile and make it inoperative. As the speed of the missile will be at a minimum value in the initial stage and also that the multiple war heads do not separate at that stage, it is essential that the missile be destroyed in its initial stage only. Laser energy can be directed at the missile from the ground, air or from satellites. There are many problems yet to be solved for the feasibility of such programmes.

Another idea from the SDI project was the *nuclear-pumped X-ray laser*. This was essentially an orbiting atomic bomb, surrounded by laser media in the form of glass rods; when the bomb exploded, the rods would be bombarded with highly-energetic gamma-ray photons, causing spontaneous and stimulated emission of X-ray photons in the atoms making up the rods. This would lead to optical amplification of the X-ray photons, producing an X-ray laser beam that would be minimally affected by atmospheric distortion and capable of destroying ICBMs in flight. The X-ray laser would be a strictly one-shot device, destroying itself on activation. Some initial tests of this concept were performed with underground nuclear testing; however, the results were not encouraging. Research into this approach to missile defense was discontinued after the SDI program was cancelled.

Target designator

Another military use of lasers is as a *laser target designator*. This is a low-power laser pointer used to indicate a target for a precision-guided munition, typically launched from an aircraft. The guided munition adjusts its flight-path to home in to the laser light reflected by the target, enabling a great precision in aiming.

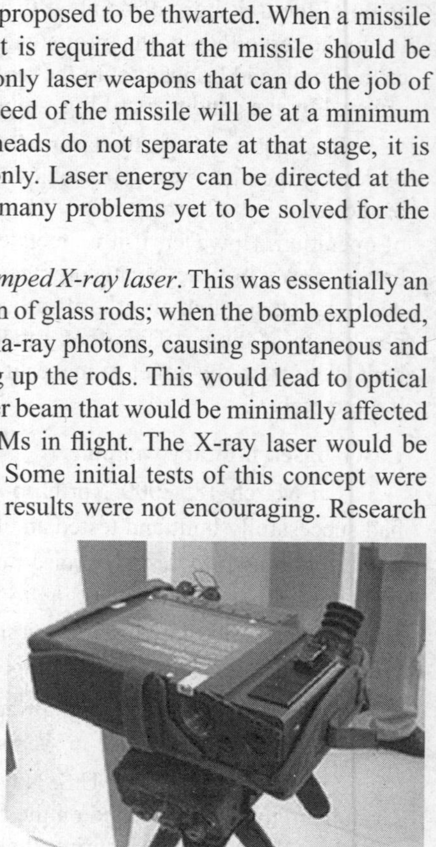

Fig.5.18 A Target designator

The beam of the laser target designator is set to a pulse rate that matches that set on the guided munitions to ensure munitions strike their designated targets and do not follow other laser beams which may be in use in the area. The laser designator can be shone onto the target by an aircraft or nearby infantry. Lasers, used for this purpose are usually infrared lasers, so the enemy cannot easily detect the guiding laser light.

Firearms Laser sight

The laser has in most firearms applications been used as a tool to enhance the targeting of other weapon systems. For example, a laser sight is a small, usually visible-light laser placed on a handgun or a rifle and aligned to emit a beam parallel to the barrel. Since a laser beam has low divergence, the laser light appears as a small spot even at long distances; the user places the spot on the desired target and the barrel of the gun is aligned (but not necessarily allowing for bullet drop, windage and the target moving while the bullet travels).

Fig. 5.19 Revolver equipped with a laser sight mounted on the trigger guard.

Most laser sights use a red laser diode. Others use an infrared diode to produce a dot invisible to the naked human eye but detectable with night vision devices. The firearms adaptive target acquisition module LLM01 laser light module combines visible and infrared laser diodes. In the late 1990s, green diode pumped solid state laser (DPSS) laser sights (532 nm) became available. Modern laser sights are small and light enough for attachment to the firearms.

In 2007, LaserMax, a company specializing in manufacturing lasers for military and police firearms, introduced the first mass-production green laser available for small arms. This laser mounts to the underside of a handgun or long arm on the accessory rail. The green laser is supposed to be more visible than the red laser in bright lighting conditions because, for the same wattage, green light appears brighter than red light.

Disorientation

Some weapons simply use a laser to disorient a person. One such weapon is the Thales Green Laser Optical Warner.

Measurement of speed of vehicles

In addition to the applications that crossover with military applications, a widely known law enforcement use of lasers is for lidar to measure the speed of enemies' vehicles.

Eye-targeted lasers

A non-lethal laser weapon was developed by the U.S. Air Force to temporarily impair an adversary's ability to fire a weapon or to otherwise threaten enemy forces. This unit illuminates an opponent with harmless low-power laser light and can have the effect of dazzling or disorienting the subject or causing him to flee. Several types of dazzlers are now available, and some have been used in combat.

There remains the possibility of using lasers to blind, since this requires much lower power levels and is easily achievable in a man-portable unit. However, most nations regard the deliberate permanent blinding of the enemy as forbidden by the rules of war (Protocol on Blinding Laser Weapons).

5.16 MEASUREMENT OF DISTANCE

5.7.1. Interferometric Methods

Modern engineering requires large-size items manufactured to very high accuracy. It involves measurement of distances from a metre to several metres accurate to within a few microns. Optical methods are used for such measurements, where Michelson interferometer or a variation of this instrument is employed. The method was limited by the coherence limitations of the conventional light sources, and distances of a few centimeters only could be measured. With the invention of lasers, sources with coherence lengths of many metres became available opening up the possibility of measuring larger distances to a very high accuracy. A frequency stabilized He-Ne laser is used as the source. The Michelson interferometer (Fig. 5.20) consists of a beam splitter, two plane mirrors and an observing telescope or a fringe counter. The distance of separation between fixed mirror M_1 and the beam splitter is kept constant to within l/10. The laser beam is divided by the beam splitter into two beams :

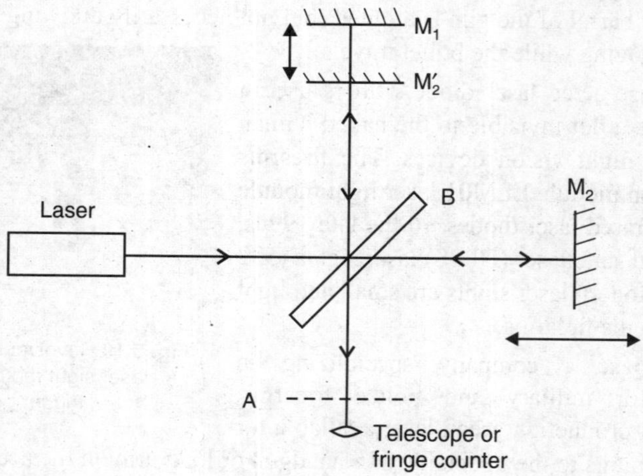

(*a*) Michelson inteferometer : M_1 is a fixed mirror, M_2 is movable and M'_2 is its image in the beamsplitter B. Interference fringes can be seen throught the telescope.

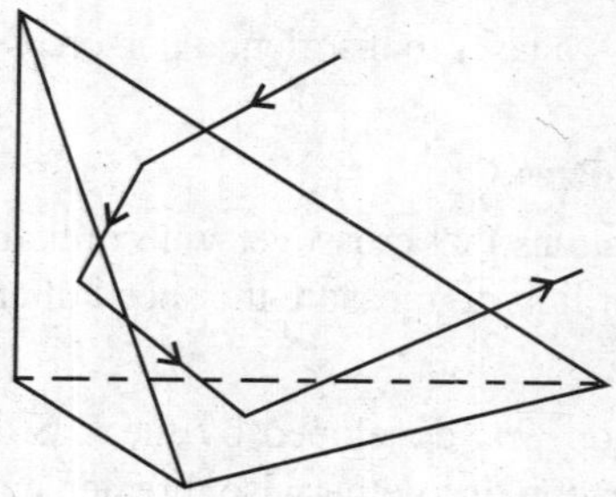

(*b*) Schematic diagram of a cube corner reflector. The incident light reflects once off each side face and emerges along a path almost exactly parallel to that of the incident light.

Fig. 5.20

a measurement beam and a reference beam. The reference beam is reflected by the fixed mirror M_1. The measurement beam is reflected by a retro-reflector fixed to the object being measured. The two reflected beams combine and produce interference pattern. If the position of the object and hence the retro-reflector is moved, a variation in intensity due to fringe pattern occurs which is detected by a photodetector and counted with a digital counter. If the object moves through a l/2 distance, one fringe will pass a reference point in the field of view of the telescope, and the signal will go from a maximum through a minimum and back to a maximum again. Therefore, we can measure the distance moved by the object in terms of fringe shifts, that is in terms of the wavelength λ of the light used.

The retroreflector is essentially a truncated triangular prism where an incident ray is internally reflected twice and is redirected along a direction parallel to its incidence path. An accuracy of 0.1 μm can be achieved and this technique can be used for a distance upto 100m.

This technique is widely used for calibration, testing and measurements of machine tools. It controls tool motions, automatically compensating for the errors due to wear, facilitates marking-off of points of engagement for a cutting tool on the surface to be machined, precisely aligns fixtures in aircraft engine manufacturing and facilitates a number of similar jobs.

Lasers can measure enormous distances with great accuracy. A laser beam travels at a constant speed (the speed of light). The time it takes a laser beam to travel from its source, reflect off an object, and return to the source, will indicate the exact distance between the source and the object. The potential of these devices to give precise figures for very long distances was shown in 1969 when the Apollo 11 astronauts became the first men ever to walk on the moon. Before blasting off on their return flight they left behind a bizarre-looking mirror. A short time later scientists on Earth claimed that the strange mirror had revealed to them the distance from Earth to the moon, a figure that was accurate to within the length of a person's finger. This moon mirror was neither mysterious nor magical, though it would have seemed so to many people only a few years before. In reality, National Aeronautics and Space Administration (NASA) scientists had instructed the astronauts on exactly how to position the mirror as part of a plan to measure the Earth-to-moon distance with a laser beam.

The scientists already knew the speed of light, so they knew what the rate of travel was. When the beam bounced off the mirror, it returned to Earth and registered on special sensors. These recorded how long the beam took to make a round trip, and scientists then knew the time factor in the equation. After some simple multiplication, they finally had the most accurate measurement of the Earth-to-moon distance possible. Knowing this has enabled them to learn much about the relationship between Earth and its natural satellite. For instance, researchers have repeated this laser-mirror experiment every year since 1969. They have found that the moon is moving away from Earth at a rate of about four centimeters a year.

5.16.2. Laser Rangers

The distances that can be measured using interferometric technique are limited to a few hundred metres. Measuring distance by means of lasers and mirrors works just as well on Earth as it does in space. Every day, surveyors use lasers to measure the distances between houses, roads, and mountains. A device called a *range finder* utilizes the same principle as the moon mirror; a surveyor aims a laser beam at a reflective target and the beam bounces back to the range finder, which records the time of the round-trip and uses this figure to calculate the exact distance to the target. This method is more accurate and also much faster than older surveying methods, which required many calculations with poles and telescopes that had to be lined up with one another. Erecting skyscrapers, excavating tunnels and canals, laying pipelines, drilling wells, leveling farmland (making it flatter and easier to exploit) are only a few of the many other projects made easier by precise laser measurements. Such measurements also have led to more exact and reliable maps; using lasers, mapmakers have now charted almost every square Km of Earth's surface,

The most advanced method for measuring larger distances is the use of *amplitude modulated CW laser beam*. The beam from a He-Ne or Ga As laser is modulated by a sinusoidal wave. As a result, the amplitude of the beam varies along the direction of beam propagation. The modulated beam is directed to the object whose distance is to be measured. The light reflected from the object is received by a detector. The phase of the modulation of the reflected beam is different from that of the emitted beam because of the finite time taken by the beam to travel to the object and return to the detector. The phase shift is given by

$$\phi = \frac{2\pi}{\lambda}(2\mu L) \tag{5.1}$$

where L is distance of the object and μ is refractive index of the atmosphere. An acuracy of 1 part in 10^6 is obtained in measuring distances of several kilometres. This technique is used for measuring large structures such as dams and bridges, for geodetic surveying etc. applications.

5.17 LIDAR (Light Detection And Ranging)

Large distances can be measured using pulse echo techniques. The acronym LIDAR is an optical remote sensing technology that can measure the distance to, or other properties of a target by illuminating the target with light, often using pulses from a laser. LIDAR technology has application in geomatics*, archaeology, geography, geology, geomorphology, seismology, forestry, remote sensing and atmospheric physics. The acronym LADAR (Laser Detection and Ranging) is often used in military contexts.

LIDAR uses ultraviolet, visible, or near infrared light to image objects and can be used with a wide range of targets, including non-metallic objects, rocks, rain, chemical compounds, aerosols, clouds and even single molecules. A narrow laser beam can be used to map physical features with very high resolution.

LIDAR has been used extensively for atmospheric research and meteorology. Down ward-looking LIDAR instruments fitted to aircraft and satellites are used for surveying and mapping. A recent example being the NASA Experimental Advanced Research Lidar.

Wavelengths in a range from about 10 micrometers to the UV (λ = 250 nm) are used to suit the target. Typically light is reflected via backscattering. Different types of scattering are used for different LIDAR applications, most common are Rayleigh scattering, Mie scattering and Raman scattering as well as fluorescence. Based on different kinds of backscattering, the LIDAR can be accordingly called Rayleigh LIDAR, Mie LIDAR, Raman LiDAR and Na/Fe/K Fluorescence LIDAR and so on.

Design: A basic LIDAR system involves a laser range finder reflected by a rotating mirror (top). The laser is scanned around the scene being digitised, in one or two dimensions (middle), gathering distance measurements at specified angle intervals (bottom).

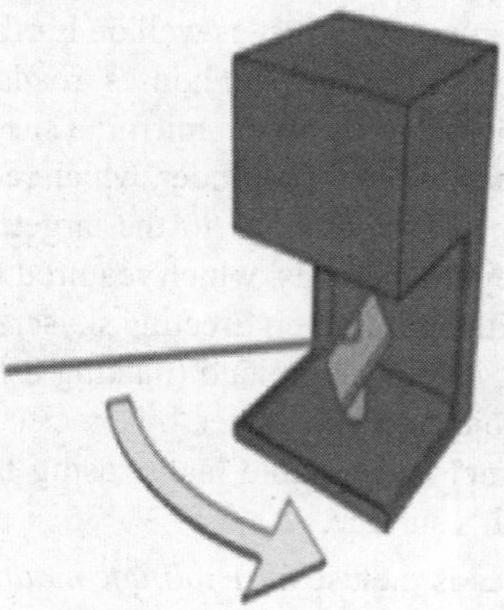

Fig. 5.21 Design of LIDAR

Fig. 5.22 A FASOR used at the Starfire Optical Ranges for LIDAR

* Geomatics is a scientific term comprising of geodesy and geoinformatics. It includes the tools and techniques used in land surveying, geography, remote sensing, cartography, Geographic Information Systems (GIS), Global Navigation Satellite Systems (GPS, GLONASS, Galileo, Compass), photogrammetry, and related forms of earth mapping.

There are several major components to a LIDAR system:

1. **Laser:** 600-1000 nm lasers are most common for non-scientific applications. They are inexpensive but since they can be focused and easily absorbed by the eye the maximum power is limited by the need to make them eye-safe. Eye-safety is often a requirement for most applications. A common alternative 1550nm lasers are eye-safe at much higher power levels since this wavelength is not focused by the eye, but the detector technology is less advanced and so these wavelengths are generally used at longer ranges and lower accuracies. They are also used for military applications as 1550 nm is not visible in night vision goggles unlike the shorter 1000 nm infrared laser. Airborne topographic mapping LIDARS generally use 1064 nm diode pumped YAG lasers, while bathymetric systems generally use 532 nm frequency doubled diode pumped YAG lasers because 532 nm penetrates water with much less attenuation than does 1064 nm. Laser settings include the laser repetition rate (which controls the data collection speed). Pulse length is generally an attribute of the laser cavity length, the number of pulses required through the gain material (YAG, YLF, etc.), and Q-switch speed. Better target resolution is achieved with shorter pulses, provided the LIDAR receiver detectors and electronics have sufficient bandwidth.
2. **Scanner and optics:** How fast images can be developed is also affected by the speed at which it can be scanned into the system. There are several options to scan the azimuth and elevation, including dual oscillating plane mirrors, a combination with a polygon mirror, a dual axis scanner. Optic choices affect the angular resolution and range that can be detected. A hole mirror or a beam splitter are options to collect a return signal.
3. **Photodetector and receiver electronic:** Two main photodetector technologies are used in LIDARS: solid state photodetectors, such as silicon avalanche photodiodes, or photomultipliers. The sensitivity of the receiver is another parameter that has to be balanced in a LIDAR design.
4. **Position and navigation systems:** LIDAR sensors that are mounted on mobile platforms such as airplanes or satellites require instrumentation to determine the absolute position and orientation of the sensor. Such devices generally include a Global Positioning System receiver and an Inertial Measurement Unit (IMU).

3D imaging can be achieved using both scanning and non-scanning systems. "*3D gated viewing laser radar*" is a non-scanning laser ranging system that applies a pulsed laser and a fast gated camera. Imaging LIDAR can also be performed using arrays of high speed detectors and modulation sensitive detectors arrays typically built on single chips using CMOS and hybrid CMOS/CC'D fabrication techniques. High resolution 3D LIDAR cameras use homodyne detection with an electronic CCD or CMOS shutter. A coherent Imaging LIDAR uses Synthetic array heterodyne detection to enables a staring single element receiver to act as though it were an imaging array avoiding the need for a gated camera and all ranges from all pixels are simultaneously available in the image.

Applications of LIDAR

Other than those applications listed above, there are a wide variety of applications of LIDAR, as often mentioned in National LIDAR Dataset. These are

1. **Agriculture**

   Agricultural Research Service scientists have developed a way to incorporate LIDAR with yield rates on agricultural fields. This technology will help farmers direct their resources toward the high-yield sections of their land. LIDAR also can be used to help farmers determine which areas of their fields to apply costly fertilizer. LIDAR can create a topological map of the fields and reveals the slopes and sun exposure of the farm land.

2. **Archaeology**

LIDAR helps in the planning of field campaigns, mapping features beneath forest canopy. LIDAR can also provide archaeologists with the ability to create high-resolution digital elevation models (DEMs) of archaeological sites that can reveal micro-topography that are otherwise hidden by vegetation. Features that could not be distinguished on the ground or through aerial photography can be identified by LIDAR. For example, discovery of Fort Beausejour - Fort Cumberland National Historic Site, Canada built in 1 755, which previously remained undiscovered.

3. **Surveying**

LIDAR has also found many applications in surveying. A laser beam is used as a reference to measure the angle of deflection from required alignment. A low power single mode He-Ne laser is used as a source. Forestry, canopy heights, biomass measurements, and leaf area can all be studied using airborne LIDAR systems. Similarly, LIDAR is also used by many industries, including Energy and Railroad, and the Department of Transportation as a faster way of surveying. Topographic maps can also be generated readily. LIDAR is also used by many industries, including Energy and Railroad, and the Department of Transportation as a faster way of surveying. Other surveying applications include pipe laying, level grading, tunnel bearing, fredging, alignment in hazardous environment like coal mines etc. LIDAR also used in oceanography, to survey generally biomass in the surface layers of the ocean. LIDAR have led to significant advances in geomorphology.

5.18 METEOROLOGY AND ENVIRONMENTAL STUDIES

Initially based on ruby lasers, LIDAR for meteorological applications was constructed shortly after the invention of the laser and represent one of the first applications of laser technology. The first LIDAR systems were used for studies of atmospheric pollutants, structure, clouds, and aerosols which includes N_2, CO_2, SO_2, NO_2 etc, gases and particulate matter such as dust, smoke and flyash. The importance of LIDAR, the laser based technique lies in the fact that they do not require sample collection or any kind of chemical processing. They yield real time data and are extremely suitable for environmental monitoring.

(*i*) **Elastic backscatter:** The LIDAR (optical radar) technique employs a pulsed laser as a source of light and light back-scattered is detected by a photo detector. The distance to particulate matter and the concentrations of particulate matter is obtained in this method. However, the method can't provide any information regarding the nature of the scattering particles. It is mainly useful in knowing the distribution of atmospheric pollutants in different vertical sections and in monitoring their variations. Environmental agencies measure concentrations of harmful gases such as SO_2 and NO_2 using this LIDAR technique. LIDAR is also used for studies of aerosols and clouds. The backscattered wavelength is identical to the transmitted wavelength, and the magnitude of the received signal at a given range depends on the backscatter coefficient of scatterers at that range and the extinction coefficients of the scatterers along the path to that range. The extinction coefficient is typically the quantity of interest.

(*ii*) **Differential Absorption LIDAR (DIAL)** is used for range-resolved measurements of a particular gas in the atmosphere, such as ozone, carbon dioxide, or water vapor. The LIDAR transmits two wavelengths: an "on-line" wavelength that is absorbed by the gas of interest and an off-line wavelength that is not absorbed. The differential absorption between the two wavelengths is a measure of the concentration of the gas as a function of range. DIAL LIDARs are essentially dual-wavelength backscatter LIDARS.

(*iii*) **Raman LIDAR** is also used for measuring the concentration of atmospheric gases, but can also be used to retrieve aerosol parameters as well. Raman LIDAR exploits inelastic scattering to single out the gas of interest from all other atmospheric constituents. A small portion of the energy of the transmitted light is deposited in the gas during the scattering process, which shifts the scattered light to a longer wavelength by an amount that is unique to the species of interest. The higher the concentration of the gas, the stronger the magnitude of the backscattered signal.

(*iv*) **Doppler LIDAR** is used to measure wind speed along the beam by measuring the frequency shift of the backscattered light. Scanning LIDARs, such as NASA's HARLIE LIDAR, have been used to measure atmospheric wind velocity in a large three dimensional cone. A doppler LIDAR system was used in the 2008 Summer Olympics to measure wind fields during the yacht competition. Doppler LIDAR systems are also now beginning to be successfully applied in the renewable energy sector to acquire wind speed, turbulence, wind veer and wind shear data. Both pulsed and continuous wave (CW) systems are being used. Pulsed systems using signal timing to obtain vertical distance resolution, whereas continuous wave (CW) systems rely on detector focusing.

5.19 VELOCITY MEASUREMENT

It is well know that when a light beam gets scattered by a moving object, the frequency of the scattered wave is different from that of the incident wave; the shift in the frequency depends on the velocity of the object. If ν is the frequency of light and if v denotes the velocity of the moving object which is moving at an angle θ with respect to the incident light beam, then the change in frequency Δν between the incident and reflected beams is given by

$$\frac{\Delta\nu}{\nu} = \frac{2\upsilon}{c}\cos\theta$$

The change in frequency, Δν is directly proportional to the velocity v of the moving object. This is known as the *Doppler shift*. This method has been successfully used for velocity determination of many types of materials from about 10 mm/min to about 150m/min. Using the above principle, portable velocity measuring meters have been fabricated to measure speeds in the range of 10-80 miles/hr. such meters are used by traffic police.

Laser Doppler velocimeters are used for measuring fluid flow rates.

The beam from a CW laser (He-Ne laser) is split by a beam splitter ; one of the components is reflacted back from a fixed mirror and the other component undergoes scattering from the moving object. The two beams are then combined and made to interfere. The change in frequency leads to beating. The beat frequency is a direct measure of the velocity of motion of the object.

5.20 HOLOGRAPHY

It is a lens-less photography. It requires a laser source for producing and viewing the image. Here the image is in the form of interference pattern.

Principle:

The working of holography is based on the principle of interference. Hence the light waves of high degree coherence is very much essential for its realization.

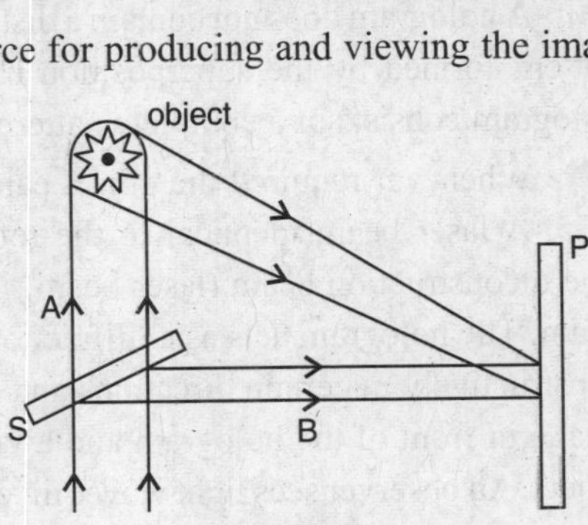

Fig. 5.23

A laser beam is splitted using a splitter S into two beams A and B. Beam A recognizes the object O whose holograph is to be recorded and a part of light scattered by the object impinges on a photographic plate P. The reflected beam B, also called as reference beam, falls on the same photog raphic plate.

The superposition of these two beams produces an interference pattern that is recorded on the plate. The plate is the developed. The developed plate is known as Holograph. The pattern is very fine and the spacing between the fringes are of the order of 0.001 mm.

Method :

In conventional photography a negative is made first and using it a positive print is produced later. The positive print is only a two dimensional record of light intensity received from a three dimensional object. It contains information about the square of the amplitude of the light wave that produced the image but information about the phase of the light wave is not recorded and which is lost. In 1947 Dennis Gabor the English physicist outlined a radically new technique of photographing objects. He called this technique *wavefront construction.* According to this technique both the phase and intensity attributes of the wave are recorded and when viewed the photograph shows a three dimensions image of the object. This technique is named *holography.* Fig. 5.24 illustrates the principle of holography. A weak but broad beam of laser light is split into two beams namely a *reference beam* and *object beam*. The reference beam is allowed to reach the photographic plate directly while the object beam illuminates the object. Part of the light scattered by the object travels towards the photographic plate and interferes with the reference beam and produces an interference pattern on the photographic plate. The photographic plate carrying the interference pattern is called a *hologram*. Holos means complete in Greek and "gramma" means writing. Thus a hologram means complete recording. Like any ordinary photographic plate, a hologram is developed, fixed and stored.

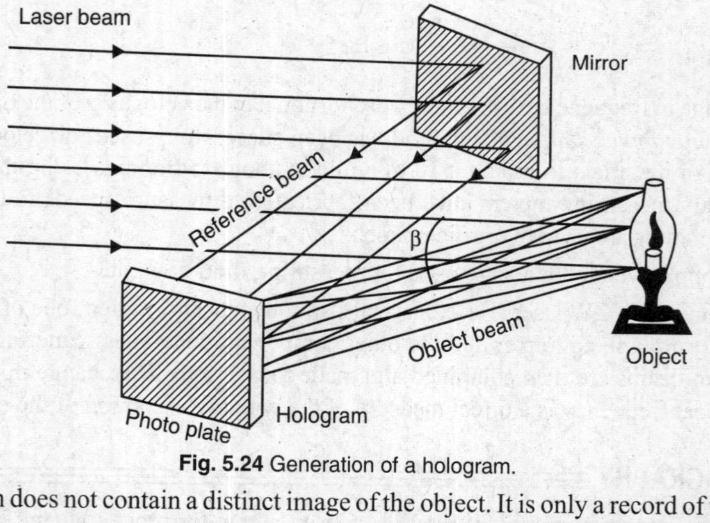

Fig. 5.24 Generation of a hologram.

A hologram does not contain a distinct image of the object. It is only a record of the interference pattern formed by the superposition of two coherent light beams. The interference pattern on a hologram consists of a complete pattern of alternate regions of dark and bright fringes.

Whenever required the object can be viewed by illuminating the hologram as shown in Fig. 5.25. A laser beam identical to the reference beam is used for the reconstruction of the object. The reconstruction beam (laser beam) illuminates the hologram at the same angle as the reference beam. The hologram acts as a diffraction grating and secondary waves from the hologram interfere constructively in certain directions and interfere destructively in other directions. They form a real image in front of the hologram and a virtual image behind the hologram at the original site of the object. An observer sees light waves diverging from the virtual image. An image of the object appears where the object once stood and that image is identical to what our eyes would have perceived in all its details. If the observer tilts his head other objects behind the first one or new details of the object which were not noticed earlier would be observed.

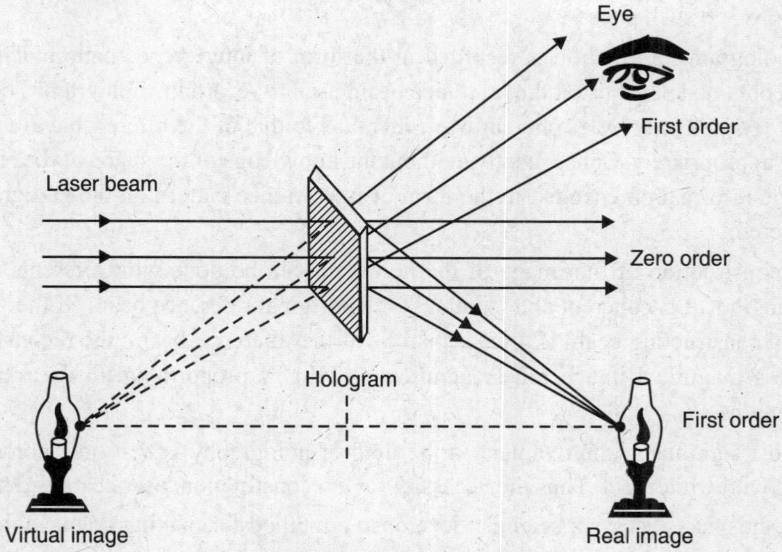

Fig. 5.25 Viewing of hologram.

Holography is thus a two stage process. In the first stage a hologram is recorded in the form of interference pattern. In the second stage the hologram acts as a diffraction grating for the reconstruction beam and the image of the object is reconstructed from the hologram.

It should be appreciated that an interference pattern can be obtained only when the reference beam and the object beam are coherent. The beams can be coherent only when they belong to the same original wave train. As the two beams travel over large distances it would be difficult to fulfil this condition with ordinary sources of light. Interference pattern can be obtained when the optical path difference between two beams, Δ is very small compared to the length of the wave train, i.e. coherence length, l_{coh}. Then only, the two parts of the wave train in the two beams will be coherent at the point of observation. For conventional monochromatic sources of light, the coherence length is found to be of the order of a few centimeters and a hologram cannot be obtained with them. Therefore holography could not be realised in practice for a long time though the principle was formulated in 1947. The invention of lasers made holography possible. Laser beams have coherence length of the order of a few kilometers and therefore the interference of split rays with a large path difference could be achieved. The first practical hologram was made using a laser beam by two American investigtors, E.N. Leith and Y. Uptneiks in 1964.

Advantages and Applications :

1. In an ordinary photograph each region contains a separate and individual part of the original object. Therefore, destruction of a portion of a photographic image leads to an irreparable loss of information corresponding to the destroyed part. On the other hand, in a hologram each part contains information about the entire object. Therefore, destruction of a part of a hologram does not cause a loss of information about the object. From even a small part of the hologram the entire image can be reconstructed if only with a reduced clarity and definition of the image. Therefore a hologram is a reliable medium for data storage.
2. It is not useful to record several images on a single photo film. Such a record cannot give information about any of the individual images. On the other hand, several images can be

recorded on a hologram. Therefore the information holding capacity of a hologram is extremely high. While a 6x9 mm photograph can hold one printed page, a hologram of the same size can store upto 300 such pages.

3. On a hologram information is recorded in the form of interference pattern. The type of the pattern obtained depends on the reference beam used to record the hologram. The information can be decoded only by a coherent wave identical to that of the reference wave which can be chosen appropriately. Consequently, without the knowledge of the shape of the reference wave front the information encoded in the form of interference pattern on the hologram cannot be deciphered.
4. The reconstruction of the image of the hologram can be done with reference beam of any wavelength if it is coherent and identical to the original reference beam. If the wave length λ of the reconstructing beam is greater than λ_0 of the reference beam, the reconstructed image will be a magnified image. The magnification will be proportional to the ratio of the two wavelengths.
5. The most significant improvement in the field of holography is that now holograms can be viewed with white light. Thus the necessity for a reconstruction laser beam is dispensed away.
6. Yu.N.Denisyuk, a Russian scientist developed a method for making colour holograms based on Lippman's colour photographic process.

The first holograms had to be viewed as well as created by laser light. Later advances made it possible to view laser-created hologram in white light. Careful choice of the way in which the hologram is recorded let the white light serve as the reference beam. The rainbow holograms are now a familiar sight seen on all types of products which are which mass-produced by stamping in metallized plastic.

Holography is extensively used in the field of nondestructive testing using interferometric methods, in the technique called *double exposure holographic* interferometry, the object under study is recorded on a photographic plate through the interference of the object wave and the reference wave. Then, without changing the object position, a stress is induced in the object. The deformed object is recorded on the same hologram with the same former reference beam. When this hologram is illuminated with a reconstruction wave, it will yield two object waves, one due to the undeformed and the other due to the deformed object. These waves will interfere and produce interference fringes covering the image of the object. They are characteristic of the strain suffered by the object. A quantitative study of the fringe pattern gives the degree and nature of deformation in different parts of the object.

Holographic interferometry is the standard technique which is routinely employed to assess the quality of aircraft tyres and many other high performance components. In modern manufacturing processes, precise control of parameters like dimensions, shape, load critical areas, stress concentrators etc is required and conventional methods cannot be used during manufacture as they disturb the fabrication process. Such measurements are possible through holographic interferometry.

A hologram can be designed to function as a lens. In military aircrafts it is often difficult for the pilot to look down at the dials and controls in the midst of a battle. It would be very convenient if their image is projected in front of his eyes so that he need not change his attention. Such image projection cannot be done using glass lenses in view of the short focal lengths and large areas required. On the other hand, the holographic lens can project an image of the control panel in the air in front of the pilot.

5.21 CONSUMER ELECTRONIC INDUSTRY

5.21.1 Supermarket Scanners (The Bar Code Reader)

A barcode also known as the Universal Product Code (UPC) is a series of lines at various widths that is a representation of information, which is read by an optic scanner called a barcode reader or scanned for an image by special software. Initially, barcodes contained the data within the widths and spacing of the parallel lines, today that's no longer true. Today, the information is concealed in patterns of dots, concentric circles, and hidden in images. The information can be price, destination, or other vital information concerning the product.

Bar codes are printed nowadays on all consumer items. The code contains information regarding the item and its manufacturer. A low power (He-Ne) laser beam is used to read the code. As the item is moved across the laser beam, dark areas of the code absorb light and the bright areas reflect it. The presence or absence of reflected light is registered by a photosensitive light detector. In this way, the supermarket scanner converts the pattern of bar code into a series of electrical signals. The information passes onto a central computer which identifies the product, looks up the price and prints the details on the bill.

Fig. 5.26 A typical universal product code on prepackaged items. The zero to the left of the code bar indicates item. The next five digits indicate the manufacturer and the next five digits identify the product.

Bar codes are adopted nowadays for a variety of purposes such as identifying materials in inventory of big concerns, to maintain data on borrowed books in libraries, to quickly process books and magazines by wholesale distributors etc.

Advantages and Applications:

Barcodes such as the UPC have become a ubiquitous element of modern civilization. Some modern applications of barcodes include:

1. Almost every item other than fresh produce from a grocery store, department store, and mass merchandiser has a UPC barcode on it. This helps track items and also reduces instances of shoplifting involving price tag swapping, although shoplifters can now print their own barcodes.
2. Barcodes are widely used in shop floor control applications software where employees can scan work orders and track the time spent on a job.
3. Retail chain membership cards (issued mostly by grocery stores and specialty "big box" retail stores such as sporting equipment, office supply, or pet stores) use bar codes to uniquely identify a consumer. Retailers can offer customized marketing and greater understanding of individual consumer shopping patterns. At the point of sale, shoppers can get product discounts or special marketing offers through the address or e-mail address provided at registration.
4. Barcodes can be used to track inventory, luggage, and even mail.
5. Warehouses, courier services can use barcodes for organizing and providing accurate status reports for individual packages or shipments.
6. Post production companies use barcodes to track videotapes, film reels, as well as other products.
7. When used on patient identification (Example of barcode on a patient identification wristband).

8. Barcodes permit clinical staff to instantly access patient data, including medical history, drug allergies, etc.
9. Document Management tools often allow for barcoded sheets to facilitate the separation and indexing of documents that have been imaged in batch scanning applications.
10. The tracking of item movement, including rental cars, airline luggage, nuclear waste, mail, express mail and parcels.
11. Tracking the organization of species in biology.
12. The barcode assigned based on the blood donor, blood acceptor, gene species allocation, etc.
13. Since 2005, airlines use an IATA-standard 2D barcode on boarding passes (BCBP), and since 2008, 2D barcodes sent to mobile phones enable electronic boarding passes.
14. Recently, researchers placed tiny barcodes on individual bees to track the insects' mating habits.
15. Barcode system is adopted in examination system in India, in order to observe the tight secrecy.
16. Barcoded entertainment event tickets allow the holder to enter sports arenas, cinemas, theatres, fairgrounds, transportation, etc. This can allow the proprietor to identify duplicate or fraudulent tickets more easily.
17. They can track the arrival and departure of vehicles from rental facilities.
18. Barcodes can integrate with in-motion checkweighers to identify the item being weighed in a conveyor line for data collection.
19. Some 2D barcodes embed a hyperlink to a web page. A capable cellphone might be used to read the barcode and browse the linked website, which can help a shopper find the best price for an item in the vicinity.
20. In 1992 the Veterans Health Administration developed Bar Code Medication Administration system (BCMA).
21. In the 21st century many artists started using barcodes in art, such as Scott Blake's Barcode Jesus, as part of the post-modernism movement.
22. Today, barcodes are issued by GS1, the most widely used supply chain standards system in the world.
23. Barcode systems can provide detailed up-to-date information on the business, accelerating decisions and with more confidence. From the point-of-sale management:
    - Fast-selling items can be identified.
    - Allowing fast-moving, more profitable items to occupy the best space.
    - Slow-selling items can be identified, preventing inventory build-up.
    - Historical data can be used to predict seasonal fluctuations very accurately.
24. Besides sales and inventory tracking, barcodes are very useful in logistics.

5.21.2. Compact Disc (CD)

Compact Discs (CDs) are similar to phonograph records in their appearance. Music is recorded in analog form on phonograph records, while it is recorded in digital form on the compact discs. As in the case of phonograph records, the *master* is first prepared. An optical flat glass disc is coated with a photoresist material layer of about A/4 thickness. A He-Cd laser whose intensity is modulated by the digitized music is directed onto the photoresist layer keeping the glass disc rotating. The encoding pattern follows a continuous, spiral path covering the entire disc surface and extending from the innermost track to the outermost track (Fig. 5.27). The data is stored on the disc with a laser (while writing), and can be accessed when the data path is illuminated (while reading) with a laser diode in an optical disc drive which spins the disc at speeds of about 200 rpm up to 4000 rpm or more depending on the drive type, disc format, and the distance of the read head from the center of the disc (inner tracks are read at a faster disc speed). For both, writing and reading purpose, optical lens is used as shown in Fig. 5.28.

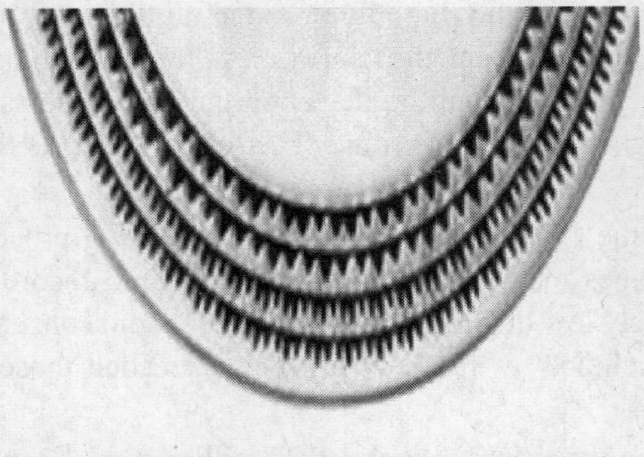

Fig. 5.27 Spiral paths of increasing radii with pits

Fig. 5.28 Optical lens of a Compact Disc drive

The pits or bumps distort the reflected laser light, hence most optical discs (except the black discs of the original PlayStation video game console) characteristically have an iridescent appearance created by the grooves of the reflective layer. The reverse side of an optical disc usually has a printed label, generally made of paper but sometimes printed or stamped onto the disc itself. This side of the disc contains the actual data and is typically coated with a transparent material, usually lacquer. Unlike the 3½-inch floppy disk, most optical discs do not have an integrated protective casing and are therefore susceptible to data transfer problems due to scratches, fingerprints, and other environmental problems.

After recording, the disc is developed which causes hardening of the unexposed area of the photoresist layer. Then the photoresist layer is etched to remove the exposed area. This process leaves a pattern of pits in the photoresist. The surface is then coated with silver and electroplated with nickel. Master is thus ready. It is used to transfer the pit pattern to a plaslic disc by injecting moulding. The plastic disc is then typically coated with a transparent material, usually lacquer.

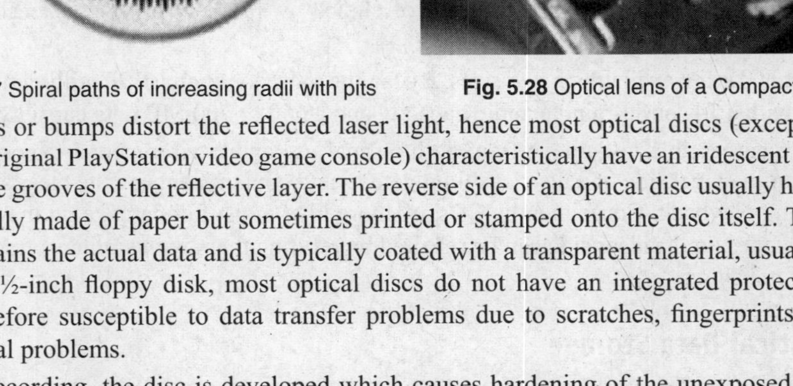

Fig 5.29 Schematic of a compact disc optical storage device. Readout mode. (Not drawn to scale.) The reflected beams in (b) are drawn under an angle. The base and bump areas covered by the probing light have to be of equal size in order that destructive interference can occur.

The music is now contained on the CD in the form of bumps. The height of these bumps is $\lambda/4$. GaAlAs laser having a wavelength of 7800 Å is mounted on the playing head of CD player. The beam size at the surface of the disc is relatively large. The light reflected from a bump and a pit is cancelled because of destructive interference. Absence of light corresponds to a *zero* in binary code, where as presence of light constitutes a *one*. The information on a CD of 12 cm size is about 7×10^9

k bits and provides a playback time of about 1 hr. The main advantages of CDs is the crisp and clean digital sound and long life. As there is no physical contact between the disc and the read-out device, mechanical wear-out of disc does not occur, thus imparting it a long useful life.

As originally developed for audio systems, the sound signal is sampled at a rate of 44,100 times a second (44.1 kHz), then each sample is measured and digitally encoded on the 4-3/4 inch (12 cm) disc as a series of microscopic pits on an otherwise polished surface. As the focused laser beam sweeps over the pits, it reproduces the binary numbers in the detection circuitry. The same function as the "pits" can be accomplished by magnetooptical recording. The digital signal is then reconverted to analog form by a D/A converter.

Since nothing touches the encoded portion, the CD is not worn out by the playing process. Introduced in 1982, the CD offered other advantages over the phonograph record and recording tape— smaller size, greater dynamic range, extremely low distortion—and met with rapid consumer acceptance, the CD became the music carrier of choice by 1991, when sales exceeded those of audiocassettes.

Standard CDs are available in two sizes. By far, the most common is 120 millimetres (4.7 inch) in diameter, with a 74- or 80-minute audio capacity and a 650 or 700 MB data capacity.

CDs are susceptible to damage from both normal use and environmental exposure. Pits are much closer to the label side of a disc, enabling defects and contaminants on the clear side to be out of focus during playback. Consequently, CDs are more likely to suffer damage on the label side of the disk. Scratches on the clear side can be repaired by refilling them with similar refractive plastic, or by careful polishing.

5.21.3 Optical Data Storage

Optical discs which are used for playback of music are also useful for high volume digital data storage. We have seen that a beam of laser focused to a high intensity spot causes a localized change in a photosensitive medium to record a 'bit' (either logic 0 logic 1) which is an essential part of digitized information data. High capacity, short access time, high reliability, and absence of mechanical motion make optical storage an attractive alternative to the conventional magnetic disc memories. The optical discs known as CD ROMs are random access, read only memory devices used in computer based applications.

The optical discs known as CD ROMs are random access, read only memory devices used in computer based application. Information is recorded in much the same way as is done in case of compact discs. The information is encoded in the length of the pit and the spacing of pits along the track. The distance between two adjacent tracks is 1.6 μm. The width of a pit is equal to a laser spot size of 0.5 μm to 0.7 μm. The light source is GaAlAs laser diode having a wavelength of 0.78 μm to 0.83 μm. The spot size is larger than the width of a pit but a single spot does not cover two tracks. Both the pit and land are coated with high reflectivity material such as aluminium. When there is no pit, the light is fully reflected from the land to the detector. When there is a pit, the focussed beam covers both the pit and the surrounding area and so the light is reflected from both of them. The depth of the pit is made such that the phase difference between the light reflected from a pit and a land is π. Consequent to 180^0 phase difference, destructive interference occurs at the detector and negligible light intensity is detected. The detector interprets no light as a *zero* in binary code while large light is interpreted as a *one*.

A disc of 12 cm diameter holds about 600 megabytes of computer data which is equivalent to three times the text of a standard 24 volume encyclopedia.

Optical discs are usually between 7.6 and 30 cm (3 to 12 in) in diameter, with 12 cm (4.75 in) being the most common size. A typical disc is about 1.2 mm (0.05 in) thick, while the track pitch (distance from the center of one track to the center of the next) is typically 1.6 μ.m.

An optical disc is designed to support one of three recording types: read-only (e.g.: CD and CD-ROM), recordable (write-once, e.g. CD-R), or re-recordable (rewritable, e.g. CD-RW). Write-once optical discs commonly have an organic dye recording layer between the substrate and the reflective layer. Rewritable discs typically contain an alloy recording layer composed of a phase change material, most often AglnSbTe, an alloy of silver, indium, antimony and tellurium.

Optical discs are most commonly used for storing music (e.g. for use in a CD player), video (e.g. for use in a DVD player), or data and programs for personal computers (PC). The Optical Storage Technology Association (OSTA) promotes standardized optical storage formats. Although optical discs are more durable than earlier audio-visual and data storage formats, they are susceptible to environmental and daily-use damage. Libraries and archives enact optical media preservation procedures to ensure continued usability in the computer's optical disc drive or corresponding disc player.

For computer data backup and physical data transfer, optical discs such as CDs and DVDs are gradually being replaced with faster, smaller, and more reliable solid-state devices, especially the USB flash drive. This trend is expected to continue as USB flash drives continue to increase in capacity and drop in price. Similarly, personal portable CD players have been supplanted by portable solid-state digital audio player (MP3 players), and MP3 music purchased or shared over the Internet has significantly reduced the number of audio CDs sold annually.

Development of Generations:

First Generation: Optical disc was invented in 1958. In 1961 and 1969, David Paul Gregg registered a patent for the analog optical disc for video recording. This form of optical disc was a very early form of the DVD. It is of special interest that U.S. Patent issued in 1990, generated royalty income for Pioneer Corporation's DVA until 2007 —then compassing the CD, DVD, and Blu-ray Disc systems.

Later, in the Netherlands in 1969, Philips Research physicists began their first optical videodisc experiments at Eindhoven. In 1975, Philips and MCA began to work together, and in 1978, commercially much too late, they presented their long-awaited Laserdisc in Atlanta. MCA delivered the discs and Philips the players. However, the presentation was a technical and commercial failure and the Philips/MCA cooperation ended.

In Japan and the U.S., Pioneer succeeded with the videodisc until the advent of the DVD; In 1979, Philips and Sony, in consortium, successfully developed the audio compact disc in 1983.

Second Generation: In the mid-1990s, a consortium of manufacturers developed the second generation of the optical disc, the DVD.

Third Generation: The third generation optical disc was developed in 2000-2006, and was introduced as Blu-ray Disc. Developed by the Blu-ray Disc Association (BDA), a group of the world's leading consumer electronics, personal computer and media manufacturers (including Apple, Dell, Hitachi, HP, JVC, LG, Mitsubishi, Panasonic, Pioneer, Philips, Samsung, Sharp, Sony, TDK and Thomson). The format was developed to enable recording, rewriting and playback of high-definition video (HD), as well as storing large amounts of data. The format offers more than five times the storage capacity of traditional DVDs and can hold up to 25 GB on a single-layer disc and 50 GB on a dual-layer disc. This extra capacity combined with the use of advanced video and audio techniques will offer consumers an unprecedented HD experience.

While current optical disc technologies such as DVD, DVD±R, DVD±RW, and DVD-RAM rely on a red laser to read and write data, the new format uses a blue-violet laser instead, hence the name Blu-ray. Despite the different type of lasers used, Blu-ray products can easily be made backwards compatible with CDs and DVDs through the use of a BD/DVD/CD-compatible optical pickup unit. The benefit of using a blue-violet laser (405 nm) is that it has a shorter wavelength than a red laser (650 nm), which makes it possible to focus the laser spot with even greater precision. This allows data to be packed more tightly and stored in less space, so it's possible to fit more data on the disc even though it's the same size as a CD/DVD. This together with the change of numerical aperture to 0.85 is what enables Blu-ray Discs to hold 25 GB/50 GB. Recent development by Pioneer has pushed the storage capacity to 500 GB on a single disc by using 20 layers. First movies on Blu-ray discs were released in June 2006. Blu-ray eventually prevailed in a high definition optical disc format war over a competing format, the HD DVD. A standard Blu-ray disc can hold about 25 GB of data, a DVD about 4.7 GB, and a CD about 700 MB.

5.22 OPTICAL COMMUNICATION USING LASER

Lasers occupy a very important place in communications. Open atmosphere communications are restricted only to line of sight. It is required that the environment be free from fog, dust and rain. With the advent of low loss optical fibres, and semiconductor lasers, the field of communications under went a great change. Optical fibres are now widely used in communications. Reduction in size, weight and volume of cabling, availability of wide bandwidth, immunity from cross talk, electromagnetic interference etc. have made the use of optical fibres an attractive proposition for communications. The semiconductor laser is well matched to optical fibres communications. Its emitting area is only a few micrometers wide, small enough to couple light efficiently into the thin optical fibres. The glass optical fibres have lower loss at 1.3 μm and 1.55 μm. In GaAsP lasers generate radiation at these wavelengths. Indium Gallium Arsenide phosphide (InGaAsP) laser is the most recent developed phosphor and takes the advantage of 'window' in the transmission specturm of optice fibre operating in the range of 1.2 to 1.6 μm.

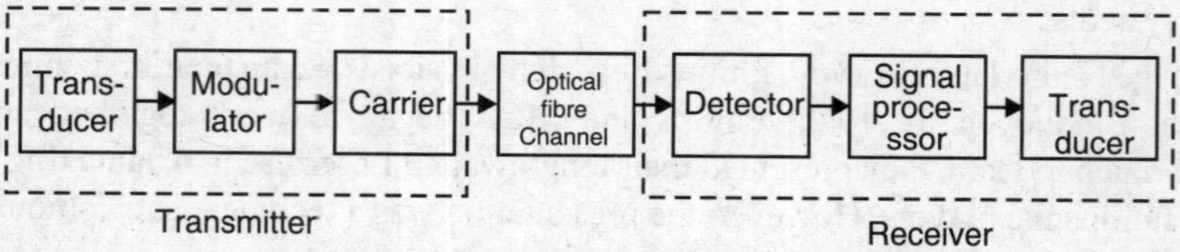

Fig. 5.30. A generalized communication system.

Satellite Frequency Bands:

The frequency used for satellite communication should be selected from bands that are most favorable in terms of power efficiencies, minimal propagation of distortion, and reduced noise and interference effects. Terrestrial systems tend to favor these same bands. So, concern for interference effect between the satellite and terrestrial systems must be made. Satellite use from space must be regulated and shared on a worldwide basis. For this reason, frequencies to be used by the satellite are established by a world body known as the International Telecommunications Union (ITU) with broadcast regulations controlled by a subgroup known as World Administrative Radio Conference (WARC). An international consultative technical committee (CCIR) provides specific recommendations on satellite frequencies under consideration by WARC. The basic objective is to allocate particular frequency bands for different types of satellite services, and also to provide international regulations in the areas of maximum radiation's level from space, co-ordination with terrestrial systems and the use of specific satellite locations in a given orbit. Within these allotments and regulations an individual country can make its own specific frequency selections based on intended uses and desired satellite services. The electromagnetic frequency spectrum allocated internationally accepted for various bands in satellite communication is as shown table 5.5.

Table 5.5 International frequency band allocation

BANDS	FREQUENCY
VHF	54 - 216 MHz
UHF	470 - 890 MHz
L	0.39 - 1.55 GHz
S	1 .55 - 5.2 GHz
C	3.9 - 6.2GHz
X	5.2 - 10.9 GHz
K	10.9 - 36 GHz
Ku	11.7 - 14.5 GHz
Ka	17 - 31 GHz
Q	36 - 46 GHz
V	46 -56 GHz

Laser Communication Process:

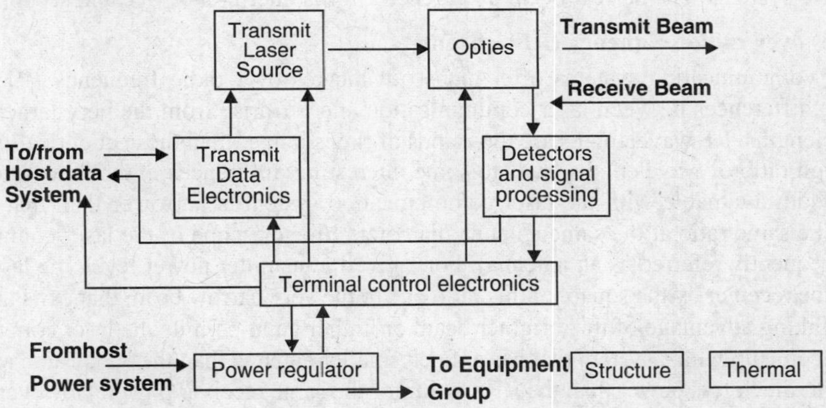

Fig. 5.31 Block diagram of Laser Communication system

Information typically in the form of digital data is input to data electronics that modulates the transmitting data source. Direct or indirect modulation techniques may be employed depending on the type of laser used. The source output passes through an optical system into the channel. The optical system typically includes transfer, beam shaping and telescope optics. The receiver beam comes in through the optical system and is passed along to the detectors and signal processing electronics. There are also terminal controlled electronics that must control the gimbals and other steering mechanism and servos to keep the acquisition and tracking system operating in the designed modes of operation.

Solid state lasers have offered higher power levels and the ability to operate in high peak power modes for the acquisition. When laser diodes are used to optically pump the lasing media graceful degradation and higher overall reliability is achieved. A variety of materials have been proposed for laser transmitters: Neodyminium doped Yttrium Aluminium Garnet (Nd:YAG) is the most widely used. Operating at wavelength 1064 nm, these lasers require an external modulator leading to a slight increase in the complexity and reliability. With the rapid development of terrestrial fiber communications, a wide area of components are available for the potential applications in space. These include detectors, lasers, multiplexers amplifiers, optical pre-amplifiers etc. operating at wavelength 1550 nm erbium doped fiber amplifiers have been developed for commercial optical fiber communications that offer levels of performance consistent with many free space communications applications.

Applications

Lasers have been considered for space communications since their realization in 1960. Specific advancements were needed in component performance and system engineering particularly for space qualified hardware. Advances in system architecture, data formatting and component technology over the past three decades have made laser communications in space not only viable but also an attractive approach into inter satellite link applications.

Information transfer is driving the requirements to higher data rates, laser cross -link technology explosions, global development activity, increased hardware, and design maturity. Most important in space laser communications has been the development of a reliable, high power, single mode laser diode as a directly modulable laser source. This technology advance offers the space laser communication system designer the flexibility to design very lightweight, high bandwidth, low-cost communication payloads for satellites whose launch costs are a very strong function of launch weigh. This feature substantially reduces blockage of fields of view of most desirable areas on satellites. The smaller antennas with diameter typically less than 30 centimeters create less momentum disturbance to any sensitive satellite sensors. Fewer on board consumables are required over the long lifetime because there are fewer disturbances to the satellite compared with heavier and larger RF systems. The narrow beam divergence affords interference free and secure operation.

Advantages over radio frequency (RF) systems

Laser communication systems offer many advantages over radio frequency (RF) systems. Most of the differences between laser communication and RF arise from the very large difference in the wavelengths. RF wavelengths are thousands of times longer than those at optical frequencies are. This high ratio of wavelengths leads to some interesting differences in the two systems. First, the beam-width attainable with the laser communication system is narrower than that of the RF system by the same ratio at the same antenna diameters (the telescope of the laser communication system is frequently referred as an antenna). For a given transmitter power level, the laser beam is brighter at the receiver by the square of this ratio due to the very narrow beam that exits the transmit telescope. Taking advantage of this brighter beam or higher gain, permits the laser communication designer to come up with a system that has a much smaller antenna than the RF system and further, need transmit much less power than the RF system for the same receiver power. However since it is much harder to point, acquisition of the other satellite terminal is more difficult.

Some advantages of laser communications over RF are smaller antenna size, lower weight, lower power and minimal integration impact on the satellite. Laser communication is capable of much higher data rates than RF. The laser beam width can be made as narrow as the diffraction limit of the optic allows. This is given by beam width = 1.22 times the wavelength of light divided by the radius of the output beam aperture. The antennae gain is proportional to the reciprocal of the beam width squared. To achieve the potential diffraction limited beam width a single mode high beam quality laser source is required; together with very high quality optical components throughout the transmitting sub system. The possible antennae gain is restricted not only by the laser source but also by the any of the optical elements. In order to communicate, adequate power must be received by the detector, to distinguish the signal from the noise. Laser power, transmitter, optical system losses, pointing system imperfections, transmitter and receiver antennae gains, receiver losses, receiver tracking losses are factors in establishing receiver power. The required optical power is determined by data rate, detector sensitivity, modulation format, noise and detection methods.

5.23 Optical Computer

Traditional electronic computers are advancing ever closer to their theoretical limits for size and speed. Some scientists believe that optical computing will one day unleash a new revolution in smaller and faster computers.

An optical computer (also called a photonic computer) is a device that uses the photons in visible light or infrared (IR) beams, rather than electric current, to perform digital computations.

An electric current flows at only about 10 percent of the speed of light. This limits the rate at which data can be exchanged over long distances, and is one of the factors that led to the evolution of optical fiber. By applying some of the advantages of visible and/or IR networks at the device and component scale, a computer might someday be developed that can perform operations 10 or more times faster than a conventional electronic computer.

Visible-light and IR beams, unlike electric currents, pass through each other without interacting. Several (or many) laser beams can be shone so their paths intersect, but there is no interference among the beams, even when they are confined essentially to two dimensions. Electric currents must be guided around each other, and this makes three-dimensional wiring necessary. Thus, an optical computer, besides being much faster than an electronic one, might also be smaller.

Photons instead of electrons

Internet connections and computers need to be ever faster and more powerful nowadays. However, conventional central processing units (CPUs) limit the performance of computers, because they produce an enormous amount of heat. The millions of transistors that switch and amplify the electronic signals in the CPUs are responsible for this. One square centimeter of CPU can emit up to 125 watts of heat, which is more than ten times as much as a square centimeter of an electric hotplate.

This is why scientists have been trying for some time to find ways to produce integrated circuits that operate on the basis of photons instead of electrons. The reason is that photons do not only generate much less heat than electrons, but they also enable considerably higher data transfer rates.

Fig. 5.32 Schematic sketch of Optical computer

Although a large part of telecomunications engineering nowadays is based on optical signal transmission, the necessary encoding of the information is generated using electronically controlled switches. Vahid Sandoghdar, Professor at the Laboratory of Physical Chemistry of ETH Zurich and his associate researchers have successfully created an **optical transistor** from a single molecule. This has brought them one step closer to an optical computer. For this, they have made use of the fact that a molecule's energy is quantized: when laser light strikes a molecule that is in its ground state, the light is absorbed. As a result, the laser beam is quenched. Conversely, it is possible to release the absorbed energy again in a targeted way with a second light beam. This occurs because the beam changes the molecule's quantum state, with the result that the light beam is amplified. This so-called stimulated emission, which Albert Einstein described over 90 years ago, also forms the basis for the principle of the laser.

Focusing on a nano scale

Jaesuk Hwang, first author of the study and a scientific member of Sandoghdar 's nano-optics group, explains that, "Amplification in a conventional laser is achieved by an enormous number of molecules." By focusing a laser beam on only a single tiny molecule, the ETH Zurich scientists have now been able to generate stimulated emission using just one molecule.

Switching light with light

By using one laser beam to prepare the quantum state of a single molecule in a controlled fashion, scientists could significantly attenuate or amplify a second laser beam. This mode of operation is identical to that of a conventional transistor, in which electrical potential can be used to modulate a second signal.

The possibility of using nonlinear crystals as switching devices opens up the possibility of building optical computers which work faster than the electronic computer. The optical switch, which forms the building block of computer, is very much similar to a laser. The two parallel end surfaces are coated with a thin film to make it semi reflecting. If monochromatic light emitted by a laser enters the crystal, part of the light bounces back and forward while part of it passes out. If the length of the crystal happens to be an integer multiple of half-wavelength of light, the light intensity within the crystal increases rapidly due to constructive interference. As a result, the intensity of the light transmitted is also intense. If the above condition is not satisfied, the transmitted light intensity will be small.

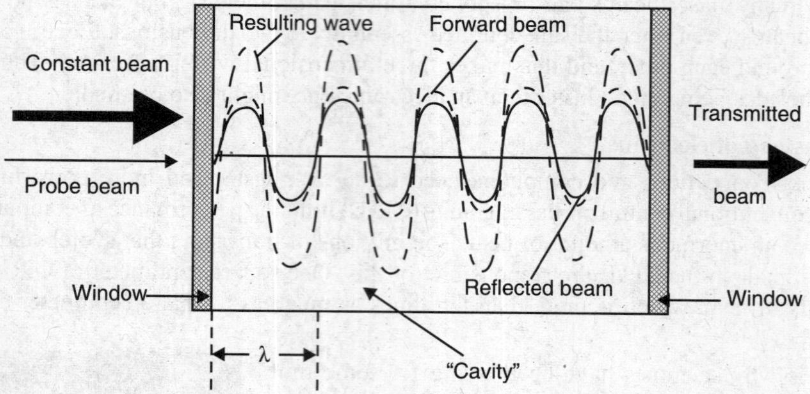

(*a*)

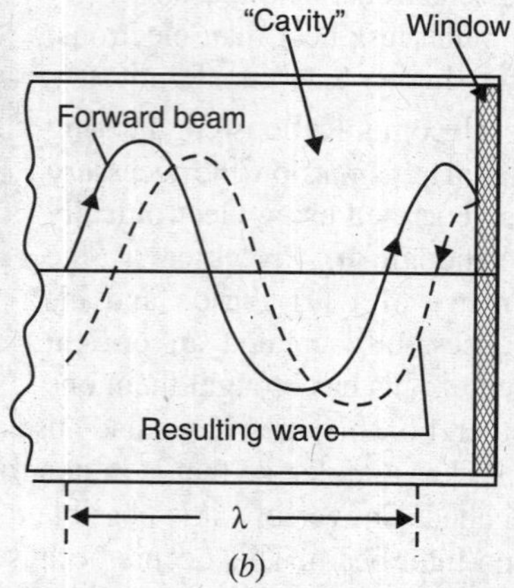

(*b*)

Fig. 5.32. Schematic representation of light waves in an optical switch. (a) Constructive interference. The length of the "cavity" equals an integer multiple of λ/2.
(b) condition (a) above is not fulfilled. The sum of many forward and reflected beams decreases the total intensity of the light.

An optical switch uses a *constant laser beam* whose intensity is not yet strong enough to trigger constructive interference. A second laser beam called *probe beam* directed onto the same spot in the crystal supplies additional energy to cause a large change in refractive index leading to constructive interference of light. As a result large output will be obtained. Thus two beams can be used to implement "AND", "OR", "NOT" gates.

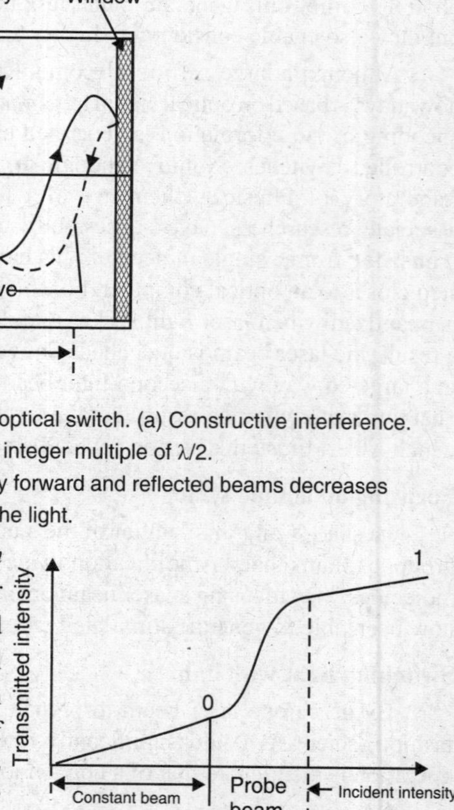

Fig. 5. 15. Schematic representation of an optical AND gate. The low transmission state may represent a "Zero" in binary logic whereas the high transmission of light may stand for a "one".

Thus component parts such as the new single molecules transistor may slao pave the way for a quantum computer. Sandoghdar says, " Many more years of research will'stil be needed before photons replace electrons in tansistors. In the meantime, scientists will learn to manipulate and control quantum systems is a targeted way, moving them closer to the dream of a quantum computer.

EXERCISE-5

1. What are the various applications of laser. **(Pune U. 2010; G.N.D.U. 2008; Garhwal U. 2004)**
2. Describe how lasers can be used in material processing. **(Pune U. 2010)**
3. How laser is useful as a 'Tool box' ?
4. Why surface treatment is needed ? How laser is used for surface treatment of a given piece of material?
5. Explain the following industrial applications of lasers:

   Cutting, Welding, Melting, Photolithography, 3-D measurements, 3-d stereo lithography.

   (B.Tech. Fourth sem., G.B.T.U. Lucknow 2010)
6. Discuss the biological hazards of a laser beam. What type of protection from such hazards can be taken? **(B.Tech. Fourth sem., G.B.T.U. Lucknow 2010)**
7. Write an essay on 'Applications of lasers in Medical Science".

   (B.Tech. Fourth sem., G.B.T.U. Lucknow 2010)
8. What are the applications of laser in military discuss in detail.

   (B.Tech. Fourth sem., G.B.T.U. Lucknow 2010)
9. What is holography ? How it differs from photography ? Write some applications of Holography.
10. Explain the use of laser in cardiology and dentistry in the medical science.
11. Write different lasers used in different medical applications.
12. Give basic principle of holography technique ? How are lasers used in recording a hologram.

    (Pune U. 2010)
13. How can lasers be used in Defence ? **(Pune 2010)**
14. Discuss briefy the application of laser for the treatment of cancer.
15. What is processessing of materials ? Write its main advantages ?
16. Explain cutting process with lasers ?
17. Explain laser welding ?
18. What is LIDAR ? Write its applications.
19. Write a short note on optical communication ?
20. What do you understand by optical computer ?
21. Explain the working of super market scanner.
22. Give advantages and disadvantages of Bar Code Reader.
23. Explain optical disc works as a data storage device ?
24. How environment monitoring is done by using laser ?
25. Discuss the use of laser in meteorology.
26. Discuss the use os LIDAR in

    (*i*) Agriculture (*ii*) Archaeology and (*iii*) Surveying.
27. Explain the use of LIDAR in velocity measurement.
28. Explain how laser can be used as defensive counter measure ?
29. How angioplasty is done using a laser beam and a ballon catheter ?
30. What is homeostasis ? Explain how laser beam can be used for bleaching the birth marks ?
31. Enunciate the applications of laser in several medical modalities.
32. What is endoscopy ? How laser beam is helpful in disease diagnosis ?

BOOKS SUGGESTED FOR FURTHER READING

1. Laser Physics and Applications by L. Tarasov
Mir Publishers, Moscow

2. Laser Age in Optics by L.V. Tarasov
Mir Publishers, Moscow

3. Understanding Lasers by Jeff Hecht
The Inst. of Electrical and Electronics Engineers Inc., New York, 1993

4. Lasers - Theory and Applications by K. Thyagarajan and A.K. Ghatak
MacMillan India Ltd., 1999

5. Laser Fundamentals by William T. Silfvast
Cambridge University Press, 1998

6. An Introduction to Lasers and Masers by A.E. Siegmann
McGraw Hill, N.Y., .1971

7. Introduction to Lasers and their Applications by D.C. O'shea, W.R. Callen and W.T. Rhides
Addison-Wesley, 1977

8. Industrial Lasers and their Applications by J.T. Luxon and D.E. Parker

9. Lasers and nonlinear optics by B.B. Laud
John Wiley & Sons, 1985

10. Optical Electronics, by A. Yariv
Holt, Rinehart, and Winston, 1985